BIOTECHNOLOGY GUIDE U.S.A.

Companies, Data and Analysis

MARK D. DIBNER

Director, Biotechnology Information Program
North Carolina Biotechnology Center

STOCKTON PRESS

To
Elaine, my partner.
She cares and understands.

©Stockton Press, 1988

Published in the United States and Canada by Stockton Press, 1988
15 East 26th Street, New York, N.Y. 10010

Library of Congress Cataloging-in-Publication Data

Dibner, Mark D.
  Biotechnology guide U.S.A. : companies, data, and analysis / Mark
D. Dibner.
     p. cm.
  Bibliography: p.
  ISBN 0-935859-40-3 : $175.00
  1. Biotechnology industries—United States.   I. Title
II. Title: Biotechnology guide USA.
HD9999.B443V63 1988
338.7'6606'0973—dc19   88-4952

Published in the United Kingdom by
MACMILLAN PUBLISHERS LTD (Journals Division), 1988
Distributed by Globe Book Services Ltd,
Brunel Road, Houndmills, Basingstoke,
Hants RG21 2XS, England

British Library Cataloguing in Publication Data

Dibner, Mark D. (Mark Douglas) 1951
  Biotechnology guide U.S.A.
companies, data and analysis.
1. United States. Biotechnology
Industries—Directories.
I. Title
338.7'6606'02573

ISBN 0-333-48551-3

Printed in the United States of America
9 8 7 6 5 4 3 2 1

CONTENTS

Contents

Contents

PREFACE

Following the explosive growth of the American biotechnology industry, there has been a demand for information about the hundreds of new companies that work with biotechnology. The need for information comes from many sectors of the industry, including clients, suppliers, partners, and competitors, as well as a variety of others wishing to monitor growth in the industry or its subsectors. Likewise, information about the industry has come from many sources.

One major source is the set of databases on the biotechnology industry maintained in our Biotechnology Information Program at the North Carolina Biotechnology Center. The databases have been used by many interested parties to gain an understanding of specific sections of the industry. For example, a client can learn about all the companies that work with biosensors or plant agricultural biotechnology in order to tap into the realm of potential clients, partners or competitors. Similarly, a client may study the U.S. corporations that have purchased equity in biotechnology companies working with therapeutics in order to identify potential sources of capital. Not only are the uses of these data manifold, but the user can also gain access to information tailored to specific needs.

This Guide, which has been developed using our databases, is not simply another in a series of directories on the U.S. biotechnology industry. It does not include a large number of companies that only have a peripheral interest in biotechnology. Rather, it lists only those companies that are actually working with the new technologies of genetic engineering (recombinant DNA), monoclonal antibodies, large-scale cell culture, and the like. As a result, we have identified 360 organizations that fit our criteria. Information has been collected by contacting companies by telephone and, in conjunction with a survey we conducted for the U.S. Congressional Office of Technology Assessment, by sending questionnaire surveys to the majority of these firms. Through an extensive search of the company and public literature, we have been able to fill our databases with current, correct and valuable information that is readily accessible.

Just as the clients of our Information Program come from a variety of institutions, a wide variety of readers should benefit from this Guide. Those in the biotechnology industry can identify potential partners, clients, competitors and sources of revenue. Those in large corporations can use this Guide to understand possible partners, suppliers of the new technology, and their competition. Government users will be able to determine industry growth, personnel demands, potential trading partners, etc. Lastly, those outsiders with an interest in the biotechnology industry can use this Guide as a source of entry, contacts and understanding.

In addition to information on the small biotechnology companies, information is included on larger corporations having significant biotechnology programs. These include the major players in the pharmaceutical, chemical and agrichemical arenas, among others. Information on the biotechnology programs in these corporations is less widely available. Often, the programs are scattered across many corporate departments and sites. Each corporation was contacted in an effort to amass this important information for our databases and for this Guide.

Finally, a significant and growing force in U.S. biotechnology is a group of more than 40 state-sponsored biotechnology centers located across the country. We have completed a study of these centers, and a section on this new and rapidly growing sector of U.S. biotechnology is also included.

Data in this Guide are presented not only as lists, but also in the form of text, tables and figures. These analyses will give the reader an understanding of the industry as a whole as well as its segments. Readers from companies will be able to ascertain how data from their firms compare to the industry as a whole or to their industry segment, such as health care or animal agriculture. In some cases, where data were not readily available or were proprietary in nature, the available data enable a snapshot of important aspects of the industry, such as research and development budgets or expected revenues for the current fiscal year. In other cases, data analysis will give the reader an overview of important issues such as who starts the new biotechnology companies or the personnel breakdown in the industry or its segments.

Since the information in this Guide is derived from computer databases that are constantly changing, this Guide will provide the reader with up-to-the-minute information. Our program constantly monitors a variety of information sources, and the databases are continuously updated. This Guide represents an excellent current source of data on an ever-changing industry.

The Biotechnology Information Program

The Biotechnology Information Program of the North Carolina Biotechnology Center was created in 1986 to compile and make available data on commercial biotechnology -- in North Carolina, in the United States, and worldwide. Through a multi-faceted approach of utilizing a commercial biotechnology-oriented library, paper datafiles, in-house computer databases (such as those used to generate the data used in this Guide) and commercially available databases, the Program is able to provide information to a variety of users. Reports from the Program's databases are made available to individuals, companies, organizations, and government agencies on a contractual, cost-recovery basis.

The North Carolina Biotechnology Center

The North Carolina Biotechnology Center was established in 1981 to catalyze economic development in North Carolina through support of biotechnology. Largely funded by the State of North Carolina, the Center is a private, non-profit corporation. In addition to the Information Program, other Center programs include grants to universities and companies, public education, special focus groups, and economic development.

Acknowledgments

The North Carolina Biotechnology Center provided the facilities and resources that made collection of most of the data in this Guide possible. I wish to thank Dr. Charles E. Hamner, Jr., President of the Center, for his generous support, and Dr. Roy E. Morse, Past Interim President of the Center, for his encouragement.

The Biotechnology Information Program staff was responsible for assisting in the collection of most of the data used in this Guide. Excellent assistance in data collection processes was provided by Nancy G. Bruce, Senior Information Specialist, and Janet Forbes, Database Assistant. Also assisting in the Program were Sharon Arnette, Pamela Dent, T.J. Foehl and Carol C. Lavrich. Janet Forbes is to be commended on keeping the databases up to date and generating reports of data, especially during the final preparation stages for this Guide.

The **Actions** and **Companies** databases were created in 1985 by the author and were maintained through 1986 in collaboration with Professor William Hamilton, Director of the Management and Technology Program of the Wharton School of the University of Pennsylvania. Collaborative studies with Professor Hamilton have contributed to text and figures in Section 9.

Some of the data on individual companies were collected in a questionnaire-based study, partially funded by a contract (No. H3-2025.0) from the U.S. Congressional Office of Technology Assessment. Data from this study were used in conjunction with newer data to generate parts of Sections 7 through 10. The study, undertaken between December 1986 and August 1987, was coordinated with Drs. Gretchen Kolsrud and Kathi E. Hanna of the OTA.

Editing and proofreading of the Guide were provided by Barry Teater with additional proofreading provided by Robin Teater and Pamela Dent.

Book design and printing were directed by Harriet Damon Shields and Associates (New York, NY). Coordination of publication was provided by Dr. Ingrid Krohn, Editor, Stockton Press.

SECTION 1   INTRODUCTION AND THE USE OF THIS GUIDE

1

INTRODUCTION AND METHODS

In parallel with the explosion of commercial biotechnology in the United States is the need for information on the companies involved. This information is important for strategic purposes: to understand competitors, clients and markets. Other users of this information include governments, academic researchers and financial analysts. The databases we have built at the North Carolina Biotechnology Center have been used for all these purposes. This Guide represents the most requested, and presumably useful, of the data, as well as analyses of trends and features.

What Is Included

There have been numerous definitions of biotechnology and what constitutes the biotechnology industry in the United States. For our database of companies, we have included those companies that are working with the new biotechnologies of genetic engineering (recombinant DNA technology), hybridoma technology (monoclonal antibodies), and cell culture of purposefully produced cells. In addition to the 360 small companies identified that fit the category, we have identified some 60 large corporations that have significant R&D efforts in these areas. This Guide focuses primarily on the small companies or biotechnology firms, and to a lesser degree on the large corporations. What are not included are companies that work in the areas ancillary to biotechnology -- areas such as equipment, media or labware. In some cases, a company that works primarily in an ancillary area has secondary efforts in the primary technologies and has been included. A company's participation in biotechnology was confirmed by direct contact with that company. A simple listing of the companies that met these criteria to be included in the database and this Guide can be found in Listing 2-1.

The data in this Guide were amassed between 1984 and the present from a variety of sources. They include contact with the companies themselves, journal and other articles, company directories, company and industry literature, and other reference and source books. Appendix D lists valuable literature resources.

The Databases

Data have been added to two databases that, in turn, are the sources of the information in this Guide. The first, **Companies**, contains information on about 360 biotechnology firms and 60 corporations in the United States. Table 1-1 describes fields of data in **Companies**. The second database, **Actions**, describes internal actions taken by companies worldwide related to biotechnology -- such as facilities, products, patents and personnel changes. The fields contained in **Actions** are listed in Table 1-2. In addition, the majority of these records describe actions taken externally to the companies -- joint ventures, licensing or marketing agreements, research contracts, equity purchases and acquisitions. The **Actions** database provides an overview not only of U.S. companies but also foreign companies.

Data have been coded for sorting by date, product involved, company and a variety of other variables. All companies and actions have been coded for industry classification, breaking down biotechnology into 31 subcategories from animal agriculture to analytical testing. Companies are classified by their primary area of work and up to four secondary areas. This classification system allows us to examine the database according to specific interests, such as to find biotechnology firms that are public corporations working with specialty chemicals, or to find joint actions between Japanese and U.S. companies in 1986 involving therapeutic products.

Questionnaires and Surveys

Questionnaires were sent to more than 300 companies in the first half of 1987 with almost all returned questionnaires being answered by company top management. These questionnaires were prepared and distributed in part under contract with the U.S. Congressional Office of Technology Assessment (OTA). In addition, more than 200 biotechnology firms and corporations were surveyed by telephone in early 1988. Other company data were obtained and updated from company and public literature, meeting with company executives, and additional telephone calls.

Hardware and Software

Data are stored in Apple Macintosh computers using the database program Omnis 3 Plus (Blythe Software). In addition to the **Companies** and **Actions** databases, another database, **States**, was created to track state-funded biotechnology centers in the United States. Data are stored on 20-megabyte hard disks. From the databases, data are generated as reports and transferred to three other software packages for preparation of this Guide: Microsoft Word 3.01 (Microsoft, Inc.) for word processing and final preparation of text and most of the listings, Microsoft Excel 1.04 (Microsoft, Inc.) for spreadsheet analysis of numbers and generation of most of the tables, and Cricket Graph 1.2 (Cricket Software) for generation of figures. All text was printed on an Apple LaserWriter II NT printer.

Limitations and Restrictions

Because of the timeliness of the data in this Guide, every effort was made to prepare the text both rapidly and accurately. Since massive data were generated from the databases for the preparation of this Guide, the listings are mostly limited to the format of data in the databases, including certain abbreviations and upper case fields. Every effort was made to assure the accuracy of the data, but since much of the data in **Companies** and almost all of the data in **Actions** come from the public literature, full accuracy in all cases cannot be assured. Also, the information is changing rapidly. For example, we have discovered that, over the last two years, the majority of biotechnology firms have changed addresses, phone numbers, names, or top management. In addition, not all data were made available by all companies and data from corporations on their often-fragmented biotechnology efforts were not readily available in many cases. As errors are identified, they will be corrected in future editions of this Guide.

Using This Guide

There are four elements of information in this Guide.

- A. introductory **text**
- B. **listings** of data
- C. **tables** of analyzed and compiled data
- D. **figures** for graphical representations

Section 1 Introduction and Use of This Guide

    This Guide is divided into 14 sections. A comprehensive directory of bio-technology firms in the United States (Listing 2-2) follows this introduction. This section is followed by sections with specific company information sorted according to subject. Thus, for example, if the reader is interested in companies in North Carolina, the listing of companies by state (Listing 3-1) would provide the names of companies to be looked up in the directory. Similarly, companies with primary or secondary interests in a specific area, such as plant agriculture (Listing 5-2) or waste disposal/treatment (Listing 5-18), can be determined by the classification listings in Section 5, and more information can be gathered on the specific companies of interest in the directory section (Listing 2-2). Section 4 covers the founding of biotechnology companies and an analysis of the people who founded the small firms. Personnel or financing information on these companies can be found in Sections 6 and 7. The subsequent three sections describe company revenues, R & D budgets, and patents and products, respectively. Section 11 describes the large corporations with significant biotechnology programs. In Section 12, partnerships of companies related to biotechnology are presented. The most recent growth area in U.S. biotechnology, state biotechnology centers, is described and analyzed in Section 13. Finally, Section 14 gives an overview of the U.S. biotechnology industry as a whole. Tables of classification codes and abbreviations follow this introduction (Tables 1-3 and 1-4; Appendix A). Companies deleted from our list of biotechnology firms for a variety of reasons are listed in Appendix B, while Appendix C describes companies that are new to our databases and who will appear in the next edition of this Guide.

    In many of the sections, data are broken down by the 31 individual industry classifications, from animal agriculture to analytical services. The Table of Contents and Index sections can be readily used to access information on individual subjects. In addition, there are many tables and figures that represent an analysis of the data. Readers should direct their attention to the Table of Contents to become familiar with the layout of this Guide and the types of information that are available.

Table 1-1 The Companies Database

TABLE 1-1 THE COMPANIES DATABASE (major fields included)

Company name

Address

Phone number

Year started

Top management (CEO, President, R&D Director)

Financing (PRIvate, PUBlic, SUBsidiary, and some details)

Primary and secondary classification codes for industry interest
(see Table 1-4)

Size of company (Small, 1-20 employees; Medium, 21-100; Large, 101 plus;
Corporation, generally 1,500 plus)

Industry of company (e.g., Biotech for the biotech firms, or Agriculture,
Pharmacology, Energy, Chemical, etc. for the larger corporations)

Personnel data -- Total employees and a breakdown of personnel, where
available

Products -- On market and in development

Sales and budget data, where available

Patent data, where available

Investor data, where available

Note: As of early 1988 there were about 360 U.S. biotechnology companies and 60
large corporations tracked using the **Companies** database.

TABLE 1-2 THE ACTIONS DATABASE (major fields included)

Company or companies involved -- up to three

Type of company (biotech firm, corporation, academic institution, government
   agency, or private institution)

Country of origin for each company

Date of action -- month, year

Type of internal action -- e.g., product, patent, facility, personnel

Type of external action -- e.g., research contract, equity purchase, licensing
   agreement

Technology involved -- e.g., rDNA, cell culture, hybridoma

Industry class of action (see code sheet, Table 1-4)

Product involved

Description of action -- one to three lines

Reference for action

Note: In early 1988 there were about 2,300 records in the **Actions** database, dating
from 1981 to the present. Almost 80 percent of the actions occured between two or
more companies.

Table 1-3   Commonly Used Abbreviations

TABLE 1-3 COMMONLY USED ABBREVIATIONS

Products

The following abbreviations are those most commonly used in the listings of company products on the pages that follow:

Abbrev.	Meaning
A B	Antibody
AG	Agriculture
ANF	Atrial natriuretic factor
BGH	Bovine growth hormone (bovine somatotropin)
BSA	Bovine serum albumin
CSF	Colony stimulating factor
CV	Cardiovascular
DX	Diagnostic(s)
EGF	Epidermal growth factor
EIA	Enzyme immunoassay
ELISA	Enzyme-linked immunosorbent assay
EPO	Erythropoietin
FGF	Fibroblast growth factor
GM-CSF	Granulocyte macrophage colony stimulating factor
HCG	Human chorionic gonadotropin
HEP B	Hepatitis B
HGH	Human growth hormone
HSA	Human serum albumin
HSV	Herpes simplex virus
IF	Interferon (IF-A=interferon-alpha, etc.)
IL	Interleukin (IL-2=interleukin-2, etc.)
KPA	Kidney plasminogen activator
MAB	Monoclonal antibody
PTA	Parathyroid hormone
RE	Restriction endonuclease; restriction enzyme
RIA	Radioimmunoassay
RX	Therapeutic(s), Drug(s)
SOD	Superoxide dismutase
TNF	Tumor necrosis factor
TPA	Tissue plasminogen activator
V X	Vaccine(s)

Financing Type

The following abbreviations are used to indicate the type of company financing:

Abbrev.	Meaning
PRI	Privately held
PUB	Public stock
SUB	Subsidiary

TABLE 1-4 CLASSIFICATION CODES FOR COMPANIES

Code	Industry Classification
A	Agriculture, Animal
B	Agriculture, Plant
C	Biomass Conversion
D	Biosensors/Bioelectronics
E	Bioseparations
F	Biotechnology Equipment
G	Biotechnology Reagents
H	Cell Culture, General
I	Chemicals, Commodity
J	Chemicals, Specialty (includes proteins and enzymes)
K	Diagnostics, Clinical Human
L	Energy
M	Food Production/Processing
N	Mining
O	Production/Fermentation
P	Therapeutics
Q	Vaccines
R	Waste Disposal/Treatment
S	Aquaculture
T	Marine Natural Products (includes algae)
U	Consulting
V	Veterinary (all animal health care)
W	Research
X	Immunological Products (non-pharmaceutical)
Y	Toxicology
Z	Venture Capital/Financing
1	Biomaterials
2	Fungi
3	Drug Delivery Systems
4	Medical Devices
5	Testing/Analytical Services

SECTION 2 DIRECTORY OF COMPANIES

Section 2 Directory of Companies

INTRODUCTION

A quick-scan listing of all of the 360 companies that were deemed to be U.S. biotechnology companies for our databases and this Guide appears in Listing 2-1 that follows. These are small companies, founded on average in 1981 and having from 2 to 1,500 employees. The quick list is a four-page overview of all companies that have been included for data display and analysis in Sections 2 through 10.

More detailed information on the U.S. biotechnology firms follows in Listing 2-2. Location, top management, telephone number, products, and year of founding are given. In addition, the financing of companies is given as PRI, PUB, or SUB, indicating private companies, public companies or subsidiary companies, respectively. In addition, the classification code for the primary industry focus of the company is given. The code sheet is presented in Table 1-4 that precedes this section. State location of the companies can be found in Listing 3-1. A breakdown of the private, public and subsidiary companies appears in Listings 3-2 through 3-4, respectively. A directory of companies listed by focus and product is presented in Section 5, Listings 5-1 to 5-29.

Listing 2-1 Quick List: Biotechnology Companies

LISTING 2-1 QUICK LIST: BIOTECHNOLOGY COMPANIES

A.M. BIOTECHNIQUES, INC.
A/G TECHNOLOGY CORP.
ABC RESEARCH CORP.
ABN
ADVANCED BIOTECHNOLOGIES
ADVANCED GENETIC SCIENCES
ADVANCED MAGNETICS, INC.
ADVANCED MINERAL TECHNOLOGIES
AGDIA, INC.
AGRACETUS
AGRI-DIAGNOSTICS ASSOCIATES
AGRIGENETICS CORP.
AGRITECH SYSTEMS, INC.
ALFACELL CORP.
ALLELIC BIOSYSTEMS
ALPHA I BIOMEDICALS, INC.
AMBICO, INC.
AMERICAN BIOCLINICAL
AMERICAN BIOGENETICS CORP.
AMERICAN BIOTECHNOLOGY CO.
AMERICAN DIAGNOSTICA, INC.
AMERICAN LABORATORIES, INC.
AMERICAN QUALEX INTERNATIONAL, INC.
AMGEN
AMTRON
AN-CON GENETICS
ANGENICS
ANTIBODIES, INC.
ANTIVIRALS, INC.
APPLIED BIOSYSTEMS, INC.
APPLIED BIOTECHNOLOGY, INC.
APPLIED DNA SYSTEMS, INC.
APPLIED GENETICS LABS, INC.
APPLIED MICROBIOLOGY, INC.
APPLIED PROTEIN TECHNOLOGIES, INC.
AQUASYNERGY, LTD.
ASTRE CORPORATE GROUP
ATLANTIC ANTIBODIES
AUTOMEDIX SCIENCES, INC.
BACHEM BIOSCIENCE, INC.
BACHEM, INC.
BEHRING DIAGNOSTICS
BEND RESEARCH, INC.
BENTECH LABORATORIES
BERKELEY ANTIBODY, INC.
BETHYL LABS, INC.
BINAX, INC.
BIO HUMA NETICS
BIO TECHNIQUES LABS, INC.
BIO-RECOVERY, INC.
BIO-RESPONSE, INC.
BIO-TECHNICAL RESOURCES, INC.
BIOCHEM TECHNOLOGY, INC.

BIOCONSEP, INC.
BIOCONTROL SYSTEMS
BIODESIGN, INC.
BIOGEN
BIOGENEX LABORATORIES
BIOGROWTH, INC.
BIOKYOWA, INC.
BIOMATERIALS INTERNATIONAL, INC.
BIOMATRIX, INC.
BIOMED RESEARCH LABS, INC.
BIOMEDICAL TECHNOLOGIES
BIOMERICA, INC.
BIONETICS RESEARCH
BIONIQUE LABS, INC.
BIOPOLYMERS, INC.
BIOPROBE INTERNATIONAL, INC.
BIOPRODUCTS FOR SCIENCE, INC.
BIOPURE
BIOSCIENCE MANAGEMENT, INC.
BIOSEARCH, INC.
BIOSPHERICS, INC.
BIOSTAR MEDICAL PRODUCTS, INC.
BIOSYSTEMS, INC.
BIOTECH RESEARCH LABS, INC.
BIOTECHNICA AGRICULTURE
BIOTECHNICA INTERNATIONAL
BIOTECHNOLOGY DEVELOPMENT CORP.
BIOTECHNOLOGY GENERAL CORP.
BIOTEST DIAGNOSTICS CORP.
BIOTHERAPEUTICS, INC.
BIOTHERAPY SYSTEMS, INC.
BIOTHERM
BIOTICS RESEARCH CORP.
BIOTROL, INC.
BIOTX
BMI, INC.
BOEHRINGER MANNHEIM DIAGNOSTICS
BRAIN RESEARCH, INC.
CALGENE, INC.
CALIFORNIA BIOTECHNOLOGY, INC.
CALIFORNIA INTEGRATED DIAGNOSTICS
CALZYME LABORATORIES, INC.
CAMBRIDGE BIOSCIENCE CORP.
CAMBRIDGE RESEARCH BIOCHEMICALS
CARBOHYDRATES INTERNATIONAL, INC.
CEL-SCI CORP.
CELLMARK DIAGNOSTICS
CELLULAR PRODUCTS, INC.
CENTOCOR
CENTRAL BIOLOGICS, INC.
CETUS CORP.
CHARLES RIVER BIOTECH. SERVICES
CHEMGENES

LISTING 2-1 QUICK LIST (Cont.)

CHEMICAL DYNAMICS CORP.
CHEMICON INTERNATIONAL, INC.
CHIRON CORP.
CIBA-CORNING DIAGNOSTIC CORP.
CISTRON BIOTECHNOLOGY, INC.
CLINETICS CORP.
CLINICAL SCIENCES, INC.
CLONTECH LABORATORIES, INC.
COAL BIOTECH CORP.
CODON CORP.
COLLABORATIVE RESEARCH, INC.
COLLAGEN CORP.
CONSOLIDATED BIOTECHNOLOGY, INC.
COOPER DEVELOPMENT CO.
COORS BIOTECH PRODUCTS CO.
COVALENT TECHNOLOGY CORP.
CREATIVE BIOMOLECULES
CROP GENETICS INTERNATIONAL CORP.
CYANOTECH CORP.
CYGNUS RESEARCH CORP.
CYTOGEN CORP.
CYTOTECH, INC.
CYTOX CORP.
DAKO CORP.
DAMON BIOTECH, INC.
DEKALB-PFIZER GENETICS
DELTOWN CHEMURGIC CORP.
DIAGNON CORP.
DIAGNOSTIC PRODUCTS CORP.
DIAGNOSTIC TECHNOLOGY, INC.
DIAMEDIX, INC.
DIGENE DIAGNOSTICS, INC.
DNA PLANT TECHNOLOGY CORP.
DNAX RESEARCH INSTITUTE
E-Y LABORATORIES, INC.
EARL-CLAY LABORATORIES, INC.
ECOGEN, INC.
ELANEX PHARMACEUTICALS, INC.
ELCATECH
ELECTRO-NUCLEONICS, INC.
EMBREX, INC.
EMTECH RESEARCH
ENDOGEN, INC.
ENDOTRONICS, INC.
ENVIRONMENTAL DIAGNOSTICS, INC.
ENZO-BIOCHEM, INC.
ENZON
ENZON, INC.
ENZYME BIO-SYSTEMS LTD.
ENZYME CENTER, INC.
ENZYME TECHNOLOGY CORP.
EPITOPE, INC.
ESCAGENETICS CORP.

EXOVIR, INC.
FERMENTA ANIMAL HEALTH
FLOW LABORATORIES, INC.
FORGENE, INC.
GAMETRICS, LTD.
GAMMA BIOLOGICALS, INC.
GEN-PROBE, INC.
GENE-TRAK SYSTEMS
GENELABS, INC.
GENENCOR, INC.
GENENTECH
GENESIS LABS, INC.
GENETIC DIAGNOSTICS CORP.
GENETIC ENGINEERING, INC.
GENETIC SYSTEMS CORP.
GENETIC THERAPY, INC.
GENETICS INSTITUTE, INC.
GENEX, INC.
GENTRONIX LABORATORIES, INC.
GENZYME CORP.
GRANADA GENETICS CORP.
HANA BIOLOGICS, INC.
HAWAII BIOTECHNOLOGIES CO.
HAZLETON BIOTECHNOLOGIES CO.
HOUSTON BIOTECHNOLOGY, INC.
HUNTER BIOSCIENCES
HYBRITECH, INC.
HYCLONE LABORATORIES, INC.
HYGEIA SCIENCES
IBF BIOTECHNICS, INC.
ICL SCIENTIFIC, INC.
IDEC PHARMACEUTICAL
IDETEK, INC.
IGEN
IGENE BIOTECHNOLOGY, INC.
IMCERA BIOPRODUCTS, INC.
IMCLONE SYSTEMS, INC.
IMMUCELL CORP.
IMMUNETECH PHARMACEUTICALS
IMMUNEX CORP.
IMMUNEX, INC.
IMMUNO CONCEPTS, INC.
IMMUNO MODULATORS LABS, INC.
IMMUNOGEN, INC.
IMMUNOMED CORP.
IMMUNOMEDICS, INC.
IMMUNOSYSTEMS, INC.
IMMUNOTECH CORP.
IMMUNOVISION, INC.
IMRE CORP.
IMREG, INC.
INCELL CORP.
INCON CORP.

Listing 2-1 Quick List: Biotechnology Companies

LISTING 2-1 QUICK LIST (Cont.)

INFERGENE CO.
INGENE, INC.
INTEGRATED CHEMICAL SENSORS
INTEGRATED GENETICS, INC.
INTEK DIAGNOSTICS, INC.
INTELLIGENETICS, INC.
INTER-AMERICAN RESEARCH ASSOCIATES
INTER-CELL TECHNOLOGIES, INC.
INTERFERON SCIENCES, INC.
INTERNATIONAL BIOTECHNOLOGIES, INC.
INTERNATIONAL ENZYMES, INC.
INTERNATIONAL PLANT RESEARCH INST.
INVITRON CORP.
JACKSON IMMUNORESEARCH LABS, INC.
KALLESTAD DIAGNOSTICS
KARYON TECHNOLOGY, LTD.
KIRIN-AMGEN
KIRKEGAARD & PERRY LABS, INC.
LABSYSTEMS, INC.
LEE BIOMOLECULAR RESEARCH LABS
LIFE SCIENCES
LIFE TECHNOLOGIES, INC.
LIFECODES CORP.
LIFECORE BIOMEDICAL, INC.
LIPOSOME CO., INC.
LIPOSOME TECHNOLOGY, INC.
LUCKY BIOTECH CORP.
LYPHOMED, INC.
MAIZE GENETIC RESOURCES
MARCOR DEVELOPMENT CO.
MARICULTURA, INC.
MARINE BIOLOGICALS, INC.
MARTEK
MAST IMMUNOSYSTEMS, INC.
MEDAREX, INC.
MELOY LABORATORIES, INC.
MICROBE MASTERS
MICROBIO RESOURCES
MICROBIOLOGICAL ASSOCIATES, INC.
MICROGENESYS, INC.
MICROGENICS
MOLECULAR BIOLOGY RESOURCES, INC.
MOLECULAR BIOSYSTEMS, INC.
MOLECULAR DIAGNOSTICS, INC.
MOLECULAR GENETIC RESOURCES
MOLECULAR GENETICS, INC.
MONOCLONAL ANTIBODIES, INC.
MONOCLONAL PRODUCTION INT'L
MONOCLONETICS INTERNATIONAL, INC.
MULTIPLE PEPTIDE SYSTEMS, INC.
MUREX CORP.
MYCOGEN CORP.
MYCOSEARCH, INC.

NATIONAL GENO SCIENCES
NATIVE PLANTS, INC.
NEOGEN CORP.
NEORX CORP.
NEUREX CORP.
NEUSHUL MARICULTURE
NITRAGIN CO.
NORDEN LABS
NORTH COAST BIOTECHNOLOGY, INC.
NYGENE CORP.
O.C.S. LABORATORIES, INC.
OCEAN GENETICS
OMNI BIOCHEM, INC.
ONCOGENE SCIENCE, INC.
ONCOR, INC.
ORGANON TEKNIKA CORP.
PAMBEC LABORATORIES
PEL-FREEZ BIOLOGICALS, INC.
PENINSULA LABORATORIES, INC.
PETROFERM, INC.
PETROGEN, INC.
PHARMACIA LKB BIOTECHNOLOGY, INC.
PHARMAGENE
PHYTOGEN
PLANT GENETICS, INC.
POLYCELL, INC.
PRAXIS BIOLOGICS
PROBIOLOGICS INTERNATIONAL, INC.
PROGENX
PROMEGA CORP.
PROTATEK INTERNATIONAL, INC.
PROTEINS INTERNATIONAL, INC.
PROVESTA CORP.
QUEST BIOTECHNOLOGY, INC.
QUEUE SYSTEMS, INC.
QUIDEL
R & A PLANT/SOIL, INC.
RECOMTEX CORP.
REPAP TECHNOLOGIES
REPLIGEN CORP.
RESEARCH & DIAGNOSTIC ANTIBODIES
RHOMED, INC.
RIBI IMMUNOCHEM RESEARCH, INC.
RICERCA, INC.
SCOTT LABORATORIES
SCRIPPS LABORATORIES
SEAPHARM, INC.
SENETEK PLC
SEPRACOR, INC.
SERAGEN, INC.
SERONO LABS
SIBIA
SOUTHERN BIOTECHNOLOGY ASSOC.

13

LISTING 2-1 QUICK LIST (Cont.)

SPHINX BIOTECHNOLOGIES
STRATAGENE, INC.
SUMMA MEDICAL CORP.
SUNGENE TECHNOLOGIES CORP.
SYNBIOTICS CORP.
SYNERGEN, INC.
SYNGENE PRODUCTS
SYNTHETIC GENETICS, INC.
SYNTRO CORP.
SYVA CO.
T CELL SCIENCES, INC.
TECHNICLONE INTERNATIONAL, INC.
TECHNOGENETICS, INC.
TEKTAGEN
THREE-M (3M) DIAGNOSTIC SYSTEMS
TOXICON
TRANSFORMATION RESEARCH, INC.
TRANSGENIC SCIENCES, INC.
TRITON BIOSCIENCES, INC.
UNIGENE LABORATORIES, INC.
UNITED AGRISEEDS, INC.

UNITED BIOMEDICAL, INC.
UNITED STATES BIOCHEMICAL CORP.
UNIVERSITY GENETICS CO.
UNIVERSITY MICRO REFERENCE LAB
UPSTATE BIOTECHNOLOGY, INC.
VECTOR LABORATORIES, INC.
VEGA BIOTECHNOLOGIES, INC.
VERAX CORP.
VIAGENE, INC.
VIRAGEN
VIROSTAT, INC.
VIVIGEN, INC.
WASHINGTON BIOLAB
WELGEN MANUFACTURING, INC.
WESTBRIDGE RESEARCH GROUP
XENOGEN
XOMA CORP.
ZOECON CORP.
ZYMED LABORATORIES, INC.
ZYMOGENETICS, INC.

Listing 2-2 Directory of Companies

LISTING 2-2 DIRECTORY OF COMPANIES

A

A.M. BIOTECHNIQUES, INC.
P.O. BOX 873
BELTON, TX 76513
Telephone: 817-939-7778
Products: VET VX FOR PARVOVIRUS,
CORONAVIRUS

CEO: ANDRES MENCHU
President: ANDRES MENCHU
R&D Dir: DR. MITCHEL JAGER
Financing: PRI
Started: 1980
Class Code: V

A/G TECHNOLOGY CORP.
34 WEXFORD ST.
NEEDHAM, MA 02194
Telephone: 617-449-5774
Products: GAS SEPARATION; ULTRA & MICRO
FILTRATION

CEO: DR. ARYE GOLLAN
President: DR. ARYE GOLLAN
R&D Dir: MILES KLEPER
Financing: PRI
Started: 1981
Class Code: E

ABC RESEARCH CORP.
3437 SW 24TH AVE.
GAINESVILLE, FL 32607
Telephone: 904-373-0436
Products: CONTRACT RESEARCH; CUSTOM STARTER
CULTURES; BACTERIAL PROTEIN; ANTIOXIDANTS

CEO: DR. WILLIAM L. BROWN
President: DR. WILLIAM L. BROWN
R&D Dir: DR. WALTER HARGRAVES
Financing: PRI
Started: 1967
Class Code: W

ABN
21377 CABOT BLVD.
HAYWARD, CA 94545
Telephone: 415-732-9000
Products: DX, REAGENTS, BT INSTRUMENTS,
DNA PROBES, BLOTTING SYSTEMS

CEO: MARTIN E. MARKS
President: FRANK RUDERMAN
R&D Dir: DR. DARRYL RAY
Financing: PUB
Started: 1981
Class Code: K

ADVANCED BIOTECHNOLOGIES
12150 TECH RD.
SILVER SPRING, MD 20904
Telephone: 301-622-5212
Products: MEDIA, SERA, GROWTH FACTORS, DNA,
RESEARCH SERVICES

President: DR. JAMES WHITMAN, JR.
R&D Dir: DR. JAMES WHITMAN, JR.
Financing: PRI
Started: 1982
Class Code: G

ADVANCED GENETIC SCIENCES
6701 SAN PABLO AVE.
OAKLAND, CA 94608
Telephone: 415-547-2395
Products: PLANT & AG RESEARCH, CHEMICALS,
SEEDS, FROSTBAN, SNOMAX

President: JOSEPH BOUCKAERT
R&D Dir: DR. JOHN R. BEDBROOK
Financing: PUB
Started: 1979
Class Code: B

ADVANCED MAGNETICS, INC.
61 MOONEY ST.
CAMBRIDGE, MA 02138
Telephone: 617-497-2070
Products: DX RIA KITS, MAGNETIC SEPARATION
REAGENTS

President: JEROME GOLDSTEIN
R&D Dir: DR. ERNEST GROMAN
Financing: PUB
Started: 1981
Class Code: K

ADVANCED MINERAL TECHNOLOGIES, INC.
5920 MCINTYRE ST.
GOLDEN, CO 80403
Telephone: 303-279-6982
Products: WASTE TREATMENT & METAL
RECOVERY

CEO: DAVID J. SCHOONMAKER
President: DAVID J. SCHOONMAKER
R&D Dir: DR. JAMES A. BRIERLEY
Financing: PRI
Started: 1982
Class Code: R

AGDIA, INC.
1901 N. CEDAR ST.
MISHAWAKA, IN 46545
Telephone: 219-255-2817
Products: PLANT AND FOOD TOX DX FOR VIRUSES

CEO: DR. CHESTER L. SUTULA
President: DR. CHESTER L. SUTULA
Financing: PRI
Started: 1981
Class Code: B

AGRACETUS
8520 UNIVERSITY AVE.
MIDDLETON, WI 53562
Telephone: 608-836-7300
Products: TERMITICIDE, VET RX, ANIMAL VX,
MICROBIAL INOCULATIONS, PESTICIDES

President: ROBERT A. FILDES
R&D Dir: DR. WINSTON J. BRILL
Financing: PRI
Started: 1981
Class Code: B

AGRI-DIAGNOSTICS ASSOCIATES
2611 BRANCH PIKE
CINNAMINSON, NJ 08077
Telephone: 609-829-0110
Products: TURF & CROP DISEASE DX KITS, FOOD
PATHOGEN DX

President: STEVE BANEGAS
R&D Dir: DR. DAVID GROTHAUS
Financing: PRI
Started: 1983
Class Code: B

AGRIGENETICS CORP.
35575 CURTIS BLVD., STE. 300
EASTLAKE, OH 44094
Telephone: 216-942-2210
Products: SEED AND PLANT GENETICS,
CORN, CROP PLANTS

CEO: JOHN A. STUDEBAKER
President: JOHN A. STUDEBAKER
R&D Dir: JOSEPH F. SCHWER
Financing: SUB
Started: 1975
Class Code: B

AGRITECH SYSTEMS, INC.
100 FORE ST.
PORTLAND, ME 04101
Telephone: 207-774-4334
Products: IMMUNOASSAYS, DX; VET DX; TESTS
FOR FEED/FOOD CONTAMINANTS

President: DAVID SHAW
R&D Dir: IRWIN WORKMAN
Financing: PRI
Started: 1984
Class Code: V

ALFACELL CORP.
225 BELLEVILLE AVE.
BLOOMFIELD, NJ 07003
Telephone: 201-748-8082
Products: TUMOR TOXINS, REAGENTS,
SERA

CEO: DR. KUSLIMA SHOGEN
President: DR. KUSLIMA SHOGEN
R&D Dir: DR. STANLEY MIKULSKI
Financing: PUB
Started: 1981
Class Code: G

ALLELIC BIOSYSTEMS
RT.1, BOX 230
KEARNEYSVILLE, WV 25430
Telephone: 304-725-5255
Products: VET TOX TESTS, PROBES, OLIGO-
NUCLEOTIDES, REAGENTS

President: STEVAN R. PHELPS
R&D Dir: DR. WILLIAM SCHILL
Financing: PRI
Started: 1984
Class Code: V

16

Listing 2-2 Directory of Companies

ALPHA I BIOMEDICALS, INC.
777 14TH ST., NW, SUITE 410
WASHINGTON, DC 20005
Telephone: 202-628-9898
Products: PEPTIDES & VX FOR DX/RX; AIDS RX, VX;
DEVELOPING THYMOSIN

CEO: J.J. FINKELSTEIN
President: J.J. FINKELSTEIN
R&D Dir: A.L. GOLDSTEIN
Financing: PUB
Started: 1982
Class Code: P

AMBICO, INC.
P.O. BOX 522, ROUTE 2
DALLAS CENTER, IA 50063
Telephone: 515-992-3747
Products: VET VX

President: DR. C. JOSEPH WELTER
R&D Dir: GERALD R. FITZGERALD
Financing: PRI
Started: 1974
Class Code: V

AMERICAN BIOCLINICAL
4432 S.E. 16TH AVE.
PORTLAND, OR 97202
Telephone: 800-547-3686
Products: RX; REAGENTS; DX KITS FOR
HORMONES

CEO: RAY R. ROGERS
President: RAY R. ROGERS
R&D Dir: DR. ROBERT BUCK
Financing: PRI
Started: 1977
Class Code: G

AMERICAN BIOGENETICS CORP.
19732 MACARTHUR BLVD.
IRVINE, CA 92715
Telephone: 714-851-7733
Products: PROCESS DESIGN MICROORGANISMS;
CHEMICALS

R&D Dir: DR. WESLEY HATFIELD
Financing: PRI
Started: 1984
Class Code: I

AMERICAN BIOTECHNOLOGY CO.
7658 STANDISH PL., SUITE 107
ROCKVILLE, MD 20855
Telephone: 301-294-9553
Products: GROWTH FACTORS, IMMUNO-
MODULATORS

CEO: DR. ROBERT W. VELTRI
President: DR. ROBERT W. VELTRI
R&D Dir: DR. MAXIM
Financing: PRI
Started: 1984
Class Code: X

AMERICAN DIAGNOSTICA, INC.
49 E. 68TH ST.
NEW YORK, NY 10021
Telephone: 212-249-2222
Products: DX; TPA ELISA, TPA ACTIVE
ASSAY KITS

CEO: DR. RICHARD HART
President: DR. RICHARD HART
R&D Dir: DR. RICHARD HART
Financing: PRI
Started: 1983
Class Code: K

AMERICAN LABORATORIES, INC.
4410 SOUTH 102 ST.
OMAHA, NE 63127
Telephone: 402-339-2494
Products: RAW MATERIALS OF BIOTECHNOLOGY
PRODUCTS

President: JEFFREY JACKSON
R&D Dir: JACK JACKSON
Financing: PRI
Started: 1962
Class Code: P

AMERICAN QUALEX INTERNATIONAL, INC.
14620 E. FIRESTONE BLVD.
LA MIRADA, CA 90638
Telephone: 714-521-3753
Products: CUSTOM MABS; ENZYMES;
DNA PROBES; INSTRUMENTS

President: DANIEL MOOTHART
R&D Dir: A. MOOTHART
Financing: PRI
Started: 1981
Class Code: G

AMGEN

1900 OAK TERRACE LANE
THOUSAND OAKS, CA 91320
Telephone: 805-499-5725
Products: RX; RESEARCH REAGENTS, EGF, IGF, IF,
EPO, IL-2, CSF, BGH, HGH

CEO: DR. GEORGE B. RATHMANN
President: DR. GEORGE B. RATHMANN
R&D Dir: DR. DANIEL VAPNEK
Financing: PUB
Started: 1980
Class Code: P

AMTRON

701 EAST BAY ST.
CHARLESTON, SC 29403
Telephone: 803-577-2931
Products: ANIMAL DISEASE PREVENTATIVES,
IMMUNOLOGICAL PRODUCTS

President: ROY SMITH
R&D Dir: GREGORY WILSON
Financing: PRI
Started: 1981
Class Code: V

AN-CON GENETICS

1 HUNTINGTON QUADRANGLE
MELVILLE, NY 11747
Telephone: 516-694-8470
Products: EQUIPMENT; INSTRUMENTS

President: ABRAHAM CONCOOL
R&D Dir: ROBERT LEDLEY
Financing: SUB
Started: 1982
Class Code: F

ANGENICS

100 INMAN ST.
CAMBRIDGE, MA 02139
Telephone: 617-876-6468
Products: SCREENING TESTS FOR DRUG
ABUSE AND MILK

CEO: ANDRE DE BRUIN
President: ANDRE DE BRUIN
R&D Dir: DR. BRUCE P. NERI
Financing: PRI
Started: 1980
Class Code: Y

ANTIBODIES, INC.

P.O. BOX 1560
DAVIS, CA 95617
Telephone: 800-824-8540
Products: ANTISERA, MABS, VET DX, HUMAN DX

President: DR. JAMES HILLMAN
R&D Dir: PATRICIA ADCOCK
Financing: PRI
Started: 1961
Class Code: X

ANTIVIRALS, INC.

249 S.W. AVERY
CORVALLIS, OR 97333
Telephone: 503-753-3635
Products: NEU-GENES (TM) FOR DX & RX USE

President: DR. JAMES SUMMERTON
R&D Dir: DR. EUGENE STIRCHAK
Financing: PRI
Started: 1980
Class Code: K

APPLIED BIOSYSTEMS, INC.

850 LINCOLN CENTER DR.
FOSTER CITY, CA 94404
Telephone: 415-570-6667
Products: DNA SYNTHESIZER; PEPTIDE SEQUENCER;
REAGENTS

CEO: ANDRE MARION
President: ANDRE MARION
R&D Dir: MICHAEL HUNKAPILLER
Financing: PUB
Started: 1981
Class Code: F

APPLIED BIOTECHNOLOGY, INC.

80 ROGERS ST.
CAMBRIDGE, MA 02142
Telephone: 617-492-7289
Products: VET VX; AIDS, TB VX; CANCER DX, RX

President: STEVEN PELTZMAN
R&D Dir: BRYAN ROBERTS
Financing: PRI
Started: 1982
Class Code: V

Listing 2-2 Directory of Companies

APPLIED DNA SYSTEMS, INC.
1450 BROADWAY
NEW YORK, NY 10018
Telephone: 212-302-7000
Products: HUMAN COLLAGENASE; MABS;
OXYRASE; IN VITRO CHEMOSENSITIVITY ASSAY

CEO: DONALD M. BACHMANN
R&D Dir: DR. KENNETH BLACKMAN
Financing: PUB
Started: 1982
Class Code: P

APPLIED GENETICS LABS, INC.
1335 GALEWAY DR., SUITE 2001
MELBOURNE, FL 32901
Telephone: 305-768-2048
Products: GENETIC TOXICITY TEST,
DNA PROBES

CEO: JOHN C. HOZIER
President: JOHN C. HOZIER
R&D Dir: DR. MARIA H. LUGO
Financing: PRI
Started: 1984
Class Code: Y

APPLIED MICROBIOLOGY, INC.
BROOKLYN NAVY YARD #5
BROOKLYN, NY 11205
Telephone: 718-852-4676
Products: RX FOR BOVINE MASTITIS, TOXIC SHOCK
SYNDROME DX, B. SUBTILIS PROD'N., STAPHYLOCIDE

President: DR. DAVID GUTTMAN
R&D Dir: DR. PETER BLACKBURN
Financing: PUB
Started: 1983
Class Code: V

APPLIED PROTEIN TECHNOLOGIES, INC.
103 BROOKLINE ST.
CAMBRIDGE, MA 02139
Telephone: 617-868-6085
Products: PEPTIDE & PROTEIN SYNTHESIZERS,
CHEMICALS, REAGENTS

CEO: MARCUS J. HORN
President: MARCUS J. HORN
R&D Dir: MARCUS J. HORN
Financing: PRI
Started: 1984
Class Code: F

AQUASYNERGY, LTD.
RT. 9, BOX 81
KINSTON, NC 28501
Telephone: 919-527-6730
Products: CO2 GENERATOR FOR AQUACULTURE PLANT
PRODUCTION; AQUATIC PLANT/ANIMAL CROPS

President: DAVID A. NUTTLE
R&D Dir: DAVID A. NUTTLE
Financing: PRI
Started: 1986
Class Code: T

ASTRE CORPORATE GROUP
1610-C QUAIL RUN
CHARLOTTESVILLE, VA 22905
Telephone: 804-973-8511
Products: AMMONIA DEGRADER; BIOPESTICIDES;
LIPASE-PRODUCING MICROBE

CEO: DR. ROY A. ACKERMAN
R&D Dir: J. TOBEY
Financing: PRI
Started: 1967
Class Code: R

ATLANTIC ANTIBODIES
10 NONESUCH RD.
SCARBOROUGH, ME 04074
Telephone: 800-343-3430
Products: DX TEST KITS; DX; MABS;
REAGENTS

President: ORWIN CARTER
R&D Dir: DENNY BARRANTES
Financing: SUB
Started: 1972
Class Code: K

AUTOMEDIX SCIENCES, INC.
19401 S. VERMONT AVE., UNIT J100
TORRANCE, CA 90502
Telephone: 213-327-1112
Products: RX PRODUCTS IN DEVELOPMENT

President: ROBERT MURTFELDT
Financing: PUB
Started: 1978
Class Code: P

B

BACHEM BIOSCIENCE, INC.
3700 MARKET ST.
PHILADELPHIA, PA 19104
Telephone: 215-387-0011
Products: AMINO ACIDS & DERIVATIVES; BIOACTIVE
PEPTIDES; ENZYME SUBSTRATES; VX EPITOPES

CEO: PETER GROGG
President: PETER GROGG
R&D Dir: DR. RONALD PEPIN
Financing: PRI
Started: 1987
Class Code: J

BACHEM, INC.
3132 KASHIWA ST.
TORRANCE, CA 90505
Telephone: 213-539-4171
Products: SYNTHETIC BIOACTIVE PEPTIDES;
GROWTH FACTORS; CRF

President: DR. RAO MAKINENI
R&D Dir: DR. NAGANA A. GOUD
Financing: PRI
Started: 1971
Class Code: J

BEHRING DIAGNOSTICS
10933 N. TORREY PINES RD.
LA JOLLA, CA 92037
Telephone: 800-854-9256
Products: ANTIBODIES; DX;
REAGENTS (CALBIOCHEM)

President: DR. WERNER SHAEFER
R&D Dir: F. BEHRENS
Financing: SUB
Started: 1952
Class Code: K

BEND RESEARCH, INC.
64550 RESEARCH ROAD
BEND, OR 97701
Telephone: 503-382-4100
Products: SEPARATION; BIOSENSORS;
IMMOBILIZED ENZYME

CEO: HAROLD LONSDALE
President: CHRIS BABCOCK
R&D Dir: DR. PAUL VAN EIKREN
Financing: PRI
Started: 1975
Class Code: E

BENTECH LABORATORIES
14350 S.E. INDUSTRIAL WAY
CLACKAMAS, OR 97015
Telephone: 503-652-2333
Products: CROP YIELD ENHANCING AGENT;
ANTIFUNGALS; WOUND HEALING

President: DR. CHRIS BENTLEY
R&D Dir: BOB LEWIS
Financing: PRI
Started: 1984
Class Code: B

BERKELEY ANTIBODY COMPANY, INC.
4131 LAKESIDE DR., SUITE B
RICHMOND, CA 94806
Telephone: 415-222-4940
Products: CUSTOM ANTIBODIES, MABS &
POLYCLONALS

President: DR. THOMAS R. ANDERSON
R&D Dir: DR. THOMAS R. ANDERSON
Financing: PRI
Started: 1983
Class Code: X

BETHYL LABS, INC.
P.O. BOX 850
MONTGOMERY, TX 77356
Telephone: 409-597-6111
Products: ANTISERA & CUSTOM ANTISERA SERVICE

President: DR. HENRY F. CARWILE
R&D Dir: JAMES BILLINGS
Financing: PRI
Started: 1977
Class Code: X

Listing 2-2 Directory of Companies

BINAX, INC.
95 DARLING AVE.
SOUTH PORTLAND, ME 04106
Telephone: 800-323-3199
Products: MABS, DX TEST KITS,
VET DX

CEO: ROGER PIASIO
President: ROGER PIASIO
R&D Dir: DR. BRUCE WATKINS
Financing: PRI
Started: 1986
Class Code: K

BIO HUMA NETICS
201 ROOSEVELT AVE.
CHANDLER, AZ 85226
Telephone: 602-961-1220
Products: SOIL CONDITIONERS; FEED
SUPPLEMENTS; WASTE TREATMENT

President: DEL BENTZ
R&D Dir: DR. JORDAN C. SMITH
Financing: PRI
Started: 1984
Class Code: B

BIO TECHNIQUES LABS, INC.
15555 N.E. 33 ST., BIOTECH RD.
REDMOND, WA 98052
Telephone: 206-883-9518
Products: FEED ADDITIVES

President: WILLIAM ST. JOHN
R&D Dir: DR. RICHARD MOORE
Financing: PRI
Started: 1982
Class Code: A

BIO-RECOVERY, INC.
P.O. 38, 193 PARIS AVE.
NORTHVALE, NJ 07647
Telephone: 201-784-9396
Products: FILTRATION MEMBRANE SYSTEMS,
LARGE SCALE PRODUCTION

President: PAUL HELLMAN
R&D Dir: LAWRENCE NIESEN
Financing: PRI
Started: 1983
Class Code: F

BIO-RESPONSE, INC.
1978 W. WINTON AVE.
HAYWARD, CA 94545
Telephone: 415-786-9744
Products: NON-RDNA TPA; MASS CELL CULTURE;
DX KITS; MABS

CEO: VAUGHN H. J. SHALSON
President: ALFRED G. DANIEL
R&D Dir: SAMUEL ROSE
Financing: PUB
Started: 1972
Class Code: P

BIO-TECHNICAL RESOURCES, INC.
1035 SOUTH 7TH ST.
MANITOWOC, WI 54220
Telephone: 414-684-5518
Products: YEAST/BACTERIAL PRODS IN FOOD,
BREWING, CHEMICALS

President: MICHAEL R. SFAT
R&D Dir: THOMAS SKATRUD
Financing: PRI
Started: 1962
Class Code: M

BIOCHEM TECHNOLOGY, INC.
66 GREAT VALLEY PKWY.
MALVERN, PA 19355
Telephone: 215-647-8610
Products: SENSORS; COMPUTER PRODUCTS &
SOFTWARE; NADH DETECTOR

President: DR. WILLIAM ARMIGER
R&D Dir: DR. LUKE JU
Financing: PRI
Started: 1977
Class Code: D

BIOCONSEP, INC.
RD3 HOMESTEAD RD, BLD 5 UNIT 9
BELLEMEAD, NJ 08502
Telephone: 201-359-6886
Products: BIOLOGICAL SEPARATIONS, R&D
AND EQUIPMENT

President: THOMAS BRADSHAW
R&D Dir: JAMES WILLIS
Financing: PRI
Started: 1983
Class Code: E

BIOCONTROL SYSTEMS
19805 NORTH CREEK PARKWAY
BOTHELL, WA 98011
Telephone: 206-395-3300
Products: TESTS FOR FOOD, WATER; BIOSENSORS

President: MAX LYON
R&D Dir: ROBERT WARD
Financing: PRI
Started: 1985
Class Code: R

BIODESIGN, INC.
432 BEACHWOOD AVE.
KENNEBUNKPORT, ME 04046
Telephone: 207-967-4173
Products: IMMUNOLOGICAL REAGENTS; BIODESIGN
ANTISERA, PURIFIED BIOLOGICAL COMPOUNDS

CEO: HOLLY SCRIBNER
President: HOLLY SCRIBNER
R&D Dir: HOLLY SCRIBNER
Financing: PRI
Started: 1987
Class Code: G

BIOGEN
14 CAMBRIDGE CENTER
CAMBRIDGE, MA 02142
Telephone: 617-864-8900
Products: RX; DX; TNF, IL-2, CSF, TPA, HSA, FACTOR
VIII, IF-G, HEP B ANTIGEN

CEO: DR. JAMES L. VINCENT
R&D Dir: RICHARD A. FLAVELL
Financing: PUB
Started: 1979
Class Code: P

BIOGENEX LABORATORIES
4600 NORRIS CANYON RD.
SAN RAMON, CA 94583
Telephone: 800-421-4149
Products: RIA DX TEST; IMMUNOHISTOCHEMICALS;
KITS; REAGENTS

President: KRISHAN KALRA
R&D Dir: DR. V. KAMALAKANNAN
Financing: PRI
Started: 1981
Class Code: K

BIOGROWTH, INC.
3065 ATLAS RD., SUITE 117
RICHMOND, CA 94806
Telephone: 415-222-2084
Products: WOUND/BONE RX; SEPARATION; PROTEIN
CHARACTERIZATION; HPLC/FPLC EQUIPMENT

CEO: STUART W. BOLINGER
President: STUART W. BOLINGER
R&D Dir: DR. CAROL VERSER
Financing: PRI
Started: 1985
Class Code: P

BIOKYOWA, INC.
975 NASH RD., P.O. BOX 1550
CAPE GIRARDEAU, MO 63701
Telephone: 314-335-4849
Products: L-LYSINE FEED ADDITIVE FOR
POULTRY/SWINE, DEVELOPING RX

CEO: SUMIO KURIHARA
President: SUMIO KURIHARA
R&D Dir: KIYOGI HATTORI
Financing: SUB
Started: 1983
Class Code: A

BIOMATERIALS INTERNATIONAL, INC.
PO BOX 8852, 420 CHIPETA WAY, SUITE 160
SALT LAKE CITY, UT 84108
Telephone: 801-583-8444
Products: RAMAN TECHNOL. FOR RESPIRATORY
APPLICATIONS; GAS MONITOR

CEO: GEORGE SIMS
President: GEORGE SIMS
R&D Dir: RICHARD VAN WAGENEN
Financing: PRI
Started: 1981
Class Code: D

BIOMATRIX, INC.
P.O. BOX 530, 65 RAILROAD AVE.
RIDGEFIELD, NJ 07657
Telephone: 201-945-9550
Products: MATERIAL FOR ARTIFICIAL ORGANS;
COSMETICS; OPHTHALMICS

President: ENDRE A. BALAZS
R&D Dir: JANET L. DENLINGER
Financing: PRI
Started: 1981
Class Code: 1

Listing 2-2 Directory of Companies

BIOMED RESEARCH LABS, INC.
1115 E. PIKE ST.
SEATTLE, WA 98122
Telephone: 206-324-0380
Products: FISH BIOLOGICS, VET PRODUCTS

President: DR. JOHN J. MAJNARICH
R&D Dir: THOMAS GOODRICH
Financing: PRI
Started: 1974
Class Code: V

BIOMEDICAL TECHNOLOGIES
378 PAGE STREET
STOUGHTON, MA 02072
Telephone: 617-344-9942
Products: CELL GROWTH FACTORS; CELL CULTURE
PRODUCTS; RIA KITS; MAB; POLYCLONALS; PROTEINS

CEO: MAURICE LAMARQUE
President: MAURICE LAMARQUE
R&D Dir: DR. RONALD FORAND
Financing: PRI
Started: 1981
Class Code: J

BIOMERICA, INC.
1533 MONROVIA AVE.
NEWPORT BEACH, CA 92663
Telephone: 714-645-2111
Products: PREGNANCY, ALLERGY, OCCULT
BLOOD DX; LAB PRODUCTS; VET DX & RX

President: JOSEPH H. IRANI
R&D Dir: WAI SAI MA
Financing: PUB
Started: 1971
Class Code: K

BIONETICS RESEARCH
1330-A PICCARD DR.
ROCKVILLE, MD 20850
Telephone: 301-258-5200
Products: AIDS DX, SALMONELLA EIA, RADIOLABELED
COMPOUNDS, LIPOPROTEINS, MELATONIN

CEO: BEN VAN DOMMELIN
President: BEN VAN DOMMELIN
R&D Dir: DR. MICHAEL HANNA
Financing: SUB
Started: 1985
Class Code: K

BIONIQUE LABS, INC.
BLOOMINGDALE RD., RT. #3
SARANAC LAKE, NY 12983
Telephone: 518-891-2356
Products: CELL CULTURE EQUIPMENT; MYCOPLASMA
DETECTION SYSTEM; CELL PROTECTIVE DEVICE; DX

President: DR. MICHAEL GABRIDGE
R&D Dir: DR. STEVEN GEARY
Financing: PRI
Started: 1983
Class Code: H

BIOPOLYMERS, INC.
74 NORTHWEST DR.
FARMINGTON, CT 06032
Telephone: 203-793-9651
Products: CELL & TISSUE, OPHTHALMIC; MOLLUSK
ADHESIVES

CEO: THOMAS M. BENEDICT
President: THOMAS M. BENEDICT
R&D Dir: DR. CHRISTINE BENEDICT
Financing: PRI
Started: 1984
Class Code: 1

BIOPROBE INTERNATIONAL, INC.
2842 WALNUT AVE., SUITE C
TUSTIN, CA 92680
Telephone: 714-544-4035
Products: SPECIALIZED GELS; SEPARATION DEVICES

President: RAPHAEL WONG
R&D Dir: DR. THAT L. NGO
Financing: PRI
Started: 1983
Class Code: E

BIOPRODUCTS FOR SCIENCE, INC.
P.O. BOX 29176
INDIANAPOLIS, IN 46229
Telephone: 608-274-2905
Products: ANTIBODIES; ANTIGEN MARKERS;
REAGENTS

President: HAL P. HARLAN
R&D Dir: PAUL A. COULIS
Financing: PRI
Started: 1985
Class Code: X

BIOPURE
68 HARRISON AVE.
BOSTON, MA 02111
Telephone: 617-350-7800
Products: CUSTOM FERMENTATION OF RX, MABS,
CHEMICALS

CEO: CARL RAUSCH
President: BRIAN MCGUINN
R&D Dir: DR. BING L. WONG
Financing: PRI
Started: 1984
Class Code: O

BIOSCIENCE MANAGEMENT, INC.
BFTC-SOUTH MOUNTAIN DR.
BETHLEHEM, PA 18015
Telephone: 215-861-0291
Products: BACTERIAL CULTURES; INOCULANTS;
INSTRUMENTS

CEO: THOMAS G. ZITRIDES
President: THOMAS G. ZITRIDES
R&D Dir: DR. M.P. PIRNIK
Financing: PRI
Started: 1984
Class Code: J

BIOSEARCH, INC.
2980 KERNER BLVD.
SAN RAFAEL, CA 94901
Telephone: 800-227-2624
Products: REAGENT CHEMICALS AND GENE/PEPTIDE
INSTRUMENTS

President: DAVID HARNDEN
R&D Dir: RON COOK
Financing: SUB
Started: 1977
Class Code: G

BIOSPHERICS, INC.
12051 INDIAN CREEK COURT
BELTSVILLE, MD 20705
Telephone: 301-369-3900
Products: WASTEWATER PROCESS & MONITORS
& PILOT PLANT

President: DR. GILBERT V. LEVIN
R&D Dir: DR. GILBERT V. LEVIN
Financing: PUB
Started: 1967
Class Code: R

BIOSTAR MEDICAL PRODUCTS, INC.
5766 CENTRAL AVE.
BOULDER, CO 80301
Telephone: 303-447-1605
Products: DX & TEST KITS FOR AUTOIMMUNE
DISEASES, INSTRUMENTATION

CEO: TIMOTHY W. STARZL
President: TIMOTHY W. STARZL
R&D Dir: DR. MICHAEL D. ROPER
Financing: PRI
Started: 1983
Class Code: K

BIOSYSTEMS, INC.
762 U.S. HIGHWAY 78
LOGANVILLE, GA 30249
Telephone: 404-466-1511
Products: MABS; GENERAL MICROBIAL R&D

President: DR. PAUL BATRA
Financing: PRI
Started: 1976
Class Code: R

BIOTECH RESEARCH LABS, INC.
1600 EAST GUDE DRIVE
ROCKVILLE, MD 20850
Telephone: 301-251-0800
Products: MABS FOR DX, AIDS DX, CANCER DX;
AIDS VX, IL-2

CEO: THOMAS M. LI
President: THOMAS M. LI
R&D Dir: ROBERT C.Y. TING
Financing: PUB
Started: 1973
Class Code: K

BIOTECHNICA AGRICULTURE
7300 W. 110TH ST., SUITE 540
OVERLAND PARK, KS 66210
Telephone: 913-661-0611
Products: DISEASE & PEST RESISTANT CORN, HAY,
SOYBEAN, WHEAT; HIGHER NUTRITIONAL CROPS

CEO: JOHN HUNT
President: DR. CHARLES H. BAKER
R&D Dir: DR. FRANK CANNON
Financing: SUB
Started: 1987
Class Code: B

Listing 2-2 Directory of Companies

BIOTECHNICA INTERNATIONAL
85 BOLTON ST.
CAMBRIDGE, MA 02140
Telephone: 617-864-0040
Products: INOCULANT, VX, DENTAL DX; ENZYMES, INDUSTRIAL YEAST, LIGHT BEER, FOODS

CEO: DR. NORMAN JACOBS
President: DR. NORMAN JACOBS
R&D Dir: LYNN C. KLOTZ
Financing: PUB
Started: 1981
Class Code: J

BIOTECHNOLOGY DEVELOPMENT CORP.
44 MECHANIC ST.
NEWTON, MA 02164
Telephone: 617-965-7255
Products: AIDS VX; EQUIPMENT; REAGENTS; DRUG DELIVERY SYSTEMS

CEO: IRWIN J. GRUVERMAN
President: IRWIN J. GRUVERMAN
R&D Dir: DR. JOHN J. KING
Financing: PUB
Started: 1982
Class Code: F

BIOTECHNOLOGY GENERAL CORP.
375 PARK AVE.
NEW YORK, NY 10152
Telephone: 212-319-8944
Products: SOD, BGH, HGH, RX, SURFACTANTS, FUNGICIDE

President: DR. SIM FASS
R&D Dir: DR. MARIAN GORECKI
Financing: PUB
Started: 1980
Class Code: P

BIOTEST DIAGNOSTICS CORP.
6 DANIEL RD. EAST
FAIRFIELD, NJ 07006
Telephone: 201-575-4500
Products: REAGENTS AND MAB MARKERS

CEO: ROBERT W. PERRY
President: ROBERT W. PERRY
Financing: SUB
Started: 1946
Class Code: G

BIOTHERAPEUTICS, INC.
357 RIVERSIDE DR.
FRANKLIN, TN 37064
Telephone: 615-794-4797
Products: CANCER RX & DX; BIOLOGICAL RESPONSE MODIFIERS

President: LOUIS P. BERNEMAN
R&D Dir: DR. ROBERT K. OLDHAM
Financing: PUB
Started: 1984
Class Code: P

BIOTHERAPY SYSTEMS, INC.
291 N. BERNARDO AVE.
MOUNTAIN VIEW, CA 94043
Telephone: 415-940-1200
Products: MABS FOR CANCER THERAPY

President: DR. WILLIAM RASTTETER
R&D Dir: DR. RICHARD A. MILLER
Financing: PUB
Started: 1984
Class Code: P

BIOTHERM
(NEW NAME: CARDIOVASCULAR DIAGNOSTICS)
P.O. BOX 13417
RESEARCH TRIANGLE PARK, NC 27709
Telephone: 919-544-2952
Products: DX, CARDIOVASCULAR DX

CEO: LEO M. STOREY, JR.
President: LEO M. STOREY, JR.
R&D Dir: DR. BRUCE OBERHARDT
Financing: PRI
Started: 1985
Class Code: K

BIOTICS RESEARCH CORP.
4850 WRIGHT RD., SUITE 150
STAFFORD, TX 77047
Telephone: 713-789-9020
Products: NUTRITIONAL SUPPLEMENTS

President: DENIS DELUCA
R&D Dir: DR. LUKE BUCCI
Financing: PRI
Started: 1975
Class Code: M

BIOTROL, INC.
11 PEAVEY RD.
CHASKA, MN 55318
Telephone: 612-448-2515
Products: PROPRIETARY BACTERIA; BIOREACTORS
FOR ENVIRONMENTAL SERVICES; DEGRADERS

CEO: DR. M.B. BURTON
President: DR. M.B. BURTON
R&D Dir: DR. DENNIS CHILCOTE
Financing: PRI
Started: 1985
Class Code: R

BIOTX (BIOSCIENCES CORP. OF TEXAS)
4900 FANNIN ST.
HOUSTON, TX 77004
Telephone: 713-526-0550
Products: DX; CANCER, NUCLEAR ANTIGEN DX KIT,
ABS TO ONCOGENES & GROWTH FACTORS

CEO: PETER G. ULRICH
President: PETER G. ULRICH
R&D Dir: DR. ROBERT L. PARDUE
Financing: PRI
Started: 1985
Class Code: K

BMI, INC.
23361 VIA LINDA, SUITE A
MISSION VIEJO, CA 92691
Telephone: 714-768-7415
Products: DX; RX; OTC PREG. TEST, ENZYME IMMUNO-
ASSAYS, ALLERGY TESTS, BOVINE PROGESTERONE RX

President: KIMBERLY KORTSCHAK
Financing: PRI
Started: 1984
Class Code: K

BOEHRINGER MANNHEIM DIAGNOSTICS
7941 CASTLEWAY DR.
INDIANAPOLIS, IN 46250
Telephone: 800-428-5433
Products: DX; RX; REAGENTS; DIGOXIN; ANIMAL
CELL TISSUE CULTURE, PLANT DX

CEO: J. VAN KAMPEN
Financing: SUB
Started: 1975
Class Code: K

BRAIN RESEARCH, INC.
46 E. 91 ST.
NEW YORK, NY 10028
Telephone: 212-831-6645
Products: CANCER DX

President: SAMUEL S. BOGOCH
Financing: PRI
Started: 1968
Class Code: K

C

CALGENE, INC.
1920 5TH ST.
DAVIS, CA 95616
Telephone: 916-753-6313
Products: TOMATOES, COTTON, CORN, SUNFLOWER,
RAPESEED, FOODS; PESTICIDES, PLANTS, TREES

CEO: ROGER SALQUIST
President: ROGER SALQUIST
R&D Dir: DR. ROBERT GOODMAN
Financing: PUB
Started: 1980
Class Code: B

CALIFORNIA BIOTECHNOLOGY, INC.
2450 BAYSHORE FRONTAGE RD.
MOUNTAIN VIEW, CA 94043
Telephone: 415-966-1550
Products: RX; ANF, EPO, SURFACTANT, HGH,
RENIN, PTH; VET RX, NASAL DRUG DELIVERY

CEO: RICHARD L. CASEY
President: RICHARD L. CASEY
R&D Dir: JOHN A. LEWICKI
Financing: PUB
Started: 1981
Class Code: P

Listing 2-2 Directory of Companies

CALIFORNIA INTEGRATED DIAGNOSTICS
1440 FOURTH ST.
BERKELEY, CA 94710
Telephone: 415-527-9400
Products: DX KITS FOR SEXUALLY TRANSMITTED
DISEASES; MABS

President: BERNIE DIDARIO
R&D Dir: SINSU TZENG
Financing: SUB
Started: 1981
Class Code: K

CALZYME LABORATORIES, INC.
3443 MIGUELITO CT.
SANLUIS OBISPO, CA 93401
Telephone: 805-541-5754
Products: ENZYMES, BIOCHEMICALS,
PROTEINS

CEO: DR. MUZAFFAR IQBAL
President: DR. MUZAFFAR IQBAL
R&D Dir: DR. JOSEPH BANNISTER
Financing: PRI
Started: 1983
Class Code: J

CAMBRIDGE BIOSCIENCE CORP.
365 PLANTATION ST.
WORCESTER, MA 01605
Telephone: 617-797-5777
Products: DX KITS FOR AIDS, ADENOVIRUS,
ROTAVIRUS; VET VX, DX

President: GERALD F. BUCK
R&D Dir: DANTE J. MARCIANI
Financing: PUB
Started: 1981
Class Code: K

CAMBRIDGE RESEARCH BIOCHEMICALS, INC.
10 EAST MERRICK RD., SUITE 202
VALLEY STREAM, NY 11580
Telephone: 516-825-1322
Products: MABS; PEPTIDES, ONCO PROTEINS;
DEVELOPING EPITOPE SCANNING PRODUCTS

President: KEN LIDDLE
R&D Dir: DR. PAUL SHEPPARD
Financing: SUB
Started: 1980
Class Code: J

CARBOHYDRATES INTERNATIONAL, INC.
150 N. MICHIGAN AVE., SUITE 1225
CHICAGO, IL 60601
Telephone: 312-781-6210
Products: INFECTIOUS DISEASE, BLOOD TYPE DX;
SUGAR STRUCTURES TO RAISE MABS

CEO: TORBJORN MOLLER
President: TORBJORN MOLLER
R&D Dir: DR. BERTIL NILSSON
Financing: SUB
Started: 1987
Class Code: J

CEL-SCI CORP.
601 WYTHE ST., SUITE 202
ALEXANDRIA, VA 22314
Telephone: 703-549-5293
Products: RX; LYMPHOKINES; AIDS VX; IL-1, IL-2

FORMERLY CALLED INTERLEUKIN-2
President: MAXIMILLIAN DE CLARA
Financing: PUB
Started: 1983
Class Code: P

CELLMARK DIAGNOSTICS
20271 GOLDENROD LANE
GERMANTOWN, MD 20874
Telephone: 301-428-4980
Products: GENE PROBING, GENETIC TESTING; DNA
FINGERPRINTING

President: ROBERT H. GOPTTHEINER
R&D Dir: DR. DAVID J. GREEN
Financing: SUB
Started: 1986
Class Code: K

CELLULAR PRODUCTS, INC.
688 MAIN ST.
BUFFALO, NY 14202
Telephone: 716-842-6270
Products: MABS, DIPSTIK (TM) DX TESTS KITS;
AIDS DX; IF

President: RICHARD MONTAGNA
R&D Dir: L. PAPSIDERO
Financing: PUB
Started: 1982
Class Code: K

CENTOCOR
244 GREAT VALLEY PARKWAY
MALVERN, PA 19355
Telephone: 215-296-4488
Products: MAB RX; CANCER, HEP B DX; MABS;
AIDS VX

CEO: HUBERT SCHOEMAKER
President: JAMES E. WAVLE, JR.
R&D Dir: VINCENT R. ZURAWSKI
Financing: PUB
Started: 1979
Class Code: K

CENTRAL BIOLOGICS, INC.
5100 HUNTINGWOOD DR.
RALEIGH, NC 27606
Telephone: 919-552-5309
Products: VET VX; CUSTOM SERA; CONTRACT
RESEARCH

CEO: DR. TERRY L. NOBLE
President: DR. TERRY L. NOBLE
Financing: PRI
Started: 1982
Class Code: Q

CETUS CORP.
1400 FIFTY-THIRD ST.
EMERYVILLE, CA 94608
Telephone: 415-420-3300
Products: RX; IF; IL-2; MABS; TNF; CSF; INSULIN;
ANTIBIOTICS; VET VX

CEO: DR. ROBERT A. FILDES
President: DR. RONALD E. CAPE
R&D Dir: DR. JEFFREY S. PRICE
Financing: PUB
Started: 1971
Class Code: P

CHARLES RIVER BIOTECH. SERVICES
251 BALLARDVALE ST.
WILMINGTON, MA 01887
Telephone: 617-658-6000
Products: MABS, CELL CULTURE SYSTEMS

President: JAMES C. FOSTER
R&D Dir: DR. MARY L. NICHOLSON
Financing: SUB
Started: 1983
Class Code: X

CHEMGENES
925 WEBSTER ST.
NEEDHAM, MA 02192
Telephone: 617-449-5051
Products: INTERMEDIATES FOR DNA SYNTHESIS

President: DR. SURESH SRIVASTAVA
Financing: PRI
Started: 1981
Class Code: G

CHEMICAL DYNAMICS CORP.
P.O. BOX 395
S. PLAINFIELD, NJ 07080
Telephone: 201-753-5000
Products: DNA TECHNOLOGY REAGENTS,
RESTRICTION ENZYMES

CEO: RONALD A. VITALI
R&D Dir: DR. EDWIN LUNDELL
Financing: PRI
Started: 1972
Class Code: G

CHEMICON INTERNATIONAL, INC.
100 LOMITA ST.
EL SEGUNDO, CA 90245
Telephone: 213-322-2451
Products: MABS; GROWTH FACTORS;
IMMUNOLOGICAL CELL GROWTH ASSAY KIT

CEO: DAVE BECKMAN
President: KEIKO KOGO
R&D Dir: DAVE BECKMAN
Financing: PRI
Started: 1981
Class Code: G

CHIRON CORP.
4560 HORTON ST.
EMERYVILLE, CA 94608
Telephone: 415-655-8730
Products: AIDS DX; DNA PROBE DX; WOUND
HEALING; HEP B VX; VET VX, SOD; ENZYMES; IF

CEO: DR. EDWARD E. PENHOET
President: DR. EDWARD E. PENHOET
R&D Dir: PABLO VALENZUELA
Financing: PUB
Started: 1981
Class Code: P

Listing 2-2 Directory of Companies

CIBA-CORNING DIAGNOSTIC CORP.
63 NORTH ST.
MEDFIELD, MA 02502
Telephone: 617-359-7711
Products: DX; HUMAN
CLINICAL DX

CEO: C. WILLIAM ZADEL
President: C. WILLIAM ZADEL
R&D Dir: DR. RICHARD D. FALB
Financing: SUB
Started: 1985
Class Code: K

CISTRON BIOTECHNOLOGY, INC.
10 BLOOMFIELD AVE., BOX 2004
PINE BROOK, NJ 07058
Telephone: 201-575-1700
Products: IL-1, IL-2, DX

President: WALTER O. FREDERICKS
R&D Dir: DR. WILLIAM J. DELORBE
Financing: PUB
Started: 1982
Class Code: P

CLINETICS CORP.
2991 DOW AVE.
TUSTIN, CA 92680
Telephone: 714-544-7991
Products: PREGNANCY ASSAY KIT, DX

President: JEFFREY L. LILLARD
R&D Dir: DR. WALTER PROTZMAN
Financing: PRI
Started: 1979
Class Code: K

CLINICAL SCIENCES, INC.
30 TROY RD.
WHIPPANY, NJ 07981
Telephone: 201-386-0030
Products: BIOCHEMICALS FOR DX AND VET USES

President: MOSHE SHURIN
R&D Dir: HELLEN C. GREENBLATT
Financing: PUB
Started: 1971
Class Code: G

CLONTECH LABORATORIES, INC.
4055 FABIAN WAY
PALO ALTO, CA 94303
Telephone: 415-424-8188
Products: CLONING, TPA, BLOOD
CLOTTING FACTORS

CEO: DR. KENNETH FONG
President: DR. KENNETH FONG
Financing: PRI
Started: 1984
Class Code: K

COAL BIOTECH CORP.
800 S. MILWAUKEE, NO. 170
LIBERTYVILLE, IL 60048
Telephone: 312-362-5620
Products: MICROBIAL COAL
DESULFURIZATION

CEO: MICHAEL W. DYBEL
President: MICHAEL W. DYBEL
R&D Dir: DR. DUANE SKIDMORE
Financing: PRI
Started: 1984
Class Code: N

CODON CORP.
213 E. GRAND AVE.
S. SAN FRANCISCO, CA 94080
Telephone: 415-952-7070
Products: VET VX; HUMAN DX

President: DR. FRED CRAVES
R&D Dir: DR. JOEL HEDGPETH
Financing: PRI
Started: 1980
Class Code: V

COLLABORATIVE RESEARCH, INC.
2 OAK PARK
BEDFORD, MA 01730
Telephone: 617-275-0004
Products: DX SERVICES (DNA PROBE); IF; RENNIN;
UROKINASE; IL-2; ENZYMES; TPA; REAGENTS

CEO: DR. ORRIE FRIEDMAN
President: JOHN E. DONALDS
R&D Dir: THOMAS O. OESTERLING
Financing: PUB
Started: 1961
Class Code: G

COLLAGEN CORP.
2500 FABER PLACE
PALO ALTO, CA 94303
Telephone: 415-856-0200
Products: MEDICAL DEVICES; COLLAGEN IMPLANT;
TISSUE CULTURES

President: H. PALEFSKY
R&D Dir: DR. BRUCE PHARRIS
Financing: PUB
Started: 1975
Class Code: 4

CONSOLIDATED BIOTECHNOLOGY, INC.
1413 WEST INDIANA AVE.
ELKHART, IN 46515
Telephone: 219-295-6767
Products: ENZYMES, FOOD RELATED RESEARCH

President: DR. J. JOHN MARSHALL
R&D Dir: DR. J. JOHN MARSHALL
Financing: PRI
Started: 1983
Class Code: G

COOPER DEVELOPMENT CO.
75 WILLOW RD.
MENLO PARK, CA 94025
Telephone: 415-853-6000
Products: ALPHA-1 ANTITRYPSIN, RX, MABS

President: JOHN L. EDWARDS
R&D Dir: KENNETH D. NOONAN
Financing: PUB
Started: 1980
Class Code: P

COORS BIOTECH PRODUCTS CO.
12200 N. PECOS ST., SUITE 100
WESTMINSTER, CO 80234
Telephone: 303-451-1853
Products: INDUSTRIAL DEGREASER; FOOD
PRODUCTS; SPECIALTY CHEMICALS

President: NORMAN O. JANGAARD
R&D Dir: DR. E.W. FOSTER
Financing: SUB
Started: 1984
Class Code: M

COVALENT TECHNOLOGY CORP.
P.O. BOX 1868
ANN ARBOR, MI 48106
Telephone: 313-769-5377
Products: DX FOR HORMONE, DISEASE;
BIOWAR DX; TROPICAL DISEASE DX

President: DR. ALEXANDER GLASS
R&D Dir: DR. MICHAEL PAPPAS
Financing: SUB
Started: 1981
Class Code: K

CREATIVE BIOMOLECULES
35 SOUTH ST.
HOPKINTON, MA 01748
Telephone: 617-435-9001
Products: PEPTIDES, CONTRACT R&D, TPA, EGF

President: DR. CHARLES COHEN
R&D Dir: DR. ROBERTO CREA
Financing: PRI
Started: 1981
Class Code: J

CROP GENETICS INTERNATIONAL CORP.
7170 STANDARD DR.
HANOVER, MD 21076
Telephone: 301-621-2900
Products: DISEASE-FREE SUGARCANE SEED,
BIOINSECTICIDES

CEO: JOHN B. HENRY
President: ROBERT L. MONTGOMERY
R&D Dir: DR. PETER S. CARLSON
Financing: PUB
Started: 1981
Class Code: B

CYANOTECH CORP.
18748 142ND AVE., N.E.
WOODINVILLE, WA 98072
Telephone: 206-481-2173
Products: PHYCOBILIPROTEIN; SPIRULINA;
BETA-CAROTENE

President: ARTHUR KARUNA-KARAN
R&D Dir: DR. GERALD CYSEWSKI
Financing: PRI
Started: 1983
Class Code: M

Listing 2-2 Directory of Companies

CYGNUS RESEARCH CORP.
701 GALVESTON DR.
REDWOOD CITY, CA 94063
Telephone: 415-369-4300
Products: DRUG DELIVERY SYSTEM; CONTRACTS
TO RX COMPANIES

CEO: DAVID DEWEESE
President: DAVID DEWEESE
R&D Dir: DAVID DEWEESE
Financing: PRI
Started: 1985
Class Code: 3

CYTOGEN CORP.
201 COLLEGE RD. EAST
PRINCETON, NJ 08540
Telephone: 609-987-8200
Products: RX; MABS; DX; CANCER RX

President: DR. RONALD J. BRENNER
R&D Dir: DR. THOMAS J. MCKEARN
Financing: PUB
Started: 1980
Class Code: K

CYTOTECH, INC.
11045 ROSELLE ST.
SAN DIEGO, CA 92121
Telephone: 619-452-1556
Products: AIDS DX, REAGENTS

President: PAUL ROSINACK
R&D Dir: JOHN TAMERIUS
Financing: PRI
Started: 1982
Class Code: K

CYTOX CORP.
954 MARCON BLVD.
ALLENTOWN, PA 18103
Telephone: 215-264-8740
Products: WORK WITH WASTE, AGRICULTURE,
POLLUTION, ENERGY

CEO: DOUGLAS R. NICHOLS
President: JOHN N. ZIKOPOULOS
R&D Dir: CURTIS S. MCDOWELL
Financing: PUB
Started: 1975
Class Code: R

D

DAKO CORP.
22 NORTH MILPAS ST.
SANTA BARBARA, CA 93103
Telephone: 800-235-5743
Products: POLYCLONAL AB & MABS,
CLINICAL TEST KITS, PROTEINS

President: VIGGO F. HARBOE
R&D Dir: TOM BOENSCH
Financing: SUB
Started: 1979
Class Code: X

DAMON BIOTECH, INC.
119 FOURTH AVE.
NEEDHAM HEIGHTS, MA 02194
Telephone: 617-449-0800
Products: MABS, DRUG DELIV., CELL
ENCAPSULATION SYSTEMS; AIDS DX; TPA

CEO: DAVID I. KOSOWSKY
President: ROBERT SCHNEIDER
R&D Dir: DR. WILLIAM TERRY
Financing: SUB
Started: 1978
Class Code: P

DEKALB-PFIZER GENETICS
3100 SYCAMORE RD.
DEKALB, IL 60115
Telephone: 815-756-3671
Products: HYBRID SEEDS, CORN, SOY, ALFALFA,
CROP PLANTS

President: DR. KENT SCHULTZ
R&D Dir: CATHERINE MACKEY
Financing: PRI
Started: 1982
Class Code: B

DELTOWN CHEMURGIC CORP.

191 MASON ST.
GREENWICH, CT 06830
Telephone: 203-629-8754
Products: PEPTONES AND HYDROLASES FOR RX,
CHEMICALS

President: DR. JACK L. HERZ
R&D Dir: TIMOTHY CALLAHAN
Financing: PRI
Started: 1968
Class Code: J

DIAGNON CORP.

11 TAFT CT.
ROCKVILLE, MD 20850
Telephone: 301-252-0633
Products: DX ASSAYS FOR FERRITIN, B12/FOLATE,
HERPES; VIRAL TREATMENT; TOXIN ASSAYS

CEO: DR. JOHN LANDON
President: DR. JOHN LANDON
Financing: PUB
Started: 1981
Class Code: K

DIAGNOSTIC PRODUCTS CORP.

5700 W. 96 ST.
LOS ANGELES, CA 90045
Telephone: 213-776-0180
Products: DX KITS, MABS; REAGENTS

CEO: DR. SIGI ZIERING
President: DR. SIGI ZIERING
R&D Dir: SAID EL SHAMI
Financing: PUB
Started: 1972
Class Code: G

DIAGNOSTIC TECHNOLOGY, INC.

240 VANDERBILT MOTOR PKWY.
HAUPPAUGE, NY 11788
Telephone: 516-582-4949
Products: DX, BLOOD TEST KITS, REAGENTS

President: DR. IMRE PINTER
R&D Dir: DR. JERRY PICKERING
Financing: PRI
Started: 1980
Class Code: K

DIAMEDIX, INC.

2140 NORTH MIAMI AVE.
MIAMI, FL 33127
Telephone: 305-324-2329
Products: DX KITS FOR INFECTIOUS & AUTOIMMUNE
DISEASES

CEO: DR. JOE GIEGEL
President: DR. JOE GIEGEL
R&D Dir: DR. LIN
Financing: PRI
Started: 1986
Class Code: K

DIGENE DIAGNOSTICS, INC.

BLDG. 334, UNIV. OF MARYLAND
COLLEGE PARK, MD 20742
Telephone: 301-454-7874
Products: DNA PROBE DX, VIRAL
DETECTION KITS

CEO: DR. FLOYD TAUB
President: DR. LEON TAUB
R&D Dir: DR. FLOYD TAUB
Financing: PRI
Started: 1984
Class Code: K

DNA PLANT TECHNOLOGY CORP.

2611 BRANCH PIKE
CINNAMINSON, NJ 08077
Telephone: 609-829-0110
Products: FOODS AND PLANTS, AG PRODS.,
TOMATOES, RICE, COFFEE, SNACKS, OILS, CORN

President: RICHARD LASTER
R&D Dir: DR. WILLIAM R. SHARP
Financing: PUB
Started: 1981
Class Code: B

DNAX RESEARCH INSTITUTE

901 CALIFORNIA AVE.
PALO ALTO, CA 94304
Telephone: 415-852-9196
Products: RESEARCH; MABS; RX

CEO: J. ALLAN WAITZ
President: J. ALLAN WAITZ
R&D Dir: J. ALLAN WAITZ
Financing: SUB
Started: 1980
Class Code: P

Listing 2-2 Directory of Companies

E

E-Y LABORATORIES, INC.
107 N. AMPHLETT BLVD.
SAN MATEO, CA 94401
Telephone: 800-821-0044
Products: ANTISERA;
REAGENTS; DX

President: DR. ALBERT E. CHU
R&D Dir: PETER CHUN
Financing: PRI
Started: 1978
Class Code: X

EARL-CLAY LABORATORIES, INC.
890 LAMONT AVE.
NOVATO, CA 94945
Telephone: 415-892-8512
Products: MABS; ULTRACLONE GROWTH CHAMBERS
IN TISSUE CULTURE; LYMPHOCYTE STIM. ASSAY

CEO: DR. ROBERT E. LOVINS
President: DR. ROBERT E. LOVINS
R&D Dir: DR. O. DILE HOLTON, III
Financing: PUB
Started: 1984
Class Code: X

ECOGEN, INC.
2005 CABOT BLVD. WEST
LANGHORNE, PA 19047
Telephone: 215-757-1590
Products: BIOPESTICIDES; MICROBIAL AG
PRODUCTS; INSECTICIDES

CEO: JOHN E. DAVIES
President: JAMES LESLIE
R&D Dir: DR. BRUCE C. CARLTON
Financing: PUB
Started: 1983
Class Code: B

ELANEX PHARMACEUTICALS, INC.
22121 17TH AVE. S.E., SUITE 115
BOTHELL, WA 98021
Telephone: 206-487-3526
Products: EPO, BLOOD RELATED PRODUCTS

CEO: LAWRENCE H. THOMPSON
President: LAWRENCE H. THOMPSON
Financing: PRI
Started: 1984
Class Code: P

ELCATECH, INC.
1001 S. MARSHALL, SUITE 64
WINSTON-SALEM, NC 27101
Telephone: 919-777-3624
Products: RESEARCH; COAGULATION
ASSAY TECHNOLOGY

CEO: DR. GEORGE J. DOELLGAST
President: DR. GEORGE J. DOELLGAST
R&D Dir: DR. MARK TRISCOTT
Financing: PRI
Started: 1984
Class Code: W

ELECTRO-NUCLEONICS, INC.
350 PASSAIC AVE.
FAIRFIELD, NJ 07006
Telephone: 201-227-6700
Products: RX; DX; AIDS DX; IL-2; HTLV-III
ANTIBODY TEST; TOXIN DX

CEO: VINCENT V. ABAJIAN
President: HERBERT APPLETON
R&D Dir: DR. ALAN FISHMAN
Financing: PUB
Started: 1960
Class Code: K

EMBREX, INC.
P.O. BOX 13989
RESEARCH TRIANGLE PARK, NC 27709
Telephone: 919-941-5185
Products: BIOLOGICALS FOR POULTRY IN-OVO
APPLICATIONS

CEO: ALAN HEROSIAN
President: ALAN HEROSIAN
R&D Dir: DR. J. P. THAXTON
Financing: PRI
Started: 1985
Class Code: A

EMTECH RESEARCH

15 W. PARK DR.
MOUNT LAUREL, NJ 08054
Telephone: 609-866-2900
Products: BIOTECH R&D FOR COSMETICS AND
AGRICULTURE

President: JAMES BUTLER, SR.
R&D Dir: HOWARD WORNE
Financing: PRI
Started: 1983
Class Code: B

ENDOGEN, INC.

451 D STREET
BOSTON, MA 02210
Telephone: 617-439-3250
Products: DX ASSAYS (QUANTITATIVE); DEVELOPING
HUMAN MABS

President: OWEN A. DEMPSEY
R&D Dir: ROY A. DEMPSEY
Financing: PRI
Started: 1985
Class Code: K

ENDOTRONICS, INC.

8500 EVERGREEN BLVD.
COON RAPIDS, MN 55433
Telephone: 612-786-0302
Products: RX; DX; MABS; APS SERIES; ACUSYST-P; VX;
BIOLOGICALS; HEP B VX

CEO: EUGENE SCHUSTER
President: RICHARD SAKOWICZ
R&D Dir: BERNARD HORWATH
Financing: PUB
Started: 1981
Class Code: F

ENVIRONMENTAL DIAGNOSTICS, INC.

PO BOX 908, 2990 ANTHONY RD.
BURLINGTON, NC 27215
Telephone: 919-226-6311
Products: ENVIRONMENTAL AND TOXIN TESTING
IN FOODS, ANIMALS & HUMANS

President: DR. JAMES SKINNER
R&D Dir: CAROLE A. GOLDEN
Financing: PUB
Started: 1983
Class Code: Y

ENZO-BIOCHEM, INC.

325 HUDSON ST.
NEW YORK, NY 10013
Telephone: 212-741-3838
Products: ENZYMES; LECTINS; FILTERS; NUCLEOTIDES;
NON-RADIOACTIVE DNA PROBES; MAB; DX

President: DR. ELAZAR RABBANI
R&D Dir: DR. DEAN ENGELHARDT
Financing: PUB
Started: 1976
Class Code: K

ENZON

518 LOGUE AVE.
MOUNTAIN VIEW, CA 94043
Telephone: 415-964-7676
Products: RX; SOD FOR HUMAN AND VET USE

President: HENRY LERMAN
R&D Dir: MARK SAIFER
Financing: PUB
Started: 1965
Class Code: P

ENZON, INC.

300-C CORPORATE CT.
S. PLAINFIELD, NJ 07080
Telephone: 201-668-1800
Products: HUMAN RX; ENZYMES

President: ABRAHAM ABUCHOWSKI
R&D Dir: RICHARD STONE
Financing: PUB
Started: 1981
Class Code: P

ENZYME BIO-SYSTEMS LTD.

INTERNATIONAL PLAZA
ENGLEWOOD CLIFFS, NJ 07632
Telephone: 201-894-2320
Products: INDUSTRIAL ENZYME PRODUCTS USED
FOR STARCH, BREWING & BAKING

CEO: EDWARD A. KUSKE
President: EDWARD A. KUSKE
R&D Dir: DR. W. MARTIN TEAGUE
Financing: SUB
Started: 1983
Class Code: J

Listing 2-2 Directory of Companies

ENZYME CENTER, INC.

36 FRANKLIN ST.
MALDEN, MA 02148
Telephone: 617-322-4885
Products: LARGE SCALE FERMENTATION; ENZYME
SEPARATION & PURIFICATION

President: STANLEY CHARM
Financing: PRI
Started: 1978
Class Code: O

ENZYME TECHNOLOGY CORP.

783 US 250 E., RTE. 2
ASHLAND, OH 44805
Telephone: 419-289-3706
Products: GLUCOAMYLASE & ALPHA-AMYLASE
ENZYMES

President: LEWIS LEONARD
R&D Dir: DR. HAL R. TURNER
Financing: SUB
Started: 1981
Class Code: O

EPITOPE, INC.

15425 SOUTHWEST KOLL PKWY.
BEAVERTON, OR 97006
Telephone: 503-641-6115
Products: DX; AIDS DX; VIRAL DETECTION KITS--
WESTERN BLOT, MAB TO HIV

CEO: MICHAEL C. HUBBARD
President: MICHAEL C. HUBBARD
R&D Dir: ANDREW GOLDSTEIN
Financing: PUB
Started: 1979
Class Code: G

ESCAGENETICS CORP.

830 BRANSTEN RD.
SAN CARLOS, CA 94070
Telephone: 415-595-5335
Products: TISSUE CULTURE FLAVORS; POTATO SEEDS;
TOMATOES; MICROPROPAGATED DATE PALMS

CEO: RAYMOND J. MOSHY
President: RAYMOND J. MOSHY
R&D Dir: DR. WALTER GOLDSTEIN
Financing: PUB
Started: 1986
Class Code: B

EXOVIR, INC.

111 GREAT NECK RD., SUITE 607
GREAT NECK, NY 11021
Telephone: 516-466-2110
Products: RX, TOPICAL FOR HSV-II, OTHER SKIN RX

CEO: WILLIAM M. SULLIVAN
President: WILLIAM M. SULLIVAN
Financing: PUB
Started: 1981
Class Code: P

F

FERMENTA ANIMAL HEALTH

P.O. BOX 901350
KANSAS CITY, MO 64190
Telephone: 800-422-0333
Products: VET DX & VX; PESTICIDES

CEO: FRANK BARRY
President: HENRY BOBE
Financing: SUB
Started: 1986
Class Code: V

FLOW LABORATORIES, INC.

7655 OLD SPRING HOUSE RD.
MCLEAN, VA 22102
Telephone: 703-893-5925
Products: MEDIA, FERMENTATION EQUIPMENT,
REAGENTS

President: RICHARD HOZIK
R&D Dir: PEGGY LEVER-FISCHER
Financing: SUB
Started: 1961
Class Code: G

FORGENE, INC.

7014 FIRE TOWER RD.
RHINELANDER, WI 54501
Telephone: 715-282-5247
Products: FORESTRY APPLICATIONS-NEW GENETIC
TECHNOLOGIES FOR TREE CROPS

President: DR. NEIL NELSON
R&D Dir: DR. NEIL NELSON
Financing: PRI
Started: 1986
Class Code: B

G

GAMETRICS, LTD.

324 S. THIRD ST.
LAS VEGAS, NV 89101
Telephone: 702-384-1049
Products: VET DX; VET & HUMAN SEX SELECTION,
SPERM ISOLATION

President: DR. RONALD J. ERICSSON
R&D Dir: DR. RONALD J. ERICSSON
Financing: PRI
Started: 1974
Class Code: V

GAMMA BIOLOGICALS, INC.

3700 MANGUM RD.
HOUSTON, TX 77092
Telephone: 800-231-5655
Products: REAGENT-BASED DX TESTS FOR HUMAN
BLOOD SERA

CEO: DAVID HATCHER
President: DAVID HATCHER
R&D Dir: RON DEMARS
Financing: PUB
Started: 1969
Class Code: K

GEN-PROBE, INC.

9880 CAMPUS POINT DR.
SAN DIEGO, CA 92121
Telephone: 619-546-8000
Products: DNA PROBE DX TECHNOLOGY FOR
RESEARCH & CLINIC

CEO: THOMAS A. BOLOGNA
President: THOMAS A. BOLOGNA
R&D Dir: DR.DAVID E. KOHNE
Financing: PUB
Started: 1984
Class Code: K

GENE-TRAK SYSTEMS

31 NEW YORK AVE.
FRAMINGHAM, MA 01701
Telephone: 617-872-3113
Products: RX PROTEINS, DNA PROBE-BASED DX;
CV PROTEINS/HEMOPOIETIC GROWTH FACTORS

CEO: ROBERT J. CARPENTER
President: ROBERT J. CARPENTER
R&D Dir: DR. ALAN SMITH
Financing: PUB
Started: 1981
Class Code: J

GENELABS, INC.

505 PENOBSCOT DR.
REDWOOD CITY, CA 94063
Telephone: 415-369-9500
Products: AIDS RX; CYTOKINES; NON-A, NON-B
HEPATITIS; SOLID PHASE BLOOD TESTING

CEO: DR. FRANK KUNG
President: DR. FRANK KUNG
R&D Dir: DR. FRANK KUNG
Financing: PRI
Started: 1984
Class Code: K

GENENCOR, INC.

180 KIMBALL WAY
S. SAN FRANCISCO, CA 94080
Telephone: 415-742-7500
Products: ENZYMES IN FOOD/BEVERAGE
PROCESSING

President: ROBERT E. LEACH
R&D Dir: DR. HERBERT HEYNEKER
Financing: PRI
Started: 1982
Class Code: J

Listing 2-2 Directory of Companies

GENENTECH
460 POINT SAN BRUNO BLVD.
S. SAN FRANCISCO, CA 94080
Telephone: 415-266-1614
Products: RX; DX; VX; INSULIN, IF, TNF, RENIN,
FACTOR VIII, PROTROPIN HGH, ACTIVASE TPA

CEO: ROBERT A. SWANSON
President: G. KIRK RAAB
R&D Dir: DR. DAVID W. MARTIN, JR.
Financing: PUB
Started: 1976
Class Code: P

GENESIS LABS, INC.
5182 WEST 76TH ST.
MINNEAPOLIS, MN 55435
Telephone: 612-835-3446
Products: DX; BLOOD GLUCOSE TEST; DRUG MONITOR

CEO: LEO T. FURCHT
R&D Dir: ALAN R. DAY
Financing: PRI
Started: 1984
Class Code: K

GENETIC DIAGNOSTICS CORP.
160 COMMUNITY DR.
GREAT NECK, NY 11021
Telephone: 516-487-4711
Products: DX TEST KITS FOR ABUSE DRUGS, DISEASES

President: DR. HERBERT A. PLATT
R&D Dir: DR. HARVEY BRANDWEIN
Financing: PUB
Started: 1981
Class Code: K

GENETIC ENGINEERING, INC.
136 AVE. & N. WASHINGTON ST.
DENVER, CO 80233
Telephone: 303-457-1311
Products: ANIMAL SEX SELECTION, BULL
SEMEN LABS

CEO: CHARLES SREBNIK
President: CHARLES SREBNIK
R&D Dir: JONATHAN VAN BLERKOM
Financing: PUB
Started: 1980
Class Code: A

GENETIC SYSTEMS CORP.
3005 FIRST AVE.
SEATTLE, WA 98121
Telephone: 206-728-4900
Products: MAB BASED HUMAN DX AND RX;
AIDS DX; HERPES DX; LEGIONNAIRES DX

CEO: DR. ROBERT NOWINSKI
President: JOSEPH ASHLEY
R&D Dir: DR. ROBERT NOWINSKI
Financing: SUB
Started: 1980
Class Code: K

GENETIC THERAPY, INC.
19 FIRSTFIELD RD.
GAITHERSBURG, MD 20878
Telephone: 301-590-2626
Products: ANIMAL/HUMAN GENE THERAPY; RX;
RETROVIRAL VECTORS;TISSUE TARGETED DELIVERY

CEO: DR. M. JAMES BARRETT
President: DR. M. JAMES BARRETT
R&D Dir: DR. PAUL TOLSTOSHEV
Financing: PRI
Started: 1986
Class Code: P

GENETICS INSTITUTE, INC.
87 CAMBRIDGE PARK DR.
CAMBRIDGE, MA 02140
Telephone: 617-876-1170
Products: RX; DX; GM-CSF; EPO; TPA; HSA; AIDS RX;
IL-3; IF; FACTOR VIII; SEEDS; CORN

President: GABRIEL SCHMERGEL
R&D Dir: DR. ROBERT KAMEN
Financing: PUB
Started: 1981
Class Code: P

GENEX, INC.
16020 INDUSTRIAL DR.
GAITHERSBURG, MD 20877
Telephone: 301-258-0552
Products: COMMODITY & SPECIALTY CHEMICALS;
ENZYMES, PROTEINS, DRAIN CLEANER, VITAMIN B-12

CEO: GARY E. FRASHIER
President: GARY E. FRASHIER
R&D Dir: DR. JUDITH HAUTALA
Financing: PUB
Started: 1977
Class Code: I

GENTRONIX LABORATORIES, INC.
P.O. BOX 34244
BETHESDA, MD 20817
Telephone: 301-299-2037
Products: BIOCHIPS, BIOELECTRONICS

CEO: JOHN M. WEHRUNG
President: JOHN M. WEHRUNG
Financing: PRI
Started: 1972
Class Code: D

GENZYME CORP.
75 KNEELAND ST.
BOSTON, MA 02111
Telephone: 800-332-1042
Products: RX, DX FOR HUMAN HEALTH CARE; MABS;
EYE PRODUCTS, SKIN PRODUCTS; SURFACTANTS

CEO: SHERIDAN SNYDER
President: HENRI TERMEER
R&D Dir: DR. J.R. RASMUSSEN
Financing: PUB
Started: 1981
Class Code: P

GRANADA GENETICS CORP.
TEXAS A&M UNIV., 1 RESEARCH PARK
COLLEGE STATION, TX 77843
Telephone: 713-783-1310
Products: BULL GENETICS, SHRIMP, FOOD

President: DAVID ELLER
R&D Dir: DR. JOSEPH MASSEY
Financing: SUB
Started: 1979
Class Code: A

H

HANA BIOLOGICS, INC.
850 MARINA VILLAGE PKWY.
ALAMEDA, CA 94501
Telephone: 800-772-4262
Products: CELL IMPLANTS TO TREAT
DISEASE

CEO: CRAIG R. MCMULLEN
President: CRAIG R. MCMULLEN
R&D Dir: DR. H. FRED VOSS
Financing: PUB
Started: 1979
Class Code: P

HAWAII BIOTECHNOLOGY GROUP, INC.
99-193 AIEA HEIGHTS DR.
AIEA, HI 96701
Telephone: 808-487-5565
Products: IMMUNOTOXINS, PESTICIDE
DETECTION KIT

CEO: STEVEN K. BRAUER
President: STEVEN K. BRAUER
R&D Dir: DR. TOM HUMPHREYS
Financing: PRI
Started: 1982
Class Code: B

HAZLETON BIOTECHNOLOGIES CO.
9200 LEESBURG TURNPIKE
VIENNA, VA 22180
Telephone: 703-893-5400
Products: LAB REAGENTS AND EQUIPMENT,
MEDIA, MABS

CEO: LEWIS E.S. PARKER
President: LEWIS E.S. PARKER
R&D Dir: DR. JOAN RENER
Financing: SUB
Started: 1983
Class Code: G

HOUSTON BIOTECHNOLOGY, INC.
3606 RESEARCH FOREST DR.
THE WOODLANDS, TX 77380
Telephone: 713-363-0999
Products: OPHTHAMOLOGY AND NEUROLOGY
PRODUCTS

CEO: DR. DOMINIC MAN-KIT LAM
President: ALFRED G. SCHEID
R&D Dir: DAVID POTTER
Financing: PRI
Started: 1984
Class Code: P

Listing 2-2 Directory of Companies

HUNTER BIOSCIENCES, INC.
3208 SPRING FOREST RD.
RALEIGH, NC 27604
Telephone: 919-872-9686
Products: BIORESTORATION OF CONTAMINATED
WASTE SITES

CEO: DR. JASON CAPLAN
President: DR. JASON CAPLAN
R&D Dir: TONY LIEBERMAN
Financing: PRI
Started: 1987
Class Code: R

HYBRITECH, INC.
P.O. BOX 269006
SAN DIEGO, CA 92126
Telephone: 619-455-6700
Products: RX; DX; MABS; INSULIN; GROWTH
HORMONE; CANCER DX

President: DONALD W. GRIMM
R&D Dir: DR. DAVID S. KABAKOFF
Financing: SUB
Started: 1978
Class Code: P

HYCLONE LABORATORIES, INC.
1725 S. STATE HIGHWAY 89-91
LOGAN, UT 84321
Telephone: 801-753-4584
Products: IMMUNOCHEMICALS; ANIMAL SERA;
HYDBRIDOMAS

CEO: LELAND FOSTER
President: DR. REX S. SPENDLOVE
R&D Dir: DALE KERN
Financing: PRI
Started: 1975
Class Code: H

HYGEIA SCIENCES
330 NEVADA ST.
NEWTON, MA 02160
Telephone: 671-964-0200
Products: DX; OVULATION PREDICTOR, PREGNANCY
TEST KITS

President: DR. ELLIOT BLOCK
R&D Dir: L. EDWARD CANNON
Financing: PUB
Started: 1980
Class Code: K

I

IBF BIOTECHNICS, INC.
8510 CORRIDOR RD.
SAVAGE, MD 20763
Telephone: 800-752-5277
Products: TISSUE CULTURE REAGENTS; CHROMA -
TOGRAPHY COLUMNS; SERUM SUBSTITUTES

CEO: DR. GARY J. CALTON
President: AGNES LEMARCHAND
R&D Dir: EGISTTO BOSCHETTI
Financing: SUB
Started: 1987
Class Code: G

ICL SCIENTIFIC, INC.
11040 CONDOR AVE.
FOUNTAIN VALLEY, CA 92708
Telephone: 714-546-9581
Products: DX

President: DR. RICHARD HAMILL
R&D Dir: DR. TOM FOLEY
Financing: SUB
Started: 1981
Class Code: K

IDEC PHARMACEUTICAL
291 N. BERNARDO AVE.
MOUNTAIN VIEW, CA 92037
Telephone: 415-940-1200
Products: MABS (ANTI-IDIOTYPE)

President: DR. WILLIAM RASTETTER
R&D Dir: DR. HEINZ KOHLER
Financing: PRI
Started: 1986
Class Code: P

IDETEK, INC.
1057 SNEATH LANE
SAN BRUNO, CA 94066
Telephone: 415-952-2844
Products: DX & TOXIN TEST KITS FOR FOOD &
AGRICULTURE

CEO: DR. DIANE G. OLIVER
President: DR. DIANE G. OLIVER
R&D Dir: DR. PRITHIPAL SINGH
Financing: PRI
Started: 1983
Class Code: Y

IGEN
1530 E. JEFFERSON ST.
ROCKVILLE, MD 20852
Telephone: 301-984-7964
Products: MABS FOR HUMAN PATHOGENS;
CONTRACT RESEARCH

President: SAM WOHLSTADER
Financing: PRI
Started: 1982
Class Code: X

IGENE BIOTECHNOLOGY, INC.
9110 RED BRANCH RD.
COLUMBIA, MD 21045
Telephone: 301-997-2599
Products: MILK/EGG WHITE REPLACER, CALCIUM &
MINERAL FOOD SUPPLEMENTS, FISH AQUACULTURE

President: DR. ROBERT A. MILCH
R&D Dir: HUEI-HSIONG YANG
Financing: PUB
Started: 1981
Class Code: M

IMCERA BIOPRODUCTS, INC.
1810 FRONTAGE RD.
NORTHBROOK, IL 60062
Telephone: 800-345-3843
Products: IGF, RDNA MURINE EGF FOR LARGE SCALE
CELL CULTURE; RDNA & OTHER ADHESION PEPTIDES

CEO: CHARLES E. SEENEY
President: CHARLES E. SEENEY
R&D Dir: DR. DAVID PARKER
Financing: SUB
Started: 1986
Class Code: H

IMCLONE SYSTEMS, INC.
180 VARICK ST., 7TH FLOOR
NEW YORK, NY 10014
Telephone: 212-645-1405
Products: DX; VX; RX; AIDS DX; HEP B DX; DNA
PROBES, IMMUNODIAGNOSTICS FOR VIRAL DISEASES

CEO: DR. SAMUEL D. WAKSAL
President: DR. HARLAN WAKSAL
Financing: PRI
Started: 1984
Class Code: K

IMMUCELL CORP.
966 RIVERSIDE ST.
PORTLAND, ME 04103
Telephone: 207-797-8386
Products: HUMAN DX; REAGENTS; ANIMAL DX; VX

President: FREDERICK J. FOLEY
R&D Dir: DR. FRANK E. RUCH, JR.
Financing: PUB
Started: 1982
Class Code: A

IMMUNETECH PHARMACEUTICALS
11045 ROSELLE ST., SUITE A
SAN DIEGO, CA 92121
Telephone: 619-457-2553
Products: IMMUNE ACTIVE PEPTIDES AS RX

President: GORDON RAMSEIER
R&D Dir: DR. GARY S. HAHN
Financing: PRI
Started: 1981
Class Code: X

IMMUNEX CORP.
51 UNIVERSITY ST.
SEATTLE, WA 98101
Telephone: 206-587-0430
Products: RX; IL-2; GM-CSF; IL-1; IL-4;
LYMPHOKINES; MAF

CEO: STEPHEN A. DUZAN
President: STEPHEN A. DUZAN
R&D Dir: DR. STEVEN GILLIS
Financing: PUB
Started: 1981
Class Code: P

Listing 2-2 Directory of Companies

IMMUNEX, INC.
1445 HOT SPRING RD.
CARSON CITY, NV 89701
Telephone: 702-885-9400
Products: DX; MABS, POLYCLONAL AB TO HAPTENS;
RADIOIMMUNOASSAY KITS; REAGENTS

CEO: HEINZ A. PAULUS
President: HEINZ A. PAULUS
R&D Dir: WILLIAM DRELL
Financing: PRI
Started: 1981
Class Code: K

IMMUNO CONCEPTS, INC.
9779 BUSINESS PARK DR., SUITE I
SACRAMENTO, CA 95827
Telephone: 916-363-2649
Products: DX; ANTINUCLEAR AB TESTS; FLUORESCENT
DNA TEST; AUTOANTIBODY TEST SYSTEMS

CEO: DON W. VALENCIA
President: DON W. VALENCIA
R&D Dir: DR. HOY
Financing: PRI
Started: 1981
Class Code: K

IMMUNO MODULATORS LABS, INC.
10521 CORPORATE DR.
STAFFORD, TX 77477
Telephone: 713-240-9595
Products: RX; IF; AIDS RX; HCG; COSMETICS

President: OLIVER BRIGHT, JR.
R&D Dir: DR. JERZY GEORGIADES
Financing: PRI
Started: 1981
Class Code: P

IMMUNOGEN, INC.
148 SIDNEY ST.
CAMBRIDGE, MA 02139
Telephone: 617-661-9312
Products: CONJUGATED ABS FOR TUMORS & DX

President: DR. MITCHEL SAYARE
R&D Dir: DR. WALTER BLAETTLER
Financing: PRI
Started: 1981
Class Code: W

IMMUNOMED CORP.
5910-G BRECKENRIDGE PKWY.
TAMPA, FL 33610
Telephone: 813-621-9447
Products: VX; IMMUNOTHERAPEUTICS; RABIES VX

President: DR. KENT R. VAN KAMPEN
R&D Dir: DR. WILLIAM COX
Financing: PRI
Started: 1979
Class Code: Q

IMMUNOMEDICS, INC.
150 MT. BETHEL RD.
WARREN, NJ 07060
Telephone: 201-456-4779
Products: DX; RX; TEST KITS; CANCER DX; AUTO-
IMMUNE DISEASE ASSAY; ALLERGY PRODUCTS

CEO: DR. DAVID M. GOLDENBERG
President: GEORGE MASTERS
R&D Dir: DR. HANS HANSEN
Financing: PUB
Started: 1983
Class Code: P

IMMUNOSYSTEMS, INC.
8 LINCOLN ST., P.O. BOX AY
BIDDEFORD, ME 04005
Telephone: 207-282-4158
Products: ENZYME IMMUNOASSAYS

CEO: BRUCE S. FERGUSON
President: BRUCE S. FERGUSON
Financing: PRI
Started: 1981
Class Code: G

IMMUNOTECH CORP.
90 WINDOM ST.
BOSTON, MA 02134
Telephone: 617-787-1010
Products: IMMUNODIAGNOSTIC TEST KITS, DX

President: FRANCIS CAPITANIO
R&D Dir: DR. AMY DINGLEY
Financing: PRI
Started: 1980
Class Code: K

IMMUNOVISION, INC.
1506 FORD AVE.
SPRINGDALE, AR 72764
Telephone: 501-751-7005
Products: AB'S; DX; PURIFIED PROTEINS FOR AUTO-
IMMUNE DISEASE; BLOOD PROTEINS, ANIMAL TISSUE

CEO: PATRICIA HALE
President: PATRICIA HALE
R&D Dir: PAUL SLAYMAKER
Financing: PRI
Started: 1985
Class Code: P

IMRE CORP.
130 FIFTH AVE., N.
SEATTLE, WA 98109
Telephone: 206-448-1000
Products: IMMUNO BLOOD PURIFIER SYSTEMS;
PROTEIN A

CEO: DR. FRANK R. JONES
President: DR. FRANK R. JONES
R&D Dir: DR. HARRY W. SNYDER
Financing: PUB
Started: 1981
Class Code: E

IMREG, INC.
144 ELK PLACE, SUITE 1400
NEW ORLEANS, LA 70112
Telephone: 504-523-2875
Products: IMMUNOREGULATORS; CANCER DX; AIDS,
ARTHRITIS RX

CEO: DR. A. ARTHUR GOTTLIEB
President: DR. A. ARTHUR GOTTLIEB
R&D Dir: DR. A. ARTHUR GOTTLIEB
Financing: PUB
Started: 1981
Class Code: P

INCELL CORP.
9075 W. HEATHER AVE.
MILWAUKEE, WI 53224
Telephone: 800-545-4141
Products: PEPTIDE HORMONES; AMINO ACIDS;
REAGENTS; CHEMICALS

President: DR. PETER BAYNE
R&D Dir: DR. SANDRA NEUENDORF
Financing: PRI
Started: 1982
Class Code: J

INCON CORP.
137 WHITE OAK DR.
YOUNGSVILLE, NC 27596
Telephone: 919-556-5352
Products: R&D ON NEW PLANT/ANIMAL AQUA-
CULTURE CROPS

President: DAVID A. NUTTLE
R&D Dir: LINDA E. SHURTLEFF
Financing: PRI
Started: 1983
Class Code: S

INFERGENE CO.
433 INDUSTRIAL WAY
BENICIA, CA 94510
Telephone: 707-746-6804
Products: DX FOR HERPES, OTHERS; ENZYMES; BAKING
& BREWING STRAINS, GLUCOAMYLASE STRAIN

President: DR. BERNIE DIDARIO
R&D Dir: DR. WILLIAM SIKKEMA
Financing: PUB
Started: 1984
Class Code: J

INGENE, INC.
1545 17TH ST.
SANTA MONICA, CA 90404
Telephone: 213-829-7681
Products: RX; BONE & CARTILAGE RX; CANCER RX;
SWEETENERS; PESTICIDE

CEO: DR. GARY WILCOX
President: DR. GARY WILCOX
R&D Dir: DR. ARUP SEN
Financing: PUB
Started: 1980
Class Code: P

INTEGRATED CHEMICAL SENSORS
44 MECHANIC ST.
NEWTON, MA 02164
Telephone: 617-965-6950
Products: INSTRUMENTS AND BIOSENSORS

President: CARL M. GOOD
R&D Dir: GLEN BAASTIANS
Financing: PRI
Started: 1984
Class Code: D

Listing 2-2 Directory of Companies

INTEGRATED GENETICS, INC.
31 NEW YORK AVE.
FRAMINGHAM, MA 01701
Telephone: 617-875-1336
Products: DX; RX; AIDS DX; FOOD TOXIN DX;
CANCER DX; TPA; HCG; EPO; HEP B VX

CEO: ROBERT J. CARPENTER
President: ROBERT J. CARPENTER
R&D Dir: DR. ALAN E. SMITH
Financing: PUB
Started: 1981
Class Code: P

INTEK DIAGNOSTICS, INC.
1450 ROLLINS RD.
BURLINGAME, CA 94010
Telephone: 415-340-0530
Products: T CELL CLONES, ANTIGENS, FLOW
CYTOMETRY

President: DICK ROGERS
R&D Dir: VICTOR LIU
Financing: PRI
Started: 1983
Class Code: X

INTELLIGENETICS, INC.
700 EAST EL CAMINO REAL
MOUNTAIN VIEW, CA 94040
Telephone: 415-962-7300
Products: GENETIC ENGINEERING SOFTWARE;
DNA & PROTEIN ANALYSIS

President: DR. MICHAEL KELLY
R&D Dir: RANCE DELONG
Financing: PRI
Started: 1980
Class Code: F

INTER-AMERICAN RESEARCH ASSOCIATES
18919 PREMEIR CT.
GAITHERSBURG, MD 20879
Telephone: 301-926-0713
Products: PURIFIED ANTIGENS, ANTISERA, BLOOD
PRODUCTS

CEO: TOM KELLY
President: TOM KELLY
R&D Dir: THOMAS LANKENAU
Financing: PRI
Started: 1979
Class Code: G

INTER-CELL TECHNOLOGIES, INC.
422 ROUTE 206, SUITE 143
SOMERVILLE, NJ 08876
Telephone: 201-359-8228
Products: BIOLOGICALS; IMMUNOLOGICAL
REAGENTS; LYMPHOKINES

President: DR. DENNIS DELLA PENTA
R&D Dir: MARGARET CIPUZAK
Financing: PRI
Started: 1982
Class Code: X

INTERFERON SCIENCES, INC.
783 JERSEY AVE.
NEW BRUNSWICK, NJ 08901
Telephone: 201-249-3250
Products: IF-A; IF-G; MABS

President: DR. SAMUEL RONEL
R&D Dir: DR. DOUGLAS TESTA
Financing: SUB
Started: 1981
Class Code: P

INTERNATIONAL BIOTECHNOLOGIES, INC.
25 SCIENCE PARK
NEW HAVEN, CT 06511
Telephone: 203-786-5600
Products: NUCLEIC ACIDS, CLONING, SEQUENCING,
GENOME ANALYSES & ELECTROPHORESIS SYSTEMS

President: MARTIN J. MATTESSICH
R&D Dir: DOUGLAS VIZARD
Financing: SUB
Started: 1982
Class Code: G

INTERNATIONAL ENZYMES, INC.
1667 S. MISSION RD.
FALLBROOK, CA 92028
Telephone: 619-728-5205
Products: ENZYMES; ANTISERA; SERUM-FREE MEDIA
SUPPLIES

President: CHARLES G. HAUGH
R&D Dir: JEFFERSON B. OFFICER
Financing: PRI
Started: 1983
Class Code: G

INTERNATIONAL PLANT RESEARCH INST.
830 BRANSTEN RD.
SAN CARLOS, CA 94070
Telephone: 415-595-5335
Products: DATE PALMS; OIL PALMS; CEREALS;
FOOD PLANTS; GENETIC ENG. REAGENTS

CEO: DR. RAYMOND J. MOSHY
President: DR. RAYMOND J. MOSHY
R&D Dir: DON DILTZ
Financing: PRI
Started: 1978
Class Code: B

INVITRON CORP.
4649 LE BOURGET DR.
ST. LOUIS, MO 63134
Telephone: 314-426-5000
Products: DX; RX; TPA, HGH, FACTOR VIII; CELL
PRODUCTS

President: DR. CHARLES V. BENTON
R&D Dir: DR. PETER REEVE
Financing: PRI
Started: 1984
Class Code: P

J

JACKSON IMMUNORESEARCH LABS, INC.
872 W. BALTIMORE DR.
WEST GROVE, PA 19390
Telephone: 800-367-5296
Products: AFFINITY PURIFICATION PROTEINS; PURE
PROTEINS

President: DR. WILLIAM J. STEGEMAN
R&D Dir: DR. CYNARA Y. KO
Financing: PRI
Started: 1982
Class Code: G

K

KALLESTAD DIAGNOSTICS
1120 CAPITAL OF TEXAS HWY., S.
AUSTIN, TX 78746
Telephone: 512-329-5555
Products: DX PRODUCTS

President: DAVID MCWILLIAMS
R&D Dir: DR. RICHARD TURNER
Financing: PRI
Started: 1967
Class Code: K

KARYON TECHNOLOGY, LTD.
333 PROVIDENCE HWY.
NORWOOD, MA 02062
Telephone: 617-769-6970
Products: TISSUE PROCESS CULTURE TECHNOLOGY

President: V.M. ESPOSITO
R&D Dir: DR. ROBERT J. BUEHLER
Financing: PRI
Started: 1984
Class Code: H

KIRIN-AMGEN
C/O AMGEN, 1900 OAK TERRACE LANE
THOUSAND OAKS, CA 91320
Telephone: 805-499-5725
Products: EPO

President: DR. GEORGE RATHMANN
R&D Dir: DANIEL VAPNEK
Financing: PRI
Started: 1984
Class Code: P

KIRKEGAARD & PERRY LABS, INC.
2 CESSNA COURT
GAITHERSBURG, MD 20879
Telephone: 301-948-7755
Products: POULTRY DX; VET DX; IMMUNOLOGICAL
REAGENTS

CEO: ALBERT PERRY
President: ALBERT PERRY
R&D Dir: LESLIE KIRKEGAARD
Financing: PRI
Started: 1979
Class Code: V

Listing 2-2 Directory of Companies

L

LABSYSTEMS, INC.
6200 W. OAKTON
MORTON GROVE, IL 60053
Telephone: 312-967-5220
Products: MICROBIOLOGY ASSAYS &
EQUIPMENT; MABS

CEO: GLENN SEYMOUR
President: GLENN SEYMOUR
R&D Dir: PAUL PARTANEN
Financing: SUB
Started: 1983
Class Code: K

LEE BIOMOLECULAR RESEARCH LABS
11211 SORRENTO VALLEY RD.
SAN DIEGO, CA 92121
Telephone: 619-452-7700
Products: IF; ANTI-IF SERA

President: DR. LEE KRONENBERG
R&D Dir: SANFORD KRONENBERG
Financing: PRI
Started: 1980
Class Code: P

LIFE SCIENCES
2900 72 ST., N.
ST. PETERSBURG, FL 33710
Telephone: 813-345-9371
Products: IF; RESEARCH ON VIRUSES, ANIMAL
STRAINS

President: SIMON SRYBNIK
R&D Dir:
Financing: PUB
Started: 1962
Class Code: W

LIFE TECHNOLOGIES, INC.
8717 GROVEMONT CIRCLE
GAITHERSBURG, MD 20877
Telephone: 301-840-8000
Products: IF; PROBE ASSAY DX; REAGENTS

President: DR M. JAMES BARRETT
R&D Dir: DR. DIETMAR RABUSSAY
Financing: PUB
Started: 1983
Class Code: K

LIFECODES CORP.
OLD SAWMILL RIVER RD.
VALHALLA, NY 10595
Telephone: 914-592-4122
Products: REFERENCE LAB DX USING DNA PROBES

President: JOHN K. WINKLER, JR.
R&D Dir: DR. JEFFREY GLASSBERG
Financing: SUB
Started: 1982
Class Code: K

LIFECORE BIOMEDICAL, INC.
1055 10TH AVE., S.E.
MINNEAPOLIS, MN 55414
Telephone: 612-379-8080
Products: HYALURONIC ACID,
MEDICAL DEVICES

CEO: DR. JAMES W. BRACKE
President: DR. JAMES W. BRACKE
R&D Dir: DR. KIPLING THACKER
Financing: PUB
Started: 1975
Class Code: J

LIPOSOME COMPANY, INC.
ONE RESEARCH WAY
PRINCETON, NJ 08540
Telephone: 609-452-7060
Products: LIPOSOMES; DRUG DELIVERY;
CANCER RX; VX

CEO: EDGAR T. MERTZ
R&D Dir: DR. ROBERT M. COHN
Financing: PUB
Started: 1981
Class Code: 3

LIPOSOME TECHNOLOGY, INC.
1050 HAMILTON CT.
MENLO PARK, CA 94025
Telephone: 415-323-9011
Products: LIPOSOMES FOR RX, DRUG DELIVERY,
VETERINARY, AGRICULTURE

CEO: DR. NICOLAS ARVANITIDIS
President: DR.NICOLAS ARVANITIDIS
R&D Dir: FRANCIS J. MARTIN
Financing: PRI
Started: 1981
Class Code: 3

LUCKY BIOTECH CORP.
4560 HORTON ST.
EMERYVILLE, CA 94608
Telephone: 415-653-0734
Products: RX; IF; IF-A; IF-B; IF-G; IL-2; HEP B DX;
PORCINE GROWTH HORMONE

President: SHIN KOO HUH
R&D Dir: DR. JOONG MYUNG CHO
Financing: SUB
Started: 1984
Class Code: P

LYPHOMED, INC.
10401 WEST TOUHY
ROSEMONT, IL 60018
Telephone: 312-390-6500
Products: RX, VITAMIN SUPPLEMENTS FOR
INTRAVENOUS USE; ANTIBIOTICS

CEO: DR. JOHN N. KAPOOR
President: DR. JOHN N. KAPOOR
R&D Dir: ABU ALAM
Financing: PUB
Started: 1981
Class Code: P

M

MAIZE GENETIC RESOURCES, INC.
421 ANGIER HWY.
BENSON, NC 27504
Telephone: 919-934-9908
Products: PROPRIETARY TECHNOL. FOR CORN
BREEDING STRESS-RESISTANT VARIETIES

CEO: DR. JAMES W. FRIEDRICH
President: DR. JAMES W. FRIEDRICH
Financing: PRI
Started: 1984
Class Code: B

MARCOR DEVELOPMENT CO.
206 PARK ST.
HACKENSACK, NJ 07601
Telephone: 201-489-5700
Products: REAGENTS FOR FERMENTATION, ENZYMES

President: CHARLES J. GARBARINI
R&D Dir: R.G. BERGER, JR.
Financing: PRI
Started: 1977
Class Code: G

MARICULTURA, INC.
P.O. DRAWER 565
WRIGHTSVILLE BEACH, NC 28480
Telephone: 919-256-5010
Products: MARINE NATURAL PRODUCTS & CHEMICALS

CEO: DR. THOMAS V. LONG, III
President: DR. THOMAS V. LONG, III
Financing: PRI
Started: 1984
Class Code: T

MARINE BIOLOGICALS, INC.
P.O. BOX 546
MARMORA, NJ 08223
Telephone: 609-390-1333
Products: MARINE BYPRODUCTS, LIMULUS
AMEBOCYTE LYSATE

President: JAMES J. FINN
R&D Dir: JAMES J. FINN
Financing: PRI
Started: 1981
Class Code: T

MARTEK
9115 GUILFORD RD.
COLUMBIA, MD 21046
Telephone: 301-490-2566
Products: STABILIZATOPE-LABELLED BIOCHEMICALS,
RADIOLABELLED LUBRICANTS, AMINO ACIDS

CEO: RICHARD RADMER
R&D Dir: JOHN COX
Financing: PRI
Started: 1985
Class Code: J

Listing 2-2 Directory of Companies

MAST IMMUNOSYSTEMS, INC.
630 CLYDE CT.
MOUNTAIN VIEW, CA 94043
Telephone: 415-961-5501
Products: ALLERGY DX; ADAPTING PATENTED
TECHNOLOGY TO INFECTIOUS DISEASES

CEO: JOHN LUCAS
President: JOHN LUCAS
R&D Dir: CHRISTOPHER BROWN
Financing: PRI
Started: 1979
Class Code: K

MEDAREX, INC.
1401 BROAD ST.
CLIFTON, NJ 07015
Telephone: 201-773-7032
Products: MAB-BASED
RX & DX

CEO: DONALD L. DRAKEMAN
President: DONALD L. DRAKEMAN
R&D Dir: DR. MICHAEL W. FANGER
Financing: PRI
Started: 1987
Class Code: P

MELOY LABORATORIES, INC.
6715 ELECTRONIC DR.
SPRINGFIELD, VA 22151
Telephone: 703-354-2600
Products: SERVICES; REAGENTS; R&D; IF

President: DR. JAMES TRETTER
R&D Dir: DR. WILLIAM DROHAN
Financing: SUB
Started: 1970
Class Code: W

MICROBE MASTERS
11814 COURSEY BLVD., SUITE 285
BATON ROUGE, LA 70816
Telephone: 504-665-1903
Products: INDUSTRIAL WASTEWATER TREATMENT
USING STRESS ACCLIMATED BACTERIAL PRODUCTS

President: J. PETER PEREZ
R&D Dir: MS. TRACEY KOENIG
Financing: PRI
Started: 1982
Class Code: R

MICROBIO RESOURCES
6150 LUSK BLVD., SUITE B105
SAN DIEGO, CA 92121
Telephone: 619-587-0600
Products: FOOD ADDITIVES & COLORANTS; SPECIAL
CHEMICALS

CEO: DEXTER W. GASTON
President: DEXTER W. GASTON
R&D Dir: DR. KENNETH G. SPENCER
Financing: PRI
Started: 1981
Class Code: J

MICROBIOLOGICAL ASSOCIATES, INC.
9900 BLACKWELL RD.
ROCKVILLE, MD 20850
Telephone: 301-738-1000
Products: TESTING SERVICE FOR COMPLIANCE
WITH REGULATORY AGENCIES (FDA/EPA)

President: LEWIS SHUSTER
R&D Dir: ANDREW LOSIKOFF
Financing: PRI
Started: 1949
Class Code: 5

MICROGENESYS, INC.
400 FRONTAGE RD.
WEST HAVEN, CT 06516
Telephone: 203-932-3203
Products: AIDS DX; AIDS VX; VIRAL INSECTICIDE;
DEVELOPING HUMAN INFECTIOUS DISEASE VX

CEO: FRANK VOLVOVITZ
President: FRANK VOLVOVITZ
R&D Dir: DR. MARK A COCHRAN
Financing: PRI
Started: 1983
Class Code: Q

MICROGENICS CORP.
2380A BISSO LANE
CONCORD, CA 94520
Telephone: 415-674-0667
Products: DX ASSAYS;
DIGOXIN ASSAY

CEO: DR. FREDERIC FELDMAN
President: DR. FREDERIC FELDMAN
R&D Dir: DR. PYARE KHANNA
Financing: PRI
Started: 1981
Class Code: K

MOLECULAR BIOLOGY RESOURCES, INC.

5520 W. BURLEIGH ST.
MILWAUKEE, WI 53210
Telephone: 414-871-7199
Products: REAGENTS; DNA & RNA MODIFYING
ENZYMES & NUCLEIC ACIDS

CEO: PETER J. SMYCZEK
President: PETER J. SMYCZEK
R&D Dir: JAMES F. WICK
Financing: PRI
Started: 1986
Class Code: G

MOLECULAR BIOSYSTEMS, INC.

10030 BARNES CANYON RD.
SAN DIEGO, CA 92121
Telephone: 619-452-0681
Products: DX; DNA & RNA PROBES, ALBUMEX,
EXTRACTOR KITS

CEO: DR. KENNETH WIDDER
President: VINCENT A. FRANK
R&D Dir: DR. JERRY RUTH
Financing: PUB
Started: 1981
Class Code: K

MOLECULAR DIAGNOSTICS, INC.

400 MORGAN LN.
WEST HAVEN, CT 06516
Telephone: 203-934-9229
Products: RX FOR DIABETES, CANCER, INFECTIOUS
DISEASES, HUMAN GENETIC DISEASES

CEO: GORDON P. POLLEY
President: GORDON P. POLLEY
Financing: SUB
Started: 1981
Class Code: K

MOLECULAR GENETIC RESOURCES

6201 JOHNS RD., SUITE 8
TAMPA, FL 33634
Telephone: 813-886-5338
Products: ENZYMES; REV. TRANSCRIPTASE; TERM.
TRANSFERASE; MRNA ANALYSIS KIT

President: DR. G.E. HOUTS
Financing: PRI
Started: 1983
Class Code: G

MOLECULAR GENETICS, INC.

10320 BREN ROAD EAST
MINNETONKA, MN 55343
Telephone: 612-935-7335
Products: DX; AG DX; BGH; VET DX; SCOURS VX

CEO: KENNETH TEMPERO
R&D Dir: DR. CHARLES MUSCOPLAT
Financing: PUB
Started: 1979
Class Code: A

MONOCLONAL ANTIBODIES, INC.

2319 CHARLESTON RD.
MOUNTAIN VIEW, CA 94043
Telephone: 415-960-1320
Products: DX TEST KITS-PREGNANCY, REPRODUCTION,
INFERTILITY; VET DX

CEO: THOMAS A. GLAZE
President: JOHN D. DIEKMAN
R&D Dir: ROB FRASER
Financing: PUB
Started: 1979
Class Code: K

MONOCLONAL PRODUCTION INT'L

TWENTIETH AND SYDNEY STREETS
FORT SCOTT, KS 66701
Telephone: 316-223-4230
Products: MABS FORMED IN SHEEP/COW UTERUS

President/CEO: WILBUR SMITH
R&D Dir: JIM CULLOR
Financing: PRI
Started: 1983
Class Code: X

MONOCLONETICS INTERNATIONAL, INC.

18333 EGRET BAY BLVD., SUITE 270
HOUSTON, TX 77058
Telephone: 713-335-1611
Products: HUMAN SCREENING TESTS & DX,
EPITOPE SCREENS

CEO: RICHARD E. WARRINGTON
President: RICHARD WARRINGTON
R&D Dir: ABBAS KHAN
Financing: PRI
Started: 1984
Class Code: G

Listing 2-2 Directory of Companies

MULTIPLE PEPTIDE SYSTEMS, INC.
10955 JOHN JAY HOPKINS DR., BLDG. 2
SAN DIEGO, CA 92121
Telephone: 619-455-3710
Products: HYDROGEN FLUORIDE CLEAVAGE EQUIP.;
PEPTIDE SYNTHESIS

CEO: SUZANNE PRATT
President: DR. C. HENDRICKSON
R&D Dir: DR. RICHARD HOUGHTEN
Financing: PRI
Started: 1986
Class Code: F

MUREX CORP.
P.O. BOX 2003
NORCROSS, GA 30071
Telephone: 404-662-0660
Products: DX ; MAB REAGENTS &
ANTIBODY KITS

CEO: JAREL R. KELSEY
President: JAREL R. KELSEY
R&D Dir: WILLIAM M. MARTIN
Financing: PRI
Started: 1984
Class Code: K

MYCOGEN CORP.
5451 OBERLIN DR.
SAN DIEGO, CA 92121
Telephone: 619-453-8030
Products: PLANT HERBICIDES/PESTICIDES BASED
ON NATURAL PATHOGENS

CEO: DR. JERRY CAULDER
President: DR. JERRY CAULDER
R&D Dir: LEO KIMI
Financing: PUB
Started: 1982
Class Code: B

MYCOSEARCH, INC.
P.O. BOX 941
CHAPEL HILL, NC 27514
Telephone: 919-968-0730
Products: CULTURES OF RARE & NEW FUNGI;
LIVE CULTURES; SPECIALTY CHEMICALS

President: DR. BARRY KATZ
R&D Dir: DR. BARRY KATZ
Financing: PRI
Started: 1979
Class Code: 2

N

NATIONAL GENO SCIENCES
22150 W. NINE MILE ROAD
SOUTHFIELD, MI 48034
Telephone: 313-552-8540
Products: LEUCOCYTE IF

CEO: DR. R. J. GANESH
Financing: PRI
Started: 1980
Class Code: P

NATIVE PLANTS, INC.
417 WAKARA WAY
SALT LAKE CITY, UT 84108
Telephone: 801-582-0144
Products: POTATO, FLOWER, VEGETABLE SEEDS;
SOIL INOCULANTS; PLANTS

CEO: PETER D. MELDRUM
R&D Dir: DR. CYRUS MCKELL
Financing: PUB
Started: 1973
Class Code: B

NEOGEN CORP.
620 LESHER PLACE
LANSING, MI 48912
Telephone: 517-372-9200
Products: FOOD, PLANT HEALTH, ANIMAL
HEALTH, DX

President: DR. JAMES L. HERBERT, JR.
R&D Dir: DR. BRINTON M. MILLER
Financing: PRI
Started: 1981
Class Code: B

NEORX CORP.
410 W. HARRISON ST.
SEATTLE, WA 98119
Telephone: 206-281-7001
Products: MELANOMA IMAGING AGENT WITH
TECHNETIUM

CEO: DR. ROBERT T. ABBOTT
President: DR. ROBERT T. ABBOTT
R&D Dir: DR. PAUL ABRAMS
Financing: PRI
Started: 1984
Class Code: P

NEUREX CORP.
3760 HAVEN AVE.
MENLO PARK, CA 94025
Telephone: 415-853-1500
Products: RX, NEUROPEPTIDES, NEUROHORMONES

CEO: THOMAS L. BARTON
R&D Dir: DR. J. RAMACHANDRAN
Financing: PRI
Started: 1986
Class Code: P

NEUSHUL MARICULTURE
475 KELLOGG WAY
GOLETA, CA 93117
Telephone: 805-964-5844
Products: ANTI-VIRAL; ENZYMES; MOLLUSCAN
SEED & CULTURE

CEO: MICHAEL NEUSHUL
President: MICHAEL NEUSHUL
R&D Dir: GEOFFRY BROSSEAU
Financing: PRI
Started: 1978
Class Code: S

NITRAGIN CO.
3101 WEST CUSTER
MILWAUKEE, WI 53029
Telephone: 414-462-7600
Products: NITROGEN FIXING BACTERIA INOCULANTS

President: TOM WINKOSKE
R&D Dir: ARTHUR A. NETHERY
Financing: SUB
Started: 1898
Class Code: B

NORDEN LABS
601 W. CORNHUSKER HWY.
LINCOLN, NB 68521
Telephone: 402-475-4541
Products: VX; VET VX & RX
HEP B VX

CEO: HENRY WENDT
President: GEORGE EBRIGHT
R&D Dir: DR. STANLEY T. CROOKE
Financing: SUB
Started: 1983
Class Code: Q

NORTH COAST BIOTECHNOLOGY, INC.
3100 E. OVERLOOK
CLEVELAND HEIGHTS, OH 44118
Telephone: 216-932-0090
Products: ENZYMES FOR RX MANUFACTURING

CEO: DR. DAVID P. NORBY
R&D Dir: DR. DAVID P. NORBY
Financing: PRI
Started: 1986
Class Code: J

NYGENE CORP.
6 EXECUTIVE PLAZA
YONKERS, NY 10701
Telephone: 914-964-8300
Products: BIOAFFINITY SEPARATION SYSTEMS;
HYBRIDOMA PRODUCTION

President: FRANK P. BRUNETTA
R&D Dir: DR. ZENON STEPLEWSKI
Financing: PRI
Started: 1985
Class Code: F

O

O.C.S. LABORATORIES, INC.
P.O. BOX 1940, RTE. 2
SANGER, TX 76266
Telephone: 817-458-4422
Products: CUSTOM MADE SYNTHETIC GENES &
PEPTIDES

President: MIGUEL M. CASTRO
R&D Dir: DEBORAH R. CASTRO
Financing: PRI
Started: 1983
Class Code: J

Listing 2-2 Directory of Companies

OCEAN GENETICS
140 DUBOIS ST.
SANTA CRUZ, CA 95060
Telephone: 408-458-5205
Products: SPECIAL CHEMICALS, RX OF MARINE ALGAE

CEO: WAYNE A. HARVEY
R&D Dir: KENNETH L. TERRY
Financing: PRI
Started: 1981
Class Code: T

OMNI BIOCHEM, INC.
2215 CLEVELAND AVE.
NATIONAL CITY, CA 92050
Telephone: 619-541-0309
Products: PEPTIDES; ENZYMES; AMINO ACIDS;
REAGENTS

CEO: GERALD L. MYRES
President: DR. K.C. BASAVA
R&D Dir: DR. K.C. BASAVA
Financing: PRI
Started: 1986
Class Code: J

ONCOGENE SCIENCE, INC.
350 COMMUNITY DRIVE
MANHASSET, NY 11030
Telephone: 516-365-9300
Products: ONCOGENE PROBES FOR DX AND RX, MABS

President: DR. JOHN R. STEPHENSON
R&D Dir: DR. JOHN R. STEPHENSON
Financing: PUB
Started: 1983
Class Code: K

ONCOR, INC.
P.O. BOX 870
GAITHERSBURG, MD 20877
Telephone: 301-963-3500
Products: AIDS DX, DX REAGENTS, DNA CANCER
PROBES, HTLV-III RNA PROBE, B & T CELL PROBES

President: STEPHEN TURNER
R&D Dir: DR. JAY GEORGE
Financing: PUB
Started: 1983
Class Code: K

ORGANON TEKNIKA CORP.
800 CAPITOLA DR.
DURHAM, NC 27713
Telephone: 919-361-1995
Products: HEALTH CARE/MEDICAL DX

CEO: J.F. SISTERMANS
President: J.F. SISTERMANS
R&D Dir: R. DRISCOLL
Financing: SUB
Started: 1978
Class Code: K

P

PAMBEC LABORATORIES
36533 N. STREAMWOOD DR.
GURNEE, IL 60031
Telephone: 414-963-8594
Products: CUSTOM/CONTRACT WORK-GENETIC ENG.
PROKARYOTES, EUKARYOTES, DNA PROBES, AIDS RX

CEO: TIM RASICO
President: JOHN BADTKE
R&D Dir: TIM RASICO
Financing: PRI
Started: 1987
Class Code: J

PEL-FREEZ BIOLOGICALS, INC.
P.O. BOX 68
ROGERS, AR 72757
Telephone: 800-643-3426
Products: VARIOUS SPECIALTY REAGENTS, MAB SERA

President: DAVID DUBBELL
R&D Dir: DR. P.K. CHUNG
Financing: PRI
Started: 1911
Class Code: G

PENINSULA LABORATORIES, INC.
611 TAYLOR WAY
BELMONT, CA 94002
Telephone: 415-592-5392
Products: PEPTIDES, ENZYMES, SPECIALTY REAGENTS

President: MEIKYO SHIMIZU
R&D Dir: BILL PETERSON
Financing: PRI
Started: 1971
Class Code: J

PETROFERM, INC.
5400 FIRST COAST HWY, SUITE 200
FERNANDINA BEACH, FL 32034
Telephone: 904-261-8286
Products: MICROBIALS FOR
IMPROVED FUELS

CEO: WILLIAM R. GALLOWAY, JR.
President: WILLIAM R. GALLOWAY, JR.
R&D Dir: DR. MICHAEL HAYES
Financing: SUB
Started: 1977
Class Code: L

PETROGEN, INC.
2452 EAST OAKTON
ARLINGTON HEIGHTS, IL 60005
Telephone: 312-952-1309
Products: MICROBES FOR IMPROVED
OIL RECOVERY

CEO: PAAL PRESTEGAARD
President: PAAL PRESTEGAARD
R&D Dir: PAAL PRESTEGAARD
Financing: PRI
Started: 1980
Class Code: L

PHARMACIA LKB BIOTECHNOLOGY, INC.
800 CENTENNIAL AVE.
PISCATAWAY, NJ 08854
Telephone: 201-457-8000
Products: CHROMATOGRAPHY CHEMICALS &
INSTS.; ENZYMES; REAGENTS; BIOCHEMICALS

President: DR. RICHARD EASTERDAY
R&D Dir: DR. ROBERT M. FLORA
Financing: SUB
Started: 1960
Class Code: E

PHARMAGENE
607 EAST KINGSTON AVE.
CHARLOTTE, NC 28203
Telephone: 704-374-1224
Products: RX; CHEMICALS; SPECIFIC SITE
RECOMBINANT PROCESSES

CEO: MAC SASSER
President: MAC SASSER
R&D Dir: MAC SASSER
Financing: PRI
Started: 1986
Class Code: P

PHYTOGEN
101 WAVERLY DR.
PASADENA, CA 91105
Telephone: 818-792-1802
Products: IMPROVED CROP PLANTS; MUNG BEANS,
COTTON, SOY

CEO: DR. DAVID ANDERSON
President: DR. DAVID ANDERSON
R&D Dir: DR. DAVID ANDERSON
Financing: PRI
Started: 1980
Class Code: B

PLANT GENETICS, INC.
1930 FIFTH ST.
DAVIS, CA 95616
Telephone: 916-753-1400
Products: POTATOES; HYBRID VEGETABLES;
ALFALFA; CORN; SEEDS

President: DR. ZACHARY S. WOCHOK
R&D Dir: DR. KEITH A. WALKER
Financing: PUB
Started: 1981
Class Code: B

POLYCELL, INC.
321 FISHER BLDG.
DETROIT, MI 48202
Telephone: 313-871-4151
Products: DX; MABS; IMAGING FOR HEART DISEASE,
OVARIAN CANCER DX

CEO: JASON PANKIN
President: JASON PANKIN
R&D Dir: DR. CHRIS READING
Financing: SUB
Started: 1983
Class Code: K

PRAXIS BIOLOGICS
30 CORPORATE WOODS, SUITE 300
ROCHESTER, NY 14623
Telephone: 716-272-7000
Products: VX; TEST KITS; RX; DX

CEO: DR. DAVID H. SMITH
President: DR. DAVID H. SMITH
R&D Dir: DR. WAYNE T. HOCKMEYER
Financing: PUB
Started: 1983
Class Code: Q

Listing 2-2 Directory of Companies

PROBIOLOGICS INTERNATIONAL, INC.
1000 PARK FORTY PLAZA, SUITE 200
RESEARCH TRIANGLE PARK, NC 27709
Telephone: 919-737-7652
Products: RX, PROBIOTICS, ANTIBIOTICS FOR
HUMAN & ANIMAL USE

CEO: DR. WALTER DOBROGOSZ
President: DR. WILLARD HAMILTON
R&D Dir: DR. IVAN A. CASAS
Financing: PRI
Started: 1987
Class Code: P

PROGENX
10955 JOHN HOPKINS DR.
SAN DIEGO, CA 92121
Telephone: 619-455-3900
Products: CANCER DX MARKERS;
CANCER RX

CEO: HOWARD C. BIRNDORF
President: HOWARD C. BIRNDORF
R&D Dir: DR. TINA BERGER
Financing: PRI
Started: 1987
Class Code: K

PROMEGA CORP.
2800 SOUTH FISH HATCHERY RD.
MADISON, WI 53711
Telephone: 608-274-4330
Products: REAGENTS FOR MOLECULAR BIOLOGY;
ENZYMES

President: WILLIAM LINTON
R&D Dir: DR. RANDALL DIMOND
Financing: PRI
Started: 1978
Class Code: G

PROTATEK INTERNATIONAL, INC.
1491 ENERGY PARK DRIVE
ST. PAUL, MN 55108
Telephone: 612-644-5391
Products: VIRUS, DX TEST KITS; RX; SPECIALIZED
MICROBES; VET VX, RX, DX

President: DR. G.H. KELLERMAN
R&D Dir: DR. K. ROGER TSANG
Financing: PRI
Started: 1984
Class Code: K

PROTEINS INTERNATIONAL, INC.
1858 STAR BATT DR.
ROCHESTER, MI 48309
Telephone: 313-852-1067
Products: IMMUNOLOGICAL DX;
CUSTOM MABS

CEO: DR. DENIS M. CALLEWAERT
President: DR. DENIS M. CALLEWAERT
R&D Dir: DR. CYNTHIA L. SEVILLA
Financing: PRI
Started: 1983
Class Code: K

PROVESTA CORP.
15 PHILLIPS BUILDING
BARTLESVILLE, OK 74004
Telephone: 918-661-5281
Products: FERMENTATION-BASED FLAVORS; ENZYMES;
YEAST FEEDS; PEST CONTROL PHEROMONES; TPA

President: DR. JOHN NORRELL
R&D Dir: GARY HAYNE
Financing: SUB
Started: 1975
Class Code: J

Q

QUEST BIOTECHNOLOGY, INC.
321 FISHER BLDG.
DETROIT, MI 48202
Telephone: 313-873-0200
Products: BISPECIFIC MAB; IN VITRO DX; CANCER
RX & IMAGING

CEO: EUGENE SCHUSTER
President: EUGENE SCHUSTER
R&D Dir: DR. WERNER H. WAHL
Financing: PUB
Started: 1986
Class Code: K

QUEUE SYSTEMS, INC.
1250 ROUTE 28
NORTH BRANCH, NJ 08876
Telephone: 201-218-0558
Products: CRYO PRESERVATION; CELL CULTURE;
CELL ENVIRONMENTAL SYSTEMS

CEO: WILLIAM GELB
President: WILLIAM GELB
Financing: PRI
Started: 1980
Class Code: H

QUIDEL
11077 N. TORREY PINES RD.
LA JOLLA, CA 92037
Telephone: 619-450-1533
Products: DX ASSAYS FOR PREGNANCY, FERTILITY,
STREP-A, ALLERGY, INFECTIOUS DISEASE

President: JOSEPH STEMLER
R&D Dir: DR. JOHN BURD
Financing: PRI
Started: 1981
Class Code: K

R

R & A PLANT/SOIL, INC.
24 PASCO-KAHLOTUS RD.
PASCO, WA 99301
Telephone: 509-545-6867
Products: ALGAL SOIL CONDITIONERS, FERTILIZERS

President: RONALD JOHNSON
R&D Dir: DR. BLAINE METTING
Financing: PRI
Started: 1978
Class Code: B

RECOMTEX CORP.
4700 S. HAGADORN RD., SUITE 290
EAST LANSING, MI 48823
Telephone: 517-332-1600
Products: GENETIC MEDICAL DX

President: ALAN P. SUITS
R&D Dir: DR. HAROLD L. SADOFF
Financing: PRI
Started: 1983
Class Code: K

REPAP TECHNOLOGIES
PO BOX 766, 2650 EISENHOWER AVE.
VALLEY FORGE, PA 19482
Telephone: 215-630-9630
Products: BIOMASS CONVERSION FOR ALCOHOLS,
SUGARS

President: DR. E. ENDALL PYE
R&D Dir: DR. MALCOLM CRONLUND
Financing: SUB
Started: 1981
Class Code: C

REPLIGEN CORP.
ONE KENDALL SQ., BLDG. 700
CAMBRIDGE, MA 02139
Telephone: 617-225-6000
Products: ENZYMES, AIDS VX, PROTEINS, PROTEIN A,
HTLVIII, PESTICIDES

President: SANDFORD D. SMITH
R&D Dir: DR. THOMAS FRASER
Financing: PUB
Started: 1981
Class Code: J

RESEARCH & DIAGNOSTIC ANTIBODIES
P.O. BOX 7653
BERKELEY, CA 94707
Telephone: 415-652-7330
Products: DX KITS, MAB REAGENTS

President: DR. ROBERT J. WEBBER
Financing: PRI
Started: 1985
Class Code: K

RHOMED, INC.
1020 TIJERAS N.E.
ALBUQUERQUE, NM 87106
Telephone: 505-764-9977
Products: DX, AB DELIVERY SYSTEM FOR CANCER
DX & RX; IMMUNOASSAYS

CEO: BUCK A. RHODES
President: BUCK A. RHODES
R&D Dir: K.D. PANT
Financing: PRI
Started: 1986
Class Code: 3

Listing 2-2 Directory of Companies

RIBI IMMUNOCHEM RESEARCH, INC.
P.O. BOX 1409
HAMILTON, MT 59840
Telephone: 406-363-6214
Products: VET ANTI-TUMOR PRODUCTS; VET DX;
VET VX

CEO: ROBERT E. IVY
President: ROBERT E. IVY
R&D Dir: J. A. RUDBACH
Financing: PUB
Started: 1981
Class Code: V

RICERCA, INC.
7528 AUBURN RD., P.O. BOX 1000
PAINESVILLE, OH 44077
Telephone: 216-357-3300
Products: FERMENTATION AND RESEARCH

President: DR. JAMES A. SCOZZIE
R&D Dir: DR. RITA MANAK
Financing: SUB
Started: 1986
Class Code: Y

S

SCOTT LABORATORIES
771 MAIN ST.
FISKVILLE, RI 40182
Telephone: 800-556-6480
Products: REAGENTS AND SPECIALTY CHEMICALS

President: DAVID MAIDMAN
R&D Dir: GARY WITMAN
Financing: SUB
Started: 1981
Class Code: G

SCRIPPS LABORATORIES
9950 SCRIPPS LAKE DRIVE
SAN DIEGO, CA 92131
Telephone: 619-566-3505
Products: BULK HUMAN PROTEINS; HORMONES;
ENZYMES; DX

President: SIMON KHOURY
R&D Dir: DR. BRIAN AUGUSTUS
Financing: PRI
Started: 1980
Class Code: P

SEAPHARM, INC.
791 ALEXANDER RD.
PRINCETON, NJ 08540
Telephone: 609-452-7140
Products: MARINE RX PRODS; TISSUE CULTURE

CEO: JUAN CARLOS TORRES
R&D Dir: DR. JACOB CLEMENT
Financing: PRI
Started: 1983
Class Code: T

SENETEK, PLC
444 CASTRO ST., SUITE 710
MOUNTAIN VIEW, CA 94011
Telephone: 415-962-0925
Products: MAB-BASED DX FOR ALZHEIMER'S DISEASE;
SKIN AGING REVERSAL RX

CEO: JOHN P. BENNETT
President: JOHN P. BENNETT
R&D Dir: NICHOLAS KOPPARD
Financing: PUB
Started: 1983
Class Code: K

SEPRACOR, INC.
33 LOCKE DRIVE
MARLBOROUGH, MA 01752
Telephone: 617-460-0412
Products: MEMBRANE COLUMNS-ANTI-FACTOR VII
& ANTI-TPA; SEPARATIONS

President: TIMOTHY J. BARBERICH
R&D Dir: DR. STEPHEN MATSON
Financing: PRI
Started: 1984
Class Code: E

SERAGEN, INC.
97 SOUTH ST.
HOPKINTON, MA 01748
Telephone: 617-435-2331
Products: CANCER RX; MABS; REAGENTS; ANTISERA

President: KARL AHRENDT
R&D Dir: DR. JEAN NICHOLS
Financing: PRI
Started: 1979
Class Code: X

SERONO LABS

280 POND ST.
RANDOLPH, MA 02368
Telephone: 617-963-8154
Products: RX AND DX FOR FERTILITY, GROWTH,
HGH, IMMUNOLOGICAL & VIROLOGICAL DISEASES

President: THOMAS G. WIGGANS
R&D Dir: RUSSEL W. PELHAM
Financing: PRI
Started: 1971
Class Code: P

SIBIA

P.O. BOX 85200
SAN DIEGO, CA 92138
Telephone: 619-452-5892
Products: RESEARCH IN ANIMAL & HUMAN HEALTH;
PLANT/AG SCIENCE, BT PROCESS DEVELOPMENT

President: FREDERIC DE HOFFMAN
R&D Dir: CHARLES COOK
Financing: PRI
Started: 1981
Class Code: W

SOUTHERN BIOTECHNOLOGY ASSOCIATES

160A OXMOOR BLVD.
BIRMINGHAM, AL 35209
Telephone: 205-945-1774
Products: COLLAGEN; COLLAGEN AB; CUSTOM
CONJUGATION AB DEVELOPMENT

CEO: CHARLES LICHTMAN
President: DR. WILLIAM GATHINGS
R&D Dir: DR. ROGER L. LALLONE
Financing: PRI
Started: 1982
Class Code: J

SPHINX BIOTECHNOLOGIES

P.O. BOX 12194
RESEARCH TRIANGLE PARK, NC 27709
Telephone: 919-541-6590
Products: CELL REGULATION THERAPEUTICS

President: DR. ROBERT M. BELL
R&D Dir: DR. CARSON R. LOOMIS
Financing: PRI
Started: 1987
Class Code: P

STRATAGENE, INC.

11099 N. TORREY PINES RD.
LA JOLLA, CA 92037
Telephone: 619-535-5400
Products: BT REAGENTS; MOLECULAR BIOLOGICALS;
GENE DX; CUSTOM RESEARCH

CEO: DR. JOSEPH SORGE
President: ANTHONY SORGE
R&D Dir: DR. WILLIAM HUSE
Financing: PRI
Started: 1984
Class Code: G

SUMMA MEDICAL CORP.

4272 BALLOON PARK RD., N.E.
ALBUQUERQUE, NM 87109
Telephone: 505-345-8891
Products: HISTOCHEMICAL MARKERS AND DX KITS,
CANCER DX

President: FRANCISCO URREA, JR.
R&D Dir: DR. PAUL O. ZAMORA
Financing: PUB
Started: 1978
Class Code: G

SUNGENE TECHNOLOGIES CORP.

2050 CONCOURSE DR.
SAN JOSE, CA 95131
Telephone: 415-856-3200
Products: PLANT BREEDING, CORN,
SUNFLOWERS, SEEDS

President: WILLIAM J. REID
R&D Dir: DR. PHILIP FILNER
Financing: PRI
Started: 1981
Class Code: B

SYNBIOTICS CORP.

11011 VIA FRONTERA
SAN DIEGO, CA 92129
Telephone: 619-451-3770
Products: VET DX; HUMAN DX; VX

President: DR. EDWARD T. MAGGIO
R&D Dir: MORTON A. VODIAN
Financing: PUB
Started: 1982
Class Code: V

Listing 2-2 Directory of Companies

SYNERGEN, INC.
1885 33RD ST.
BOULDER, CO 80301
Telephone: 303-938-6200
Products: RX; RESEARCH LUNG DISEASE; FGF; VET VX

President: DR. LARRY SOLL
R&D Dir: DR. DAVID I. HIRSCH
Financing: PUB
Started: 1981
Class Code: P

SYNGENE PRODUCTS
15 AND OAK, P.O. BOX 338
ELWOOD, KS 66024
Telephone: 800-255-6829
Products: PRODUCTS FOR ANIMAL AND
HUMAN HEALTH

President: DR. V. DAVID HEIN
R&D Dir: SCOTT WINSTON
Financing: SUB
Started: 1980
Class Code: P

SYNTHETIC GENETICS, INC.
10455 ROSELLE ST.
SAN DIEGO, CA 92121
Telephone: 619-587-0320
Products: CUSTOM & SYNTHETIC DNA
& PEPTIDES

CEO: RICHARD H. TULLIS
President: RICHARD H. TULLIS
R&D Dir: RICHARD H. TULLIS
Financing: PRI
Started: 1985
Class Code: J

SYNTRO CORP.
10655 SORRENTO VALLEY RD.
SAN DIEGO, CA 92121
Telephone: 619-453-4000
Products: ANIMAL VX & DX; ANIMAL
HEALTH PRODUCTS

President: DR. J. THOMAS PARMETER
R&D Dir: DR. RONALD D. BROWN
Financing: PUB
Started: 1981
Class Code: A

SYVA CO.
900 ARASTRADERO RD.
PALO ALTO, CA 94304
Telephone: 415-493-2200
Products: DX PRODUCTS; THERAPEUTIC DRUG
MONITORING; DRUG ABUSE/TOXINS

President: WILLIAM GOMEZ
R&D Dir: DR. TED ULLMAN
Financing: SUB
Started: 1966
Class Code: K

T

T CELL SCIENCES, INC.
840 MEMORIAL DRIVE
CAMBRIDGE, MA 02139
Telephone: 617-864-2160
Products: MABS; CANCER DX; T CELL PRODUCTS
FOR ARTHRITIS; IL-2; RX

CEO: MR JAMES D. GRANT
R&D Dir: DR. PATRICK C. KUNG
Financing: PUB
Started: 1984
Class Code: P

TECHNICLONE INTERNATIONAL, INC.
3301 S. HARBOR BLVD., SUITE 101
SANTA ANA, CA 92704
Telephone: 714-557-5913
Products: CANCER DX, TISSUE CULTURES, MABS,
MAB CULTURING EQUIPMENT

CEO: LON H. STONE
President: DR. ROBERT LUNDAK
R&D Dir: DR. ROBERT LUNDAK
Financing: PUB
Started: 1981
Class Code: K

TECHNOGENETICS, INC.
110 CHARLOTTE PLACE
ENGLEWOOD CLIFFS, NJ 07632
Telephone: 201-871-9898
Products: MABS FOR IMMUNOCYTOLOGY, HUMAN DX KITS

President: GEORGE AARON
Financing: PUB
Started: 1983
Class Code: K

TEKTAGEN
358 TECHNOLOGY DR.
MALVERN, PA 19355
Telephone: 215-640-4550
Products: SERVICE LAB FOR FDA TESTING

CEO: DR. CHARLES NAWROT
President: DR. CHARLES NAWROT
Financing: PRI
Started: 1987
Class Code: 5

THREE-M (3M) DIAGNOSTIC SYSTEMS
3380 CENTRAL EXPRESSWAY
SANTA CLARA, CA 95051
Telephone: 408-739-2200
Products: DX

President: CLINT SEVERSON
Financing: SUB
Started: 1981
Class Code: K

TOXICON
225 WILDWOOD AVE.
WOBURN, MA 01801
Telephone: 617-769-8820
Products: TOXICOLOGY SERVICE

President: LAXMAN DESAI
R&D Dir: DR. JAMES KINCH
Financing: PRI
Started: 1978
Class Code: Y

TRANSFORMATION RESEARCH, INC.
155 N. BEACON ST.
BRIGHTON, MA 02135
Telephone: 617-783-4124
Products: ANTIBODIES, GROWTH FACTORS

President: DR. JEFF YEH
Financing: PRI
Started: 1981
Class Code: X

TRANSGENIC SCIENCES, INC.
365 PLANTATION ST.
WORCESTER, MA 01604
Telephone: 617-752-4442
Products: DISEASE RESISTANT POULTRY; USE OF TRANSGENIC ANIMALS FOR RX; GENE THERAPY

President: F. DONALD HUDSON
R&D Dir: DR. C.M. WEI
Financing: PRI
Started: 1988
Class Code: V

TRITON BIOSCIENCES, INC.
1501 HARBOR BAY PARKWAY
ALAMEDA, CA 94501
Telephone: 415-769-5200
Products: RX, IF-B; DX RESEARCH REAGENTS; MAB, RESEARCH

President: RICHARD L. LOVE
R&D Dir: DR. JOHN F. COLE
Financing: SUB
Started: 1983
Class Code: P

U

UNIGENE LABORATORIES, INC.
110 LITTLE FALLS RD.
FAIRFIELD, NJ 07006
Telephone: 201-882-0860
Products: PEPTIDES, HORMONES, ENZYMES, CALCITONIN, GROWTH FACTORS

CEO: DR. WARREN P. LEVY
President: DR. WARREN P. LEVY
R&D Dir: DR. RONALD LEVY
Financing: PUB
Started: 1980
Class Code: P

Listing 2-2 Directory of Companies

UNITED AGRISEEDS, INC.
P.O. BOX 4011
CHAMPAIGN, IL 61820
Telephone: 217-373-5300
Products: PLANTS, SEEDS

President: ROD N. STACEY
R&D Dir: DR. JOHN R. SNYDER
Financing: SUB
Started: 1981
Class Code: B

UNITED BIOMEDICAL, INC.
2 NEVADA DRIVE
LAKE SUCCESS, NY 11042
Telephone: 516-354-1060
Products: DX, MABS, SPECIALTY PROTEINS, AIDS VX

President: NEAN HU
R&D Dir: DR. CHANG YI WANG
Financing: PRI
Started: 1983
Class Code: K

UNITED STATES BIOCHEMICAL CORP.
26111 MILES RD.
CLEVELAND, OH 44128
Telephone: 216-765-5000
Products: REAGENTS AND EQUIPMENT

President: THOMAS A. MANN
R&D Dir: DR. PARKE FLICK
Financing: PRI
Started: 1973
Class Code: G

UNIVERSITY GENETICS CO.
P.O. BOX 5117
WESTPORT, CT 06881
Telephone: 203-259-2829
Products: CLONED OR CULTURED PLANTS;
MONOCLONAL CELLS & EMBRYOS

CEO: DR. ALAN G. WALTON
President: RANDALL CHARLTON
R&D Dir: DR. DONALD J. SILVERT
Financing: PUB
Started: 1981
Class Code: B

UNIVERSITY MICRO REFERENCE LAB
611(P) HAMMONDS FERRY RD.
LINTHICUM, MD 21090
Telephone: 800-468-4876
Products: BACTERIAL REFERENCE STRAINS; BATCH
PROCESSING OF MICROBIAL STRAINS

President: DR. REX A. D'AGOSTINO
R&D Dir: DR. REX A. D'AGOSTINO
Financing: PRI
Started: 1980
Class Code: G

UPSTATE BIOTECHNOLOGY, INC.
89 SARANAC AVE.
LAKE PLACID, NY 12946
Telephone: 518-523-1518
Products: GROWTH FACTORS, EGF; BRAIN EXTRACTS
FROM PITUITARY; SERUM SUBSTITUTE

CEO: THOMAS TARKA, III
President: THOMAS TARKA, III
R&D Dir: THOMAS TARKA, III
Financing: PRI
Started: 1986
Class Code: G

V

VECTOR LABORATORIES, INC.
30 INGOLD RD.
BURLINGAME, CA 94010
Telephone: 415-697-3600
Products: MOLECULAR BIOLOGICAL REAGENTS
AND EQUIPMENT

President: DR. JAMES S. WHITEHEAD
R&D Dir: DR. LAURENCE MCINTYRE
Financing: PRI
Started: 1976
Class Code: G

VEGA BIOTECHNOLOGIES, INC.
1250 EAST AERO PARK BLVD.
TUCSON, AZ 85706
Telephone: 602-746-1401
Products: PEPTIDE SYNTHESIZERS; REAGENTS;
ENZYMES

President: VELIMIR CUBRILOVIC
R&D Dir: WILLIAM DEMBROWSKI
Financing: PUB
Started: 1979
Class Code: F

VERAX CORP.

HC61, BOX 6, ETNA RD.
LEBANON, NH 03766
Telephone: 603-448-4445
Products: LARGE SCALE CELL CULTURE PRODUCTION,
MAMMALIAN CELL GROWTH

President: FREDERIC HILDEBRANDT
R&D Dir: PETER W. RUNSTADLER, JR.
Financing: PRI
Started: 1978
Class Code: O

VIAGENE, INC.

11075 ROSELLE ST., SUITE B
SAN DIEGO, CA 92121
Telephone: 619-452-1288
Products: GENE TRANSFER TECHNOLOGY FOR
INFECTIONS, CANCERS, & GENE RELATED DISEASES

President: DAVID F. HALE
R&D Dir: DR. DOUGLAS JOLLY
Financing: PRI
Started: 1987
Class Code: E

VIRAGEN

2201 WEST 36TH ST.
HIALEAH, FL 33016
Telephone: 305-558-4000
Products: IF, PEPTIDES, RX

President: THOMAS LANGHAM
R&D Dir: DR. DAVID C. MUNCH
Financing: PUB
Started: 1980
Class Code: P

VIROSTAT, INC.

P.O. BOX 8522
PORTLAND, OR 04104
Telephone: 207-892-3201
Products: IMMUNOCHEMICALS FOR INFECTIOUS
DISEASE RESEARCH

CEO: DOUGLAS MCALLISTER
President: DOUGLAS MCALLISTER
R&D Dir: DOUGLAS MCALLISTER
Financing: PRI
Started: 1985
Class Code: J

VIVIGEN, INC.

435 ST. MICHAELS DR.
SANTA FE, NM 87501
Telephone: 505-988-9744
Products: GENETIC TESTING LAB, AMNIOCENTESIS
CHROMOSOME ANALYSIS

CEO: THOMAS K. REED, JR.
President: DR. GREGG L. MAYER
R&D Dir: DR. CARL GOODPASTURE
Financing: PUB
Started: 1981
Class Code: 5

W

WASHINGTON BIOLAB

IOWA STATE, ISIS CENTER
AMES, IA 50011
Telephone: 515-296-9919
Products: AIDS RX; ENZYMES, RAW MATERIALS FOR
BIOTECHNOLOGY

CEO: PAUL KI
President: PAUL KI
R&D Dir: P.F. KI
Financing: PRI
Started: 1986
Class Code: J

WELGEN MANUFACTURING, INC.

250 CENTERVILLE RD., BLDG. F
WARWICK, RI 02886
Telephone: 401-738-4280
Products: RX, TPA, IF-A

CEO: WANDERLEY RIBEIRO
President: WANDERLEY RIBEIRO
Financing: PRI
Started: 1988
Class Code: P

Listing 2-2 Directory of Companies

WESTBRIDGE RESEARCH GROUP
9920 SCRIPPS LAKE DRIVE
SAN DIEGO, CA 92131
Telephone: 619-586-7211
Products: PLANT GROWTH REGULATOR; WATER
TREATMENT

President: GERALD R. HADDOCK
R&D Dir: THEODORE A. HYMOS
Financing: PRI
Started: 1982
Class Code: J

X

XENOGEN
1734 STORRS RD.
MANSFIELD, CT 06268
Telephone: 203-871-4581
Products: RDNA RESEARCH FOR DX AND CHEMICALS;
DX FOR BREAST CANCER, GONORRHEA

CEO: DR. TODD SCHUSTER
President: DR. TODD SCHUSTER
R&D Dir: MITCHEL SAYRE
Financing: PRI
Started: 1981
Class Code: K

XOMA CORP.
2910 SEVENTH ST.
BERKELEY, CA 94710
Telephone: 415-644-1170
Products: MAB-BASED ANTICANCER RX

CEO: DR. STEVEN C. MENDELL
President: DR. PATRICK J. SCANNON
R&D Dir: DR. PATRICK J. SCANNON
Financing: PUB
Started: 1981
Class Code: P

Z

ZOECON CORP.
975 CALIFORNIA AVE., #10975
PALO ALTO, CA 94304
Telephone: 415-857-1130
Products: PLANT IMPROVEMENT; SEEDS;
AG CHEMICALS

CEO: JOHN DICKMAN
President: JOHN DICKMAN
R&D Dir: L.K. GRILL
Financing: SUB
Started: 1968
Class Code: B

ZYMED LABORATORIES, INC.
52 S. LINDEN AVE., SUITE 4
S. SAN FRANCISCO, CA 94080
Telephone: 415-871-4494
Products: IMMUNOCHEMICALS, MABS, PROTEINS,
ENZYMES

President: DR. DEAN TSAO
R&D Dir: DR. AGOSTIN BELLA, JR.
Financing: PRI
Started: 1980
Class Code: G

ZYMOGENETICS, INC.
4225 ROOSEVELT WAY, N.E.
SEATTLE, WA 98105
Telephone: 206-547-8080
Products: RX, WOUND HEALING, COAGULATION,
PROTEINS, ANTI-INFECTIVES

President: GREGORY PHELPS
R&D Dir: DR. BRUCE CARTER
Financing: PRI
Started: 1981
Class Code: P

SECTION 3   BIOTECHNOLOGY COMPANY LOCATION AND TYPE

Section 3 Company Location and Type

INTRODUCTION

This section contains two types of information. The locations of the biotechnology companies in the United States, by state, are presented in Listing 3-1. The city locations are also given to allow the reader to locate individual companies. This listing is followed by Table 3-1 analyzing the number of companies in each state. Table 3-2 gives the ranking of states by number of biotechnology companies located there.

The second half of this section divides the companies by their financing status; public, private or subsidiary. These divisions can be found in Listings 3-2 through 3-4, respectively. The year of founding and number of employees are also given for each firm. Later on in this Guide, certain data will be analyzed by all companies and further analyzed by their public/private/subsidiary status.

Listing 3-1 Location of Biotechnology Companies

LISTING 3-1 LOCATION OF BIOTECHNOLOGY COMPANIES BY STATE

Company	City
Alabama	
SOUTHERN BIOTECHNOLOGY ASSOCIATES	BIRMINGHAM
Arizona	
BIO HUMA NETICS	CHANDLER
VEGA BIOTECHNOLOGIES, INC.	TUCSON
Arkansas	
IMMUNOVISION, INC.	SPRINGDALE
PEL-FREEZ BIOLOGICALS, INC.	ROGERS
California	
ABN	HAYWARD
ADVANCED GENETIC SCIENCES	OAKLAND
AMERICAN BIOGENETICS CORP.	IRVINE
AMERICAN QUALEX INTERNATIONAL, INC.	LA MIRADA
AMGEN	THOUSAND OAKS
ANTIBODIES, INC.	DAVIS
APPLIED BIOSYSTEMS, INC.	FOSTER CITY
AUTOMEDIX SCIENCES, INC.	TORRANCE
BACHEM, INC.	TORRANCE
BEHRING DIAGNOSTICS	LA JOLLA
BERKELEY ANTIBODY CO., INC.	RICHMOND
BIO-RESPONSE, INC.	HAYWARD
BIOGENEX LABORATORIES	SAN RAMON
BIOGROWTH, INC.	RICHMOND
BIOMERICA, INC.	NEWPORT BEACH
BIOPROBE INTERNATIONAL, INC.	TUSTIN
BIOSEARCH, INC.	SAN RAFAEL
BIOTHERAPY SYSTEMS, INC.	MOUNTAIN VIEW
BMI, INC.	MISSION VIEJO
CALGENE, INC.	DAVIS
CALIFORNIA BIOTECHNOLOGY, INC.	MOUNTAIN VIEW
CALIFORNIA INTEGRATED DIAGNOSTICS	BERKELEY
CALZYME LABORATORIES, INC.	SAN LUIS OBISPO
CETUS CORPORATION	EMERYVILLE
CHEMICON INTERNATIONAL, INC.	EL SEGUNDO
CHIRON CORP.	EMERYVILLE
CLINETICS CORP.	TUSTIN
CLONTECH LABORATORIES, INC.	PALO ALTO
CODON CORP.	S. SAN FRANCISCO
COLLAGEN CORPORATION	PALO ALTO
COOPER DEVELOPMENT CO.	MENLO PARK
CYGNUS RESEARCH CORP.	REDWOOD CITY

Company	City
California (Cont.)	
CYTOTECH, INC.	SAN DIEGO
DAKO CORP.	SANTA BARBARA
DIAGNOSTIC PRODUCTS CORP.	LOS ANGELES
DNAX RESEARCH INSTITUTE	PALO ALTO
E-Y LABORATORIES, INC.	SAN MATEO
EARL-CLAY LABORATORIES, INC.	NOVATO
ENZON	MOUNTAIN VIEW
ESCAGENETICS CORP.	SAN CARLOS
GEN-PROBE, INC.	SAN DIEGO
GENELABS, INC.	REDWOOD CITY
GENENCOR, INC.	S. SAN FRANCISCO
GENENTECH	S. SAN FRANCISCO
HANA BIOLOGICS, INC.	ALAMEDA
HYBRITECH, INC.	SAN DIEGO
ICL SCIENTIFIC, INC.	FOUNTAIN VALLEY
IDEC PHARMACEUTICAL	MOUNTAIN VIEW
IDETEK, INC.	SAN BRUNO
IMMUNETECH PHARMACEUTICALS	SAN DIEGO
IMMUNO CONCEPTS, INC.	SACRAMENTO
INFERGENE CO.	BENICIA
INGENE, INC.	SANTA MONICA
INTEK DIAGNOSTICS, INC.	BURLINGAME
INTELLIGENETICS, INC.	MOUNTAIN VIEW
INTERNATIONAL ENZYMES, INC.	FALLBROOK
INTERNATIONAL PLANT RESEARCH INST.	SAN CARLOS
KIRIN-AMGEN	THOUSAND OAKS
LEE BIOMOLECULAR RESEARCH LABS	SAN DIEGO
LIPOSOME TECHNOLOGY, INC.	MENLO PARK
LUCKY BIOTECH CORP.	EMERYVILLE
MAST IMMUNOSYSTEMS, INC.	MOUNTAIN VIEW
MICROBIO RESOURCES	SAN DIEGO
MICROGENICS	CONCORD
MOLECULAR BIOSYSTEMS, INC.	SAN DIEGO
MONOCLONAL ANTIBODIES, INC.	MOUNTAIN VIEW
MULTIPLE PEPTIDE SYSTEMS, INC.	SAN DIEGO
MYCOGEN CORP.	SAN DIEGO
NEUREX CORP.	MENLO PARK
NEUSHUL MARICULTURE	GOLETA
OCEAN GENETICS	SANTA CRUZ
OMNI BIOCHEM, INC.	NATIONAL CITY
PENINSULA LABORATORIES, INC.	BELMONT
PHYTOGEN	PASADENA
PLANT GENETICS, INC.	DAVIS
PROGENX	SAN DIEGO
QUIDEL	LA JOLLA

Listing 3-1 Location of Biotechnology Companies

Company	City
California (Cont.)	
RESEARCH & DIAGNOSTIC ANTIBODIES	BERKELEY
SCRIPPS LABORATORIES	SAN DIEGO
SENETEK PLC	MOUNTAIN VIEW
SIBIA	SAN DIEGO
STRATAGENE, INC.	LA JOLLA
SUNGENE TECHNOLOGIES CORP.	SAN JOSE
SYNBIOTICS CORP.	SAN DIEGO
SYNTHETIC GENETICS, INC.	SAN DIEGO
SYNTRO CORP.	SAN DIEGO
SYVA CO.	PALO ALTO
TECHNICLONE INTERNATIONAL, INC.	SANTA ANA
THREE-M (3M) DIAGNOSTIC SYSTEMS	SANTA CLARA
TRITON BIOSCIENCES, INC.	ALAMEDA
VECTOR LABORATORIES, INC.	BURLINGAME
VIAGENE, INC.	SAN DIEGO
WESTBRIDGE RESEARCH GROUP	SAN DIEGO
XOMA CORP.	BERKELEY
ZOECON CORP.	PALO ALTO
ZYMED LABORATORIES, INC.	S. SAN FRANCISCO
Colorado	
ADVANCED MINERAL TECHNOLOGIES	GOLDEN
BIOSTAR MEDICAL PRODUCTS, INC.	BOULDER
COORS BIOTECH PRODUCTS CO.	WESTMINSTER
GENETIC ENGINEERING, INC.	DENVER
SYNERGEN, INC.	BOULDER
Connecticut	
BIOPOLYMERS, INC.	FARMINGTON
DELTOWN CHEMURGIC CORP.	GREENWICH
INTERNATIONAL BIOTECHNOLOGIES, INC.	NEW HAVEN
MICROGENESYS, INC.	WEST HAVEN
MOLECULAR DIAGNOSTICS, INC.	WEST HAVEN
UNIVERSITY GENETICS CO.	WESTPORT
XENOGEN	MANSFIELD
District of Columbia	
ALPHA I BIOMEDICALS, INC.	WASHINGTON, D.C.
Florida	
ABC RESEARCH CORP.	GAINESVILLE
APPLIED GENETICS LABS, INC.	MELBOURNE
DIAMEDIX, INC.	MIAMI
IMMUNOMED CORP.	TAMPA
LIFE SCIENCES	ST. PETERSBURG

Company	City
Florida (Cont.)	
MOLECULAR GENETIC RESOURCES	TAMPA
PETROFERM, INC.	FERNANDINA BEACH
VIRAGEN	HIALEAH
Georgia	
BIOSYSTEMS, INC.	LOGANVILLE
MUREX CORP.	NORCROSS
Hawaii	
HAWAII BIOTECHNOLOGY GROUP, INC.	AIEA
Illinois	
CARBOHYDRATES INTERNATIONAL, INC.	CHICAGO
COAL BIOTECH CORP.	LIBERTYVILLE
DEKALB-PFIZER GENETICS	DEKALB
IMCERA BIOPRODUCTS, INC.	NORTHBROOK
LABSYSTEMS, INC.	MORTON GROVE
LYPHOMED, INC.	ROSEMONT
PAMBEC LABORATORIES	GURNEE
PETROGEN, INC.	ARLINGTON HEIGHTS
UNITED AGRISEEDS, INC.	CHAMPAIGN
Indiana	
AGDIA, INC.	MISHAWAKA
BIOPRODUCTS FOR SCIENCE, INC.	INDIANAPOLIS
BOEHRINGER MANNHEIM DIAGNOSTICS	INDIANAPOLIS
CONSOLIDATED BIOTECHNOLOGY, INC.	ELKHART
Iowa	
AMBICO, INC.	DALLAS CENTER
WASHINGTON BIOLAB	AMES
Kansas	
BIOTECHNICA AGRICULTURE	OVERLAND PARK
MONOCLONAL PRODUCTION INT'L	FORT SCOTT
SYNGENE PRODUCTS	ELWOOD
Louisiana	
IMREG, INC.	NEW ORLEANS
MICROBE MASTERS	BATON ROUGE
Maine	
AGRITECH SYSTEMS, INC.	PORTLAND
ATLANTIC ANTIBODIES	SCARBOROUGH
BINAX, INC.	SOUTH PORTLAND

Listing 3-1 Location of Biotechnology Companies

Company	City
Maine (Cont.)	
BIODESIGN, INC.	KENNEBUNKPORT
IMMUCELL CORP.	PORTLAND
IMMUNOSYSTEMS, INC.	BIDDEFORD
Maryland	
ADVANCED BIOTECHNOLOGIES	SILVER SPRING
AMERICAN BIOTECHNOLOGY CO.	ROCKVILLE
BIONETICS RESEARCH	ROCKVILLE
BIOSPHERICS, INC.	BELTSVILLE
BIOTECH RESEARCH LABS, INC.	ROCKVILLE
CELLMARK DIAGNOSTICS	GERMANTOWN
CROP GENETICS INTERNATIONAL CORP.	HANOVER
DIAGNON CORP.	ROCKVILLE
DIGENE DIAGNOSTICS, INC.	COLLEGE PARK
GENETIC THERAPY, INC.	GAITHERSBURG
GENEX, INC.	GAITHERSBURG
GENTRONIX LABORATORIES, INC.	BETHESDA
IBF BIOTECHNICS, INC.	SAVAGE
IGEN	ROCKVILLE
IGENE BIOTECHNOLOGY, INC.	COLUMBIA
INTER-AMERICAN RESEARCH ASSOCIATES	GAITHERSBURG
KIRKEGAARD & PERRY LABS, INC.	GAITHERSBURG
LIFE TECHNOLOGIES, INC.	GAITHERSBURG
MARTEK	COLUMBIA
MICROBIOLOGICAL ASSOCIATES, INC.	ROCKVILLE
ONCOR, INC.	GAITHERSBURG
UNIVERSITY MICRO REFERENCE LAB	LINTHICUM
Massachusetts	
A/G TECHNOLOGY CORP.	NEEDHAM
ADVANCED MAGNETICS, INC.	CAMBRIDGE
ANGENICS	CAMBRIDGE
APPLIED BIOTECHNOLOGY, INC.	CAMBRIDGE
APPLIED PROTEIN TECHNOLOGIES, INC.	CAMBRIDGE
BIOGEN	CAMBRIDGE
BIOMEDICAL TECHNOLOGIES	STOUGHTON
BIOPURE	BOSTON
BIOTECHNICA INTERNATIONAL	CAMBRIDGE
BIOTECHNOLOGY DEVELOPMENT CORP.	NEWTON
CAMBRIDGE BIOSCIENCE CORP.	WORCESTER
CHARLES RIVER BIOTECH. SERVICES	WILMINGTON
CHEMGENES	NEEDHAM
CIBA-CORNING DIAGNOSTIC CORP.	MEDFIELD
COLLABORATIVE RESEARCH, INC.	BEDFORD
CREATIVE BIOMOLECULES	HOPKINTON
DAMON BIOTECH, INC.	NEEDHAM HEIGHTS

Company	City
Massachusetts (Cont.)	
ENDOGEN, INC.	BOSTON
ENZYME CENTER INC., THE	MALDEN
GENE-TRAK SYSTEMS	FRAMINGHAM
GENETICS INSTITUTE, INC.	CAMBRIDGE
GENZYME CORP.	BOSTON
HYGEIA SCIENCES	NEWTON
IMMUNOGEN, INC.	CAMBRIDGE
IMMUNOTECH CORP.	BOSTON
INTEGRATED CHEMICAL SENSORS	NEWTON
INTEGRATED GENETICS, INC.	FRAMINGHAM
KARYON TECHNOLOGY, LTD.	NORWOOD
REPLIGEN CORP.	CAMBRIDGE
SEPRACOR, INC.	MARLBOROUGH
SERAGEN, INC.	HOPKINTON
SERONO LABS	RANDOLPH
T CELL SCIENCES, INC.	CAMBRIDGE
TOXICON	WOBURN
TRANSFORMATION RESEARCH, INC.	BRIGHTON
TRANSGENIC SCIENCES, INC.	WORCESTER
Michigan	
COVALENT TECHNOLOGY CORP.	ANN ARBOR
NATIONAL GENO SCIENCES	SOUTHFIELD
NEOGEN CORP.	LANSING
POLYCELL, INC.	DETROIT
PROTEINS INTERNATIONAL, INC.	ROCHESTER
QUEST BIOTECHNOLOGY, INC.	DETROIT
RECOMTEX CORP.	EAST LANSING
Minnesota	
BIOTROL, INC.	CHASKA
ENDOTRONICS, INC.	COON RAPIDS
GENESIS LABS, INC.	MINNEAPOLIS
LIFECORE BIOMEDICAL, INC.	MINNEAPOLIS
MOLECULAR GENETICS, INC.	MINNETONKA
PROTATEK INTERNATIONAL, INC.	ST. PAUL
Missouri	
BIOKYOWA, INC.	CAPE GIRARDEAU
FERMENTA ANIMAL HEALTH	KANSAS CITY
INVITRON CORP.	ST. LOUIS
Montana	
RIBI IMMUNOCHEM RESEARCH, INC.	HAMILTON

Listing 3-1 Location of Biotechnology Companies

Company	City
Nebraska	
AMERICAN LABORATORIES, INC.	OMAHA
NORDEN LABS	LINCOLN
Nevada	
GAMETRICS, LTD.	LAS VEGAS
IMMUNEX, INC.	CARSON CITY
New Hampshire	
VERAX CORP.	LEBANON
New Jersey	
AGRI-DIAGNOSTICS ASSOCIATES	CINNAMINSON
ALFACELL CORP.	BLOOMFIELD
BIO-RECOVERY, INC.	NORTHVALE
BIOCONSEP, INC.	BELLEMEAD
BIOMATRIX, INC.	RIDGEFIELD
BIOTEST DIAGNOSTICS CORP.	FAIRFIELD
CHEMICAL DYNAMICS CORP.	S. PLAINFIELD
CISTRON BIOTECHNOLOGY, INC.	PINE BROOK
CLINICAL SCIENCES, INC.	WHIPPANY
CYTOGEN CORP.	PRINCETON
DNA PLANT TECHNOLOGY CORP.	CINNAMINSON
ELECTRO-NUCLEONICS, INC.	FAIRFIELD
EMTECH RESEARCH	MOUNT LAUREL
ENZON, INC.	S. PLAINFIELD
ENZYME BIO-SYSTEMS LTD.	ENGLEWOOD CLIFFS
IMMUNOMEDICS, INC.	WARREN
INTER-CELL TECHNOLOGIES, INC.	SOMERVILLE
INTERFERON SCIENCES, INC.	NEW BRUNSWICK
LIPOSOME COMPANY, INC.	PRINCETON
MARCOR DEVELOPMENT CO.	HACKENSACK
MARINE BIOLOGICALS, INC.	MARMORA
MEDAREX, INC.	CLIFTON
PHARMACIA LKB BIOTECHNOLOGY, INC.	PISCATAWAY
QUEUE SYSTEMS, INC.	NORTH BRANCH
SEAPHARM, INC.	PRINCETON
TECHNOGENETICS, INC.	ENGLEWOOD CLIFFS
UNIGENE LABORATORIES, INC.	FAIRFIELD
New Mexico	
RHOMED, INC.	ALBUQUERQUE
SUMMA MEDICAL CORP.	ALBUQUERQUE
VIVIGEN, INC.	SANTA FE

Section 3 Company Location and Type

Company	City
New York	
AMERICAN DIAGNOSTICA, INC.	NEW YORK
AN-CON GENETICS	MELVILLE
APPLIED DNA SYSTEMS, INC.	NEW YORK
APPLIED MICROBIOLOGY, INC.	BROOKLYN
BIONIQUE LABS, INC.	SARANAC LAKE
BIOTECHNOLOGY GENERAL CORP.	NEW YORK
BRAIN RESEARCH, INC.	NEW YORK
CAMBRIDGE RESEARCH BIOCHEMICALS, INC.	VALLEY STREAM
CELLULAR PRODUCTS, INC.	BUFFALO
DIAGNOSTIC TECHNOLOGY, INC.	HAUPPAUGE
ENZO-BIOCHEM, INC.	NEW YORK
EXOVIR, INC.	GREAT NECK
GENETIC DIAGNOSTICS CORP.	GREAT NECK
IMCLONE SYSTEMS, INC.	NEW YORK
LIFECODES CORP.	VALHALLA
NYGENE CORP.	YONKERS
ONCOGENE SCIENCE, INC.	MANHASSET
PRAXIS BIOLOGICS	ROCHESTER
UNITED BIOMEDICAL, INC.	LAKE SUCCESS
UPSTATE BIOTECHNOLOGY, INC.	LAKE PLACID
North Carolina	
AQUASYNERGY, LTD.	KINSTON
BIOTHERM	RES. TRIANGLE PARK
CENTRAL BIOLOGICS, INC.	RALEIGH
ELCATECH	WINSTON-SALEM
EMBREX, INC.	RES. TRIANGLE PARK
ENVIRONMENTAL DIAGNOSTICS, INC.	BURLINGTON
HUNTER BIOSCIENCES	RALEIGH
INCON CORP.	YOUNGSVILLE
MAIZE GENETIC RESOURCES, INC.	BENSON
MARICULTURA, INC.	WRIGHTSVILLE BEACH
MYCOSEARCH, INC.	CHAPEL HILL
ORGANON TEKNIKA CORP.	DURHAM
PHARMAGENE	CHARLOTTE
PROBIOLOGICS INTERNATIONAL, INC.	RES. TRIANGLE PARK
SPHINX BIOTECHNOLOGIES	RES. TRIANGLE PARK
Ohio	
AGRIGENETICS CORP.	EASTLAKE
ENZYME TECHNOLOGY CORP.	ASHLAND
NORTH COAST BIOTECHNOLOGY, INC.	CLEVELAND HEIGHTS
RICERCA, INC.	PAINESVILLE
UNITED STATES BIOCHEMICAL CORP.	CLEVELAND

Listing 3-1 Location of Biotechnology Companies

Company	City
Oklahoma	
PROVESTA CORP.	BARTLESVILLE
Oregon	
AMERICAN BIOCLINICAL	PORTLAND
BEND RESEARCH, INC.	BEND
BENTECH LABORATORIES	CLACKAMAS
EPITOPE, INC.	BEAVERTON
VIROSTAT, INC.	PORTLAND
Pennsylvania	
BACHEM BIOSCIENCE, INC.	PHILADELPHIA
BIOCHEM TECHNOLOGY, INC.	MALVERN
BIOSCIENCE MANAGEMENT, INC.	BETHLEHEM
CENTOCOR	MALVERN
ECOGEN, INC.	LANGHORNE
JACKSON IMMUNORESEARCH LABS, INC.	WEST GROVE
REPAP TECHNOLOGIES	VALLEY FORGE
TEKTAGEN	MALVERN
Rhode Island	
SCOTT LABORATORIES	FISKVILLE
WELGEN MANUFACTURING, INC.	WARWICK
South Carolina	
AMTRON	CHARLESTON
Tennessee	
BIOTHERAPEUTICS, INC.	FRANKLIN
Texas	
A.M. BIOTECHNIQUES, INC.	BELTON
BETHYL LABS, INC.	MONTGOMERY
BIOTICS RESEARCH CORP.	STAFFORD
BIOTX (BIOSCIENCES CORP. OF TEXAS)	HOUSTON
GAMMA BIOLOGICALS, INC.	HOUSTON
GRANADA GENETICS CORP.	COLLEGE STATION
HOUSTON BIOTECHNOLOGY, INC.	THE WOODLANDS
IMMUNO MODULATORS LABS, INC.	STAFFORD
KALLESTAD DIAGNOSTICS	AUSTIN
MONOCLONETICS INTERNATIONAL, INC.	HOUSTON
O.C.S. LABORATORIES, INC.	DENTON

Section 3 Company Location and Type

Company	City
Utah	
BIOMATERIALS INTERNATIONAL, INC.	SALT LAKE CITY
HYCLONE LABORATORIES, INC.	LOGAN
NATIVE PLANTS, INC.	SALT LAKE CITY
Virginia	
ASTRE CORPORATE GROUP	CHARLOTTESVILLE
CEL-SCI CORP.	ALEXANDRIA
FLOW LABORATORIES, INC.	MCLEAN
HAZLETON BIOTECHNOLOGIES CO.	VIENNA
MELOY LABORATORIES, INC.	SPRINGFIELD
Washington	
BIO TECHNIQUES LABS, INC.	REDMOND
BIOCONTROL SYSTEMS	BOTHELL
BIOMED RESEARCH LABS, INC.	SEATTLE
CYANOTECH CORP.	WOODINVILLE
ELANEX PHARMACEUTICALS, INC.	BOTHELL
GENETIC SYSTEMS CORP.	SEATTLE
IMMUNEX CORP.	SEATTLE
IMRE CORP.	SEATTLE
NEORX CORP.	SEATTLE
R & A PLANT/SOIL, INC.	PASCO
ZYMOGENETICS, INC.	SEATTLE
West Virginia	
ALLELIC BIOSYSTEMS	KEARNEYSVILLE
Wisconsin	
AGRACETUS	MIDDLETON
BIO-TECHNICAL RESOURCES, INC.	MANITOWOC
FORGENE, INC.	RHINELANDER
INCELL CORP.	MILWAUKEE
MOLECULAR BIOLOGY RESOURCES, INC.	MILWAUKEE
NITRAGIN CO.	MILWAUKEE
PROMEGA CORP.	MADISON

Table 3-1 Companies per State

TABLE 3-1 COMPANIES PER STATE

State	No. of Companies	% of Total
ALABAMA	1	0.3
ALASKA	0	0.0
ARIZONA	2	0.6
ARKANSAS	2	0.6
CALIFORNIA	96	26.7
COLORADO	5	1.4
CONNECTICUT	7	1.9
DELAWARE	0	0.0
DISTRICT OF COLUMBIA	1	0.3
FLORIDA	8	2.2
GEORGIA	2	0.6
HAWAII	1	0.3
IDAHO	0	0.0
ILLINOIS	9	2.5
INDIANA	4	1.1
IOWA	2	0.6
KANSAS	3	0.8
KENTUCKY	0	0.0
LOUISIANA	2	0.6
MAINE	6	1.7
MARYLAND	22	6.1
MASSACHUSETTS	36	10.0
MICHIGAN	7	1.9
MINNESOTA	6	1.7
MISSISSIPPI	0	0.0
MISSOURI	3	0.8
MONTANA	1	0.3
NEBRASKA	2	0.6
NEVADA	2	0.6
NEW HAMPSHIRE	1	0.3
NEW JERSEY	27	7.5
NEW MEXICO	3	0.8
NEW YORK	21	5.8
NORTH CAROLINA	15	4.2
NORTH DAKOTA	0	0.0
OHIO	5	1.4
OKLAHOMA	1	0.3
OREGON	6	1.7
PENNSYLVANIA	9	2.5
RHODE ISLAND	2	0.6
SOUTH CAROLINA	1	0.3

Section 3 Company Location and Type

State	No. of Companies	% of Total
SOUTH DAKOTA	0	0.0
TENNESSEE	1	0.3
TEXAS	11	3.1
UTAH	3	0.8
VERMONT	0	0.0
VIRGINIA	5	1.4
WASHINGTON	11	3.1
WEST VIRGINIA	1	0.3
WISCONSIN	7	1.9
WYOMING	0	0.0

Note: This table is derived from the biotechnology companies in Listing 3-1 and the directory (Listing 2-2).

Table 3-2 State Rankings

TABLE 3-2 STATE RANKINGS

State	No. of Companies	% of Total
CALIFORNIA	96	26.7
MASSACHUSETTS	36	10.0
NEW JERSEY	27	7.5
MARYLAND	22	6.1
NEW YORK	21	5.8
NORTH CAROLINA	15	4.2
TEXAS	11	3.1
WASHINGTON	11	3.1
ILLINOIS	9	2.5
PENNSYLVANIA	9	2.5
FLORIDA	8	2.2
CONNECTICUT	7	1.9
MICHIGAN	7	1.9
WISCONSIN	7	1.9
MAINE	6	1.7
MINNESOTA	6	1.7
OREGON	6	1.7
COLORADO	5	1.4
OHIO	5	1.4
VIRGINIA	5	1.4
INDIANA	4	1.1
KANSAS	3	0.8
MISSOURI	3	0.8
NEW MEXICO	3	0.8
UTAH	3	0.8
ARIZONA	2	0.6
ARKANSAS	2	0.6
GEORGIA	2	0.6
IOWA	2	0.6
LOUISIANA	2	0.6
NEBRASKA	2	0.6
NEVADA	2	0.6
RHODE ISLAND	2	0.6
ALABAMA	1	0.3
DISTRICT OF COLUMBIA	1	0.3
HAWAII	1	0.3
MONTANA	1	0.3
NEW HAMPSHIRE	1	0.3
OKLAHOMA	1	0.3
SOUTH CAROLINA	1	0.3
TENNESSEE	1	0.3

Section 3 Company Location and Type

State	No. of Companies	% of Total
WEST VIRGINIA	1	0.3
ALASKA	0	0.0
DELAWARE	0	0.0
IDAHO	0	0.0
KENTUCKY	0	0.0
MISSISSIPPI	0	0.0
NORTH DAKOTA	0	0.0
SOUTH DAKOTA	0	0.0
VERMONT	0	0.0
WYOMING	0	0.0

Note: This table is derived from the biotechnology companies in Listing 3-1 and the directory (Listing 2-2).

Listing 3-2 Public Companies

LISTING 3-2 PUBLIC COMPANIES

Company	Started	Employees
A BN	1981	45
ADVANCED GENETIC SCIENCES	1979	130
ADVANCED MAGNETICS, INC.	1981	40
ALFACELL CORP.	1981	22
ALPHA I BIOMEDICALS, INC.	1982	11
AMGEN	1980	340
APPLIED BIOSYSTEMS, INC.	1981	350
APPLIED DNA SYSTEMS, INC.	1982	7
APPLIED MICROBIOLOGY, INC.	1983	20
AUTOMEDIX SCIENCES, INC.	1978	7
BIO-RESPONSE, INC.	1972	45
BIOGEN	1979	250
BIOMERICA, INC.	1971	35
BIOSPHERICS, INC.	1967	250
BIOTECH RESEARCH LABS, INC.	1973	167
BIOTECHNICA INTERNATIONAL	1981	145
BIOTECHNOLOGY DEVELOPMENT CORP.	1982	35
BIOTECHNOLOGY GENERAL CORP.	1980	130
BIOTHERAPEUTICS, INC.	1984	120
BIOTHERAPY SYSTEMS, INC.	1984	8
CALGENE, INC.	1980	165
CALIFORNIA BIOTECHNOLOGY, INC.	1981	155
CAMBRIDGE BIOSCIENCE CORP.	1981	97
CEL-SCI CORP.	1983	2
CELLULAR PRODUCTS, INC.	1982	101
CENTOCOR	1979	246
CETUS CORP.	1971	711
CHIRON CORP.	1981	300
CISTRON BIOTECHNOLOGY, INC.	1982	30
CLINICAL SCIENCES, INC.	1971	50
COLLABORATIVE RESEARCH, INC.	1961	150
COLLAGEN CORP.	1975	225
COOPER DEVELOPMENT CO.	1980	900
CROP GENETICS INTERNATIONAL CORP.	1981	75
CYTOGEN CORP.	1980	140
DIAGNON CORP.	1981	70
DIAGNOSTIC PRODUCTS CORP.	1972	270
DNA PLANT TECHNOLOGY CORP.	1981	80
EARL-CLAY LABORATORIES, INC.	1984	7
ECOGEN, INC.	1983	75
ELECTRO-NUCLEONICS, INC.	1960	720
ENDOTRONICS, INC.	1981	45
ENVIRONMENTAL DIAGNOSTICS, INC.	1983	55

Section 3 Company Location and Type

Company	Started	Employees
ENZO-BIOCHEM, INC.	1976	110
ENZON	1965	22
ENZON, INC.	1981	31
EPITOPE, INC.	1979	60
ESCAGENETICS CORP.	1986	50
EXOVIR, INC.	1981	5
GAMMA BIOLOGICALS, INC.	1969	166
GEN-PROBE, INC.	1984	127
GENE-TRAK SYSTEMS	1981	196
GENENTECH	1976	1,500
GENETIC DIAGNOSTICS CORP.	1981	17
GENETIC ENGINEERING, INC.	1980	23
GENETICS INSTITUTE, INC.	1981	302
GENEX, INC.	1977	80
GENZYME CORP.	1981	185
HANA BIOLOGICS, INC.	1979	130
HYGEIA SCIENCES	1980	70
IGENE BIOTECHNOLOGY, INC.	1981	30
IMMUCELL CORP.	1982	17
IMMUNEX CORP.	1981	153
IMMUNOMEDICS, INC.	1983	55
IMRE CORP.	1981	25
IMREG, INC.	1981	50
INFERGENE CO.	1984	25
INGENE, INC.	1980	65
INTEGRATED GENETICS, INC.	1981	200
LIFE SCIENCES	1962	20
LIFE TECHNOLOGIES, INC.	1983	1,300
LIFECORE BIOMEDICAL, INC.	1975	20
LIPOSOME COMPANY, INC.	1981	67
LYPHOMED, INC.	1981	1,200
MOLECULAR BIOSYSTEMS, INC.	1981	55
MOLECULAR GENETICS, INC.	1979	63
MONOCLONAL ANTIBODIES, INC.	1979	110
MYCOGEN CORP.	1982	50
NATIVE PLANTS, INC.	1973	625
ONCOGENE SCIENCE, INC.	1983	65
ONCOR, INC.	1983	30
PLANT GENETICS, INC.	1981	63
PRAXIS BIOLOGICS	1983	180
QUEST BIOTECHNOLOGY, INC.	1986	9
REPLIGEN CORP.	1981	95
RIBI IMMUNOCHEM RESEARCH, INC.	1981	35
SENETEK PLC	1983	14
SUMMA MEDICAL CORP.	1978	33
SYNBIOTICS CORP.	1982	89

Listing 3-2 Public Companies

Company	Started	Employees
SYNERGEN, INC.	1981	101
SYNTRO CORP.	1981	84
T CELL SCIENCES, INC.	1984	65
TECHNICLONE INTERNATIONAL, INC.	1981	15
TECHNOGENETICS, INC.	1983	30
UNIGENE LABORATORIES, INC.	1980	30
UNIVERSITY GENETICS CO.	1981	59
VEGA BIOTECHNOLOGIES, INC.	1979	60
VIRAGEN	1980	15
VIVIGEN, INC.	1981	70
XOMA CORP.	1981	149

Note: This listing is derived from the biotechnology companies in the directory (Section 2-2).

LISTING 3-3 PRIVATE COMPANIES

Company	Started	Employees
A.M. BIOTECHNIQUES, INC.	1980	5
A/G TECHNOLOGY CORP.	1981	8
ABC RESEARCH CORP.	1967	82
ADVANCED BIOTECHNOLOGIES	1982	17
ADVANCED MINERAL TECHNOLOGIES	1982	17
AGDIA, INC.	1981	8
AGRACETUS	1981	65
AGRI-DIAGNOSTICS ASSOCIATES	1983	22
AGRITECH SYSTEMS, INC.	1984	90
ALLELIC BIOSYSTEMS	1984	7
AMBICO, INC.	1974	25
AMERICAN BIOCLINICAL	1977	24
AMERICAN BIOGENETICS CORP.	1984	10
AMERICAN BIOTECHNOLOGY CO.	1984	5
AMERICAN DIAGNOSTICA, INC.	1983	9
AMERICAN LABORATORIES, INC.	1962	26
AMERICAN QUALEX INT'L	1981	10
AMTRON	1981	16
ANGENICS	1980	50
ANTIBODIES, INC.	1961	35
ANTIVIRALS, INC.	1980	5
APPLIED BIOTECHNOLOGY, INC.	1982	45
APPLIED GENETICS LABS, INC.	1984	11
APPLIED PROTEIN TECHNOLOGIES, INC.	1984	5
AQUASYNERGY, LTD.	1986	5
ASTRE CORPORATE GROUP	1967	148
BACHEM BIOSCIENCE, INC.	1987	15
BACHEM, INC.	1971	45
BEND RESEARCH, INC.	1975	60
BENTECH LABORATORIES	1984	7
BERKELEY ANTIBODY CO., INC.	1983	27
BETHYL LABS, INC.	1977	8
BINAX, INC.	1986	17
BIO HUMA NETICS	1984	15
BIO TECHNIQUES LABS, INC.	1982	25
BIO-RECOVERY, INC.	1983	49
BIO-TECHNICAL RESOURCES, INC.	1962	25
BIOCHEM TECHNOLOGY, INC.	1977	12
BIOCONSEP, INC.	1983	6
BIOCONTROL SYSTEMS	1985	29
BIODESIGN, INC.	1987	3
BIOGENEX LABORATORIES	1981	42
BIOGROWTH, INC.	1985	16
BIOMATERIALS INTERNATIONAL, INC.	1981	30

Listing 3-3 Private Companies

Company	Started	Employees
BIOMATRIX, INC.	1981	50
BIOMED RESEARCH LABS, INC.	1974	18
BIOMEDICAL TECHNOLOGIES	1981	10
BIONIQUE LABS, INC.	1983	11
BIOPOLYMERS, INC.	1984	30
BIOPROBE INTERNATIONAL, INC.	1983	15
BIOPRODUCTS FOR SCIENCE, INC.	1985	19
BIOPURE	1984	31
BIOSCIENCE MANAGEMENT, INC.	1984	8
BIOSTAR MEDICAL PRODUCTS, INC.	1983	30
BIOSYSTEMS, INC.	1976	15
BIOTHERM	1985	10
BIOTICS RESEARCH CORP.	1975	22
BIOTROL, INC.	1985	30
BIOTX (BIOSCIENCES CORP. OF TEXAS)	1985	20
BMI, INC.	1984	10
BRAIN RESEARCH, INC.	1968	
CALZYME LABORATORIES, INC.	1983	10
CENTRAL BIOLOGICS, INC.	1982	7
CHEMGENES	1981	8
CHEMICAL DYNAMICS CORP.	1972	45
CHEMICON INTERNATIONAL, INC.	1981	8
CLINETICS CORP.	1979	10
CLONTECH LABORATORIES, INC.	1984	30
COAL BIOTECH CORP.	1984	10
CODON CORP.	1980	120
CONSOLIDATED BIOTECHNOLOGY, INC.	1983	15
CREATIVE BIOMOLECULES	1981	70
CYANOTECH CORP.	1983	25
CYGNUS RESEARCH CORP.	1985	18
CYTOTECH, INC.	1982	43
DEKALB-PFIZER GENETICS	1982	235
DELTOWN CHEMURGIC CORP.	1968	90
DIAGNOSTIC TECHNOLOGY, INC.	1980	60
DIAMEDIX, INC.	1986	35
DIGENE DIAGNOSTICS, INC.	1984	20
E-Y LABORATORIES, INC.	1978	20
ELANEX PHARMACEUTICALS, INC.	1984	10
ELCATECH	1984	4
EMBREX, INC.	1985	20
EMTECH RESEARCH	1983	30
ENDOGEN, INC.	1985	7
ENZYME CENTER, INC.	1978	30
FORGENE, INC.	1986	4
GAMETRICS, LTD.	1974	20
GENELABS, INC.	1984	90
GENENCOR, INC.	1982	130

Company	Started	Employees
GENESIS LABS, INC.	1984	29
GENETIC THERAPY, INC.	1986	4
GENTRONIX LABORATORIES, INC.	1972	6
HAWAII BIOTECHNOLOGY GROUP, INC.	1982	10
HOUSTON BIOTECHNOLOGY, INC.	1984	30
HUNTER BIOSCIENCES	1987	8
HYCLONE LABORATORIES, INC.	1975	107
IDEC PHARMACEUTICAL	1986	45
IDETEK, INC.	1983	14
IGEN	1982	60
IMCLONE SYSTEMS, INC.	1984	50
IMMUNETECH PHARMACEUTICALS	1981	45
IMMUNEX, INC.	1981	5
IMMUNO CONCEPTS, INC.	1981	20
IMMUNO MODULATORS LABS, INC.	1981	25
IMMUNOGEN, INC.	1981	11
IMMUNOMED CORP.	1979	17
IMMUNOSYSTEMS, INC.	1981	4
IMMUNOTECH CORP.	1980	15
IMMUNOVISION, INC.	1985	6
INCELL CORP.	1982	23
INCON CORP.	1983	6
INTEGRATED CHEMICAL SENSORS	1984	8
INTEK DIAGNOSTICS, INC.	1983	35
INTELLIGENETICS, INC.	1980	52
INTER-AMERICAN RESEARCH ASSOC.	1979	10
INTER-CELL TECHNOLOGIES, INC.	1982	9
INTERNATIONAL ENZYMES, INC.	1983	4
INT'L PLANT RESEARCH INST.	1978	50
INVITRON CORP.	1984	165
JACKSON IMMUNORESEARCH LABS, INC.	1982	16
KALLESTAD DIAGNOSTICS	1967	330
KARYON TECHNOLOGY, LTD.	1984	25
KIRIN-AMGEN	1984	5
KIRKEGAARD & PERRY LABS, INC.	1979	40
LEE BIOMOLECULAR RESEARCH LABS	1980	16
LIPOSOME TECHNOLOGY, INC.	1981	65
MAIZE GENETIC RESOURCES, INC.	1984	2
MARCOR DEVELOPMENT CO.	1977	6
MARICULTURA, INC.	1984	8
MARINE BIOLOGICALS, INC.	1981	2
MARTEK	1985	16
MAST IMMUNOSYSTEMS, INC.	1979	45
MEDAREX, INC.	1987	3
MICROBE MASTERS	1982	37
MICROBIO RESOURCES	1981	30
MICROBIOLOGICAL ASSOCIATES, INC.	1949	180

Listing 3-3 Private Companies

Company	Started	Employees
MICROGENESYS, INC.	1983	32
MICROGENICS	1981	70
MOLECULAR BIOLOGY RESOURCES, INC.	1986	10
MOLECULAR GENETIC RESOURCES	1983	5
MONOCLONAL PRODUCTION INT'L.	1983	4
MONOCLONETICS INT'L, INC.	1984	5
MULTIPLE PEPTIDE SYSTEMS, INC.	1986	6
MUREX CORP.	1984	180
MYCOSEARCH, INC.	1979	6
NATIONAL GENO SCIENCES	1980	4
NEOGEN CORP.	1981	40
NEORX CORP.	1984	194
NEUREX CORP.	1986	22
NEUSHUL MARICULTURE	1978	6
NORTH COAST BIOTECHNOLOGY, INC.	1986	2
NYGENE CORP.	1985	18
O.C.S. LABORATORIES, INC.	1983	12
OCEAN GENETICS	1981	35
OMNI BIOCHEM, INC.	1986	5
PAMBEC LABORATORIES	1987	6
PEL-FREEZ BIOLOGICALS, INC.	1911	175
PENINSULA LABORATORIES, INC.	1971	75
PETROGEN, INC.	1980	6
PHARMAGENE	1986	3
PHYTOGEN	1980	20
PROBIOLOGICS INTERNATIONAL, INC.	1987	5
PROGENX	1987	20
PROMEGA CORP.	1978	85
PROTATEK INTERNATIONAL, INC.	1984	37
PROTEINS INTERNATIONAL, INC.	1983	6
QUEUE SYSTEMS, INC.	1980	85
QUIDEL	1981	325
R & A PLANT/SOIL, INC.	1978	6
RECOMTEX CORP.	1983	12
RESEARCH & DIAGNOSTIC ANTIBODIES	1985	
RHOMED, INC.	1986	8
SCRIPPS LABORATORIES	1980	60
SEAPHARM, INC.	1983	50
SEPRACOR, INC.	1984	40
SERAGEN, INC.	1979	250
SERONO LABS	1971	198
SIBIA	1981	65
SOUTHERN BIOTECHNOLOGY ASSOC.	1982	12
SPHINX BIOTECHNOLOGIES	1987	3
STRATAGENE, INC.	1984	77
SUNGENE TECHNOLOGIES CORP.	1981	28
SYNTHETIC GENETICS, INC.	1985	6

Section 3 Company Location and Type

Company	Started	Employees
TEKTAGEN	1987	3
TOXICON	1978	35
TRANSFORMATION RESEARCH, INC.	1981	4
TRANSGENIC SCIENCES, INC.	1988	3
UNITED BIOMEDICAL, INC.	1983	10
UNITED STATES BIOCHEMICAL CORP.	1973	
UNIVERSITY MICRO REFERENCE LAB	1980	6
UPSTATE BIOTECHNOLOGY, INC.	1986	4
VECTOR LABORATORIES, INC.	1976	50
VERAX CORP.	1978	80
VIAGENE, INC.	1987	9
VIROSTAT, INC.	1985	1
WASHINGTON BIOLAB	1986	10
WELGEN MANUFACTURING, INC.	1989	200
WESTBRIDGE RESEARCH GROUP	1982	15
XENOGEN	1981	
ZYMED LABORATORIES, INC.	1980	20
ZYMOGENETICS, INC.	1981	95

Note: This listing is derived from the biotechnology companies in the directory (Section 2-2).

Listing 3-4 Subsidiary Companies

LISTING 3-4 SUBSIDIARY COMPANIES

Company	Started	Subsidiary of
AGRIGENETICS CORP.	1975	LUBRIZOL
AN-CON GENETICS	1982	UNIVERSITY GENETICS
ATLANTIC ANTIBODIES	1972	INCSTAR
BEHRING DIAGNOSTICS	1952	AMERICAN HOECHST CORP.
BIOKYOWA, INC.	1983	KYOWA HAKKO KOGYO (JPN)
BIONETICS RESEARCH	1985	ORGANON TEKNIKA (NETH)
BIOSEARCH, INC.	1977	NEW BRUNSWICK SCIENTIFIC
BIOTECHNICA AGRICULTURE	1987	BIOTECHNICA INT'L
BIOTEST DIAGNOSTICS CORP.	1946	BIOTEST SERUM INST. (WGER)
BOEHRINGER MANNHEIM DIAGN.	1975	BOEHR. MANNHEIM (WGER)
CALIFORNIA INTEGRATED DIAGN.	1981	INFRAGENE
CAMBRIDGE RES. BIOCHEMICALS	1980	CAMBRIDGE RES.BIOCHEM. (UK)
CARBOHYDRATES INT'L, INC.	1987	MEDICARB (SWE)
CELLMARK DIAGNOSTICS	1986	ICI AMERICAS
CHARLES RIVER BIOTECH. SERV.	1983	BAUSCH & LOMB
CIBA-CORNING DIAGNOSTIC CORP.	1985	CIBA-GEIGY & CORNING GLASS
COORS BIOTECH PRODUCTS CO.	1984	ADOLPH COORS CO.
COVALENT TECHNOLOGY CORP.	1981	KMS INDUSTRIES
DAKO CORP.	1979	DAKOPATTS A/S (DENMARK)
DAMON BIOTECH, INC.	1978	DAMON CORP.
DNAX RESEARCH INSTITUTE	1980	SCHERING-PLOUGH
ENZYME BIO-SYSTEMS LTD.	1983	CPC INTERNATIONAL
ENZYME TECHNOLOGY CORP.	1981	GREAT LAKES CHEMICAL CORP.
FERMENTA ANIMAL HEALTH	1986	FERMENTA AB (SWE)
FLOW LABORATORIES, INC.	1961	FLOW GENERAL, INC.
GENETIC SYSTEMS CORP.	1980	SUB OF BRISTOL-MYERS
GRANADA GENETICS CORP.	1979	GRANADA CORP.
HAZLETON BIOTECHNOLOGIES CO.	1983	HAZELTON LABORATORIES
HYBRITECH, INC.	1978	ELI LILLY
IBF BIOTECHNICS, INC.	1987	IBF, SA
ICL SCIENTIFIC, INC.	1981	HYCOR BIOMED
IMCERA BIOPRODUCTS, INC.	1986	INT'L MINERALS & CHEMICALS
INTERFERON SCIENCES, INC.	1981	NATIONAL PATENTS DEVEL.
INT'L BIOTECHNOLOGIES, INC.	1982	EASTMAN KODAK
LABSYSTEMS, INC.	1983	LABSYSTEMS OY (FINLAND)
LIFECODES	1982	QUANTOM CORP.
LUCKY BIOTECH CORP.	1984	LUCKY LTD. (S. KOREA)
MELOY LABORATORIES, INC.	1970	RORER GROUP, INC.
MOLECULAR DIAGNOSTICS, INC.	1981	BAYER AG (WGER)
NITRAGIN CO.	1898	ALLIED CORP.
NORDEN LABS	1983	SMITHKLINE BECKMAN
ORGANON TEKNIKA CORP.	1978	ORGANON TEKNIKA (NETH)
PETROFERM, INC.	1977	PETROLEUM FERM., NV
PHARMACIA LKB BIOTECHNOLOGY	1960	PHARMACIA AB (SWEDEN)

Section 3 Company Location and Type

Company	Started	Subsidiary of
POLYCELL, INC.	1983	QUEST BIOTECHNOLOGY
PROVESTA CORP.	1975	PHILLIPS PETROLEUM
REPAP TECHNOLOGIES	1981	REPAP ENTERPRISES (CANADA)
RICERCA, INC.	1986	FERMENTA AB (SWE)
SCOTT LABORATORIES	1981	MICROBIOLOGICAL SCIENCES
SYNGENE PRODUCTS	1980	TECHAMERICA GROUP
SYVA CO.	1966	SYNTEX
3M DIAGNOSTIC SYSTEMS	1981	3M
TRITON BIOSCIENCES, INC.	1983	SHELL OIL CO.
UNITED AGRISEEDS, INC.	1981	DOW CHEMICAL
ZOECON CORP.	1968	SANDOZ (SWI)

Note: This listing is derived from the biotechnology companies in the directory (Section 2-2).

SECTION 4 THE FOUNDING OF BIOTECHNOLOGY COMPANIES

Section 4 Company Founding

INTRODUCTION

Although the 360 biotechnology companies described in this Guide were founded between 1911 and the present, more than 93 percent were founded in the 1970's and after. In fact, over 74 percent of the companies were founded in the 1980's. The companies founded each year are listed in Listing 4-1, and Figure 4-1 demonstrates the number of companies founded each year from 1971 to the present.

Founders Study
As part of a questionnaire-based study of U.S. biotechnology companies, we inquired about the background of the company founders, as indicated by their previous positions before founding the companies. The following data were derived from responses of 121 companies representing 292 total founders. This study did not include data from subsidiary companies or companies for which the founder information was not complete. Company founders ranged in number from one to six per company, with a mean value of 2.4 founders per company and a median of three.

Founders' Backgrounds
There have been some indications that a large number of biotechnology companies were spun off from academic laboratories. However, only 38 percent of the founders came from academic positions, whereas 44 percent were from industry (Figure 4-2). An additional 4.5 percent were from government laboratories, and 4.8 percent were from financial or venture capital backgrounds. The remaining 8.5 percent of the founders came from a variety of backgrounds, ranging from private consulting to medical to legal practices. One-fifth of the companies were started by a mix of both academic and industry founders. It appears that at least half of the company founders worked in or ran scientific laboratories in their previous positions before founding their own companies.

Founders Over Time
Companies were divided into three classes, according to when they were started. **Early** companies, founded between 1971 and 1980; **Biotech Boom** companies, founded between 1981 and 1983; and **Recent** companies, founded between 1984 and 1986.   Figure 4-3 demonstrates a decrease in founders with prior positions in academia over time — 52 percent of the founders of **Early** companies were from academia while only 19 percent of **Recent** company founders had previous posts in academia. In contrast, the number of founders with industrial backgrounds steadily increased over time. Founders who came from government or finance/venture capital firms accounted for only a small percentage of total founders in all three groups of companies (9 to 11 percent) .

Prior Company Size
Founders from industry were further divided into those with **small firm** or **large corporation** (i.e., more than $100 million in annual sales and more than 1,500 employees) backgrounds. Of the founders from industry, almost an equal number were from each category. Similarly, the number of founders with **small firm** and

large corporation backgrounds were approximately equal for **Early, Biotech Boom,** and **Recent** companies.

Biotechnology Company Size

Companies were also separated by size into **Small** companies, with 2 to 20 employees; **Medium** companies, with 21 to 100 employees; and **Large** companies, those with more than 100 employees. There were no major differences in the academic/industrial background mix of founders of companies in the three size groups. As might be expected, the industrial founders of the **Small** biotechnology companies came more from **small firm** backgrounds than from **large corporation** backgrounds. In contrast, there were twice as many industrial founders of **Large** biotechnology companies with previous positions in **large corporations** than those with positions in **small firms.** None of the 109 founders of **Small** companies had a finance/venture capital background, while 11 percent of **large** company founders came from financial backgrounds.

Notes

1. This analysis provides a picture of company founders at the time of founding, and is not necessarily an indication of current top management. The shift in company management is the topic of a current study.

2. A more detailed analysis of biotechnology company founders appears in: Dibner, Mark D. : Commercial Biotech's Founding Fathers. Bio/Technology $\underline{5}$: 571 May 1987.

LISTING 4-1 COMPANIES BY YEAR OF FOUNDING

Company	Type	Location
Pre-1970		
ABC RESEARCH CORP.	PRI	GAINESVILLE, FL
AMERICAN LABORATORIES, INC.	PRI	OMAHA, NE
ANTIBODIES, INC.	PRI	DAVIS, CA
ASTRE CORPORATE GROUP	PRI	CHARLOTTESVILLE, VA
BEHRING DIAGNOSTICS	SUB	LA JOLLA, CA
BIO-TECHNICAL RESOURCES, INC.	PRI	MANITOWOC, WI
BIOSPHERICS, INC.	PUB	BELTSVILLE, MD
BIOTEST DIAGNOSTICS CORP.	SUB	FAIRFIELD, NJ
BRAIN RESEARCH, INC.	PRI	NEW YORK, NY
COLLABORATIVE RESEARCH, INC.	PUB	BEDFORD, MA
DELTOWN CHEMURGIC CORP.	PRI	GREENWICH, CT
ELECTRO-NUCLEONICS, INC.	PUB	FAIRFIELD, NJ
ENZON	PUB	MTN. VIEW, CA
FLOW LABORATORIES, INC.	SUB	MCLEAN, VA
GAMMA BIOLOGICALS, INC.	PUB	HOUSTON, TX
KALLESTAD DIAGNOSTICS	PRI	AUSTIN, TX
LIFE SCIENCES	PUB	ST. PETERSBURG, FL
MICROBIOLOGICAL ASSOCIATES, INC.	PRI	ROCKVILLE, MD
NITRAGIN CO.	SUB	MILWAUKEE, WI
PEL-FREEZ BIOLOGICALS, INC.	PRI	ROGERS, AR
PHARMACIA LKB BIOTECHNOLOGY	SUB	PISCATAWAY, NJ
SYVA CO.	SUB	PALO ALTO, CA
ZOECON CORP.	SUB	PALO ALTO, CA
1970		
MELOY LABORATORIES, INC.	SUB	SPRINGFIELD, VA
1971		
BACHEM, INC.	PRI	TORRANCE, CA
BIOMERICA, INC.	PUB	NEWPORT BEACH, CA
CETUS CORP.	PUB	EMERYVILLE, CA
CLINICAL SCIENCES, INC.	PUB	WHIPPANY, NJ
PENINSULA LABORATORIES, INC.	PRI	BELMONT, CA
SERONO LABS	PRI	RANDOLPH, MA
1972		
ATLANTIC ANTIBODIES	SUB	SCARBOROUGH, ME
BIO-RESPONSE, INC.	PUB	HAYWARD, CA
CHEMICAL DYNAMICS CORP.	PRI	S. PLAINFIELD, NJ
DIAGNOSTIC PRODUCTS CORP.	PUB	LOS ANGELES, CA
GENTRONIX LABORATORIES, INC.	PRI	BETHESDA, MD

Listing 4-1 Companies by Year

Company	Type	Location
1973		
BIOTECH RESEARCH LABS, INC.	PUB	ROCKVILLE, MD
NATIVE PLANTS, INC.	PUB	SALT LAKE CITY, UT
UNITED STATES BIOCHEMICAL CORP.	PRI	CLEVELAND, OH
1974		
AMBICO, INC.	PRI	DALLAS CENTER, IA
BIOMED RESEARCH LABS, INC.	PRI	SEATTLE, WA
GAMETRICS, LTD.	PRI	LAS VEGAS, NV
1975		
AGRIGENETICS CORP.	SUB	EASTLAKE, OH
BEND RESEARCH, INC.	PRI	BEND, OR
BIOTICS RESEARCH CORP.	PRI	STAFFORD, TX
BOEHRINGER MANNHEIM DIAGN.	SUB	INDIANAPOLIS, IN
COLLAGEN CORP.	PUB	PALO ALTO, CA
CYTOX CORP.	PUB	ALLENTOWN, PA
HYCLONE LABORATORIES, INC.	PRI	LOGAN, UT
LIFECORE BIOMEDICAL, INC.	PUB	MINNEAPOLIS, MN
PROVESTA CORP.	SUB	BARTLESVILLE, OK
1976		
BIOSYSTEMS, INC.	PRI	LOGANVILLE, GA
ENZO-BIOCHEM, INC.	PUB	NEW YORK, NY
GENENTECH	PUB	S. SAN FRANCISCO, CA
VECTOR LABORATORIES, INC.	PRI	BURLINGAME, CA
1977		
AMERICAN BIOCLINICAL	PRI	PORTLAND, OR
BETHYL LABS, INC.	PRI	MONTGOMERY, TX
BIOCHEM TECHNOLOGY, INC.	PRI	MALVERN, PA
BIOSEARCH, INC.	SUB	SAN RAFAEL, CA
GENEX, INC.	PUB	GAITHERSBURG, MD
MARCOR DEVELOPMENT CO.	PRI	HACKENSACK, NJ
PETROFERM, INC.	SUB	FERNANDINA BCH., FL
1978		
AUTOMEDIX SCIENCES, INC.	PUB	TORRANCE, CA
DAMON BIOTECH, INC.	SUB	NEEDHAM HTS, MA
E-Y LABORATORIES, INC.	PRI	SAN MATEO, CA
ENZYME CENTER, INC.	PRI	MALDEN, MA
HYBRITECH, INC.	SUB	SAN DIEGO, CA
INT'L PLANT RES. INST.	PRI	SAN CARLOS, CA
NEUSHUL MARICULTURE	PRI	GOLETA, CA
ORGANON TEKNIKA CORP.	SUB	DURHAM, NC
PROMEGA CORP.	PRI	MADISON, WI
R & A PLANT/SOIL, INC.	PRI	PASCO, WA
SUMMA MEDICAL CORP.	PUB	ALBUQUERQUE, NM

Company	Type	Location
1978 (Cont.)		
TOXICON	PRI	WOBURN, MA
VERAX CORP.	PRI	LEBANON, NH
1979		
ADVANCED GENETIC SCIENCES	PUB	OAKLAND, CA
BIOGEN	PUB	CAMBRIDGE, MA
CENTOCOR	PUB	MALVERN, PA
CLINETICS CORP.	PRI	TUSTIN, CA
DAKO CORP.	SUB	SANTA BARBARA, CA
EPITOPE, INC.	PUB	BEAVERTON, OR
GRANADA GENETICS CORP.	SUB	COLLEGE STATION, TX
HANA BIOLOGICS, INC.	PUB	ALAMEDA, CA
IMMUNOMED CORP.	PRI	TAMPA, FL
INTER-AMERICAN RESEARCH ASSOC.	PRI	GAITHERSBURG, MD
KIRKEGAARD & PERRY LABS, INC.	PRI	GAITHERSBURG, MD
MAST IMMUNOSYSTEMS, INC.	PRI	MOUNTAIN VIEW, CA
MOLECULAR GENETICS, INC.	PUB	MINNETONKA, MN
MONOCLONAL ANTIBODIES, INC.	PUB	MOUNTAIN VIEW, CA
MYCOSEARCH, INC.	PRI	CHAPEL HILL, NC
SERAGEN, INC.	PRI	HOPKINTON, MA
VEGA BIOTECHNOLOGIES, INC.	PUB	TUCSON, AZ
1980		
A.M. BIOTECHNIQUES, INC.	PRI	BELTON, TX
AMGEN	PUB	THOUSAND OAKS, CA
ANGENICS	PRI	CAMBRIDGE, MA
ANTIVIRALS, INC.	PRI	CORVALLIS, OR
BIOTECHNOLOGY GENERAL CORP.	PUB	NEW YORK, NY
CALGENE, INC.	PUB	DAVIS, CA
CAMBRIDGE RES. BIOCHEMICALS, INC.	SUB	VALLEY STREAM, NY
CODON CORP.	PRI	S. SAN FRANCISCO, CA
COOPER DEVELOPMENT CO.	PUB	MENLO PARK, CA
CYTOGEN CORP.	PUB	PRINCETON, NJ
DIAGNOSTIC TECHNOLOGY, INC.	PRI	HAUPPAUGE, NY
DNAX RESEARCH INSTITUTE	SUB	PALO ALTO, CA
GENETIC ENGINEERING, INC.	PUB	DENVER, CO
GENETIC SYSTEMS CORP.	SUB	SEATTLE, WA
HYGEIA SCIENCES	PUB	NEWTON, MA
IMMUNOTECH CORP.	PRI	BOSTON, MA
INGENE, INC.	PUB	SANTA MONICA, CA
INTELLIGENETICS, INC.	PRI	MOUNTAIN VIEW, CA
LEE BIOMOLECULAR RESEARCH LABS	PRI	SAN DIEGO, CA
NATIONAL GENO SCIENCES	PRI	SOUTHFIELD, MI
PETROGEN, INC.	PRI	ARLINGTON HTS, IL
PHYTOGEN	PRI	PASADENA, CA
QUEUE SYSTEMS, INC.	PRI	NORTH BRANCH, NJ

Listing 4-1 Companies by Year

Company	Type	Location
1980 (Cont.)		
SCRIPPS LABORATORIES	PRI	SAN DIEGO, CA
SYNGENE PRODUCTS	SUB	ELWOOD, KS
UNIGENE LABORATORIES, INC.	PUB	FAIRFIELD, NJ
UNIVERSITY MICRO REFERENCE LAB	PRI	LINTHICUM, MD
VIRAGEN	PUB	HIALEAH, FL
ZYMED LABORATORIES, INC.	PRI	S. SAN FRANCISCO, CA
1981		
A/G TECHNOLOGY CORP.	PRI	NEEDHAM, MA
ABN	PUB	HAYWARD, CA
ADVANCED MAGNETICS, INC.	PUB	CAMBRIDGE, MA
AGDIA, INC.	PRI	MISHAWAKA, IN
AGRACETUS	PRI	MIDDLETON, WI
ALFACELL CORP.	PUB	BLOOMFIELD, NJ
AMERICAN QUALEX INT'L, INC.	PRI	LA MIRADA, CA
AMTRON	PRI	CHARLESTON, SC
APPLIED BIOSYSTEMS, INC.	PUB	FOSTER CITY, CA
BIOGENEX LABORATORIES	PRI	SAN RAMON, CA
BIOMATERIALS INT'L, INC.	PRI	SALT LAKE CITY, UT
BIOMATRIX, INC.	PRI	RIDGEFIELD, NJ
BIOMEDICAL TECHNOLOGIES	PRI	STOUGHTON, MA
BIOTECHNICA INT'L	PUB	CAMBRIDGE, MA
CALIFORNIA BIOTECHNOLOGY, INC.	PUB	MOUNTAIN VIEW, CA
CALIFORNIA INTEGRATED DIAGNOS.	SUB	BERKELEY, CA
CAMBRIDGE BIOSCIENCE CORP.	PUB	WORCESTER, MA
CHEMGENES	PRI	NEEDHAM, MA
CHEMICON INT'L, INC.	PRI	EL SEGUNDO, CA
CHIRON CORP.	PUB	EMERYVILLE, CA
COVALENT TECHNOLOGY CORP.	SUB	ANN ARBOR, MI
CREATIVE BIOMOLECULES	PRI	HOPKINTON, MA
CROP GENETICS INT'L CORP.	PUB	HANOVER, MD
DIAGNON CORP.	PUB	ROCKVILLE, MD
DNA PLANT TECHNOLOGY CORP.	PUB	CINNAMINSON, NJ
ENDOTRONICS, INC.	PUB	COON RAPIDS, MN
ENZON, INC.	PUB	S. PLAINFIELD, NJ
ENZYME TECHNOLOGY CORP.	SUB	ASHLAND, OH
EXOVIR, INC.	PUB	GREAT NECK, NY
GENE-TRAK SYSTEMS	PUB	FRAMINGHAM, MA
GENETIC DIAGNOSTICS CORP.	PUB	GREAT NECK, NY
GENETICS INSTITUTE, INC.	PUB	CAMBRIDGE, MA
GENZYME CORP.	PUB	BOSTON, MA
ICL SCIENTIFIC, INC.	SUB	FOUNTAIN VLY, CA
IGENE BIOTECHNOLOGY, INC.	PUB	COLUMBIA, MD
IMMUNETECH PHARMACEUTICALS	PRI	SAN DIEGO, CA
IMMUNEX CORP.	PUB	SEATTLE, WA
IMMUNEX, INC.	PRI	CARSON CITY, NV

Company	Type	Location
1981 (Cont.)		
IMMUNO CONCEPTS, INC.	PRI	SACRAMENTO, CA
IMMUNO MODULATORS LABS, INC.	PRI	STAFFORD, TX
IMMUNOGEN, INC.	PRI	CAMBRIDGE, MA
IMMUNOSYSTEMS, INC.	PRI	BIDDEFORD, ME
IMRE CORP.	PUB	SEATTLE, WA
IMREG, INC.	PUB	NEW ORLEANS, LA
INTEGRATED GENETICS, INC.	PUB	FRAMINGHAM, MA
INTERFERON SCIENCES, INC.	SUB	NEW BRUNSWICK, NJ
LIPOSOME COMPANY, INC.	PUB	PRINCETON, NJ
LIPOSOME TECHNOLOGY, INC.	PRI	MENLO PARK, CA
LYPHOMED, INC.	PUB	ROSEMONT, IL
MARINE BIOLOGICALS, INC.	PRI	MARMORA, NJ
MICROBIO RESOURCES	PRI	SAN DIEGO, CA
MICROGENICS	PRI	CONCORD, CA
MOLECULAR BIOSYSTEMS, INC.	PUB	SAN DIEGO, CA
MOLECULAR DIAGNOSTICS, INC.	SUB	WEST HAVEN, CT
NEOGEN CORP.	PRI	LANSING, MI
OCEAN GENETICS	PRI	SANTA CRUZ, CA
PLANT GENETICS, INC.	PUB	DAVIS, CA
QUIDEL	PRI	LA JOLLA, CA
REPAP TECHNOLOGIES	SUB	VALLEY FORGE, PA
REPLIGEN CORP.	PUB	CAMBRIDGE, MA
RIBI IMMUNOCHEM RESEARCH, INC.	PUB	HAMILTON, MT
SCOTT LABORATORIES	SUB	FISKVILLE, RI
SIBIA	PRI	SAN DIEGO, CA
SUNGENE TECHNOLOGIES CORP.	PRI	SAN JOSE, CA
SYNERGEN, INC.	PUB	BOULDER, CO
SYNTRO CORP.	PUB	SAN DIEGO, CA
TECHNICLONE INT'L, INC.	PUB	SANTA ANA, CA
THREE-M (3M) DIAGNOSTIC SYSTEMS	SUB	SANTA CLARA, CA
TRANSFORMATION RESEARCH, INC.	PRI	BRIGHTON, MA
UNITED AGRISEEDS, INC.	SUB	CHAMPAIGN, IL
UNIVERSITY GENETICS CO.	PUB	WESTPORT, CT
VIVIGEN, INC.	PUB	SANTA FE, NM
XENOGEN	PRI	MANSFIELD, CT
XOMA CORP.	PUB	BERKELEY, CA
ZYMOGENETICS, INC.	PRI	SEATTLE, WA
1982		
ADVANCED BIOTECHNOLOGIES	PRI	SILVER SPRING, MD
ADVANCED MINERAL TECHNOLOGIES	PRI	GOLDEN, CO
ALPHA I BIOMEDICALS, INC.	PUB	WASHINGTON, DC
AN-CON GENETICS	SUB	MELVILLE, NY
APPLIED BIOTECHNOLOGY, INC.	PRI	CAMBRIDGE, MA
APPLIED DNA SYSTEMS, INC.	PUB	NEW YORK, NY
BIO TECHNIQUES LABS, INC.	PRI	REDMOND, WA

Listing 4-1 Companies by Year

Company	Type	Location
1982 (Cont.)		
BIOTECHNOLOGY DEVELOPMENT	PUB	NEWTON, MA
CELLULAR PRODUCTS, INC.	PUB	BUFFALO, NY
CENTRAL BIOLOGICS, INC.	PRI	RALEIGH, NC
CISTRON BIOTECHNOLOGY, INC.	PUB	PINE BROOK, NJ
CYTOTECH, INC.	PRI	SAN DIEGO, CA
DEKALB-PFIZER GENETICS	PRI	DEKALB, IL
GENENCOR, INC.	PRI	S. SAN FRANCISCO, CA
HAWAII BIOTECHNOLOGIES CO.	PRI	AIEA, HI
IGEN	PRI	ROCKVILLE, MD
IMMUCELL CORP.	PUB	PORTLAND, ME
INCELL CORP.	PRI	MILWAUKEE, WI
INTER-CELL TECHNOLOGIES, INC.	PRI	SOMERVILLE, NJ
INT'L BIOTECHNOLOGIES, INC.	SUB	NEW HAVEN, CT
JACKSON IMMUNORESEARCH LABS	PRI	WEST GROVE, PA
LIFECODES CORP.	PRI	VALHALLA, NY
MICROBE MASTERS	PRI	BATON ROUGE, LA
MYCOGEN CORP.	PUB	SAN DIEGO, CA
SOUTHERN BIOTECHNOLOGY ASSOC.	PRI	BIRMINGHAM, AL
SYNBIOTICS CORP.	PUB	SAN DIEGO, CA
WESTBRIDGE RESEARCH GROUP	PRI	SAN DIEGO, CA
1983		
AGRI-DIAGNOSTICS ASSOCIATES	PRI	CINNAMINSON, NJ
AMERICAN DIAGNOSTICA, INC.	PRI	NEW YORK, NY
APPLIED MICROBIOLOGY, INC.	PUB	BROOKLYN, NY
BERKELEY ANTIBODY CO., INC.	PRI	RICHMOND, CA
BIO-RECOVERY, INC.	PRI	NORTHVALE, NJ
BIOCONSEP, INC.	PRI	BELLEMEAD, NJ
BIOKYOWA, INC.	SUB	CAPE GIRARDEAU, MO
BIONIQUE LABS, INC.	PRI	SARANAC LAKE, NY
BIOPROBE INT'L, INC.	PRI	TUSTIN, CA
BIOSTAR MEDICAL PRODUCTS, INC.	PRI	BOULDER, CO
CALZYME LABORATORIES, INC.	PRI	SAN LUIS OBISPO, CA
CEL-SCI CORP.	PUB	ALEXANDRIA, VA
CHARLES RIVER BIOTECH. SERVICES	SUB	WILMINGTON, MA
CONSOLIDATED BIOTECHNOLOGY	PRI	ELKHART, IN
CYANOTECH CORP.	PRI	WOODINVILLE, WA
ECOGEN, INC.	PUB	LANGHORNE, PA
EMTECH RESEARCH	PRI	MOUNT LAUREL, NJ
ENVIRONMENTAL DIAGNOSTICS, INC.	PUB	BURLINGTON, NC
ENZYME BIO-SYSTEMS, LTD.	SUB	ENGLEWOOD CLIFFS, NJ
HAZLETON BIOTECHNOLOGIES CO.	SUB	VIENNA, VA
IDETEK, INC.	PRI	SAN BRUNO, CA
IMMUNOMEDICS, INC.	PUB	WARREN, NJ
INCON CORP.	PRI	YOUNGSVILLE, NC
INTEK DIAGNOSTICS, INC.	PRI	BURLINGAME, CA

Company	Type	Location
1983 (Cont.)		
INT'L ENZYMES, INC.	PRI	FALLBROOK, CA
LABSYSTEMS, INC.	SUB	MORTON GROVE, IL
LIFE TECHNOLOGIES, INC.	PUB	GAITHERSBURG, MD
MICROGENESYS, INC.	PRI	WEST HAVEN, CT
MOLECULAR GENETIC RESOURCES	PRI	TAMPA, FL
MONOCLONAL PRODUCTION INT'L	PRI	FORT SCOTT, KS
NORDEN LABS	SUB	LINCOLN, NE
O.C.S. LABORATORIES, INC.	PRI	DENTON, TX
ONCOGENE SCIENCE, INC.	PUB	MANHASSET, NY
ONCOR, INC.	PUB	GAITHERSBURG, MD
POLYCELL, INC.	SUB	DETROIT, MI
PRAXIS BIOLOGICS	PUB	ROCHESTER, NY
PROTEINS INT'L, INC.	PRI	ROCHESTER, MI
RECOMTEX CORP.	PRI	EAST LANSING, MI
SEAPHARM, INC.	PRI	PRINCETON, NJ
SENETEK PLC	PUB	MOUNTAIN VIEW, CA
TECHNOGENETICS, INC.	PUB	ENGLEWOOD CLIFFS, NJ
TRITON BIOSCIENCES, INC.	SUB	ALAMEDA, CA
UNITED BIOMEDICAL, INC.	PRI	LAKE SUCCESS, NY
1984		
AGRITECH SYSTEMS, INC.	PRI	PORTLAND, ME
ALLELIC BIOSYSTEMS	PRI	KEARNEYSVILLE, WV
AMERICAN BIOGENETICS CORP.	PRI	IRVINE, CA
AMERICAN BIOTECHNOLOGY CO.	PRI	ROCKVILLE, MD
APPLIED GENETICS LABS, INC.	PRI	MELBOURNE, FL
APPLIED PROTEIN TECHNOLOGIES,	PRI	CAMBRIDGE, MA
BENTECH LABORATORIES	PRI	CLACKAMAS, OR
BIO HUMA NETICS	PRI	CHANDLER, AZ
BIOPOLYMERS, INC.	PRI	FARMINGTON, CT
BIOPURE	PRI	BOSTON, MA
BIOSCIENCE MANAGEMENT INC.	PRI	BETHLEHEM, PA
BIOTHERAPEUTICS, INC.	PUB	FRANKLIN, TN
BIOTHERAPY SYSTEMS, INC.	PUB	MOUNTAIN VIEW, CA
BMI, INC.	PRI	MISSION VIEJO, CA
CLONTECH LABORATORIES, INC.	PRI	PALO ALTO, CA
COAL BIOTECH CORP.	PRI	LIBERTYVILLE, IL
COORS BIOTECH PRODUCTS CO.	SUB	WESTMINSTER, CO
DIGENE DIAGNOSTICS, INC.	PRI	COLLEGE PARK, MD
EARL-CLAY LABORATORIES, INC.	PUB	NOVATO, CA
ELANEX PHARMACEUTICALS, INC.	PRI	BOTHELL, WA
ELCATECH, INC.	PRI	WINSTON-SALEM, NC
GEN-PROBE, INC.	PUB	SAN DIEGO, CA
GENELABS , INC.	PRI	REDWOOD CITY, CA
GENESIS LABS, INC.	PRI	MINNEAPOLIS, MN
HOUSTON BIOTECHNOLOGY, INC.	PRI	THE WOODLANDS, TX

Listing 4-1 Companies by Year

Company	Type	Location
1984 (Cont.)		
IMCLONE SYSTEMS, INC.	PRI	NEW YORK, NY
INFERGENE CO.	PUB	BENICIA, CA
INTEGRATED CHEMICAL SENSORS	PRI	NEWTON, MA
INVITRON CORP.	PRI	ST. LOUIS, MO
KARYON TECHNOLOGY, LTD.	PRI	NORWOOD, MA
KIRIN-AMGEN	PRI	THOUSAND OAKS, CA
LUCKY BIOTECH CORP.	SUB	EMERYVILLE, CA
MAIZE GENETIC RESOURCES, INC.	PRI	BENSON, NC
MARICULTURA, INC.	PRI	WRIGHTSVILLE BCH, NC
MONOCLONETICS INT'L, INC.	PRI	HOUSTON, TX
MUREX CORP.	PRI	NORCROSS, GA
NEORX CORP.	PRI	SEATTLE, WA
PROTATEK INT'L, INC.	PRI	ST. PAUL, MN
SEPRACOR, INC.	PRI	MARLBOROUGH, MA
STRATAGENE, INC.	PRI	LA JOLLA, CA
T CELL SCIENCES, INC.	PUB	CAMBRIDGE, MA
1985		
BIOCONTROL SYSTEMS	PRI	BOTHELL, WA
BIOGROWTH, INC.	PRI	RICHMOND, CA
BIONETICS RESEARCH	SUB	ROCKVILLE, MD
BIOPRODUCTS FOR SCIENCE, INC.	PRI	INDIANAPOLIS, IN
BIOTHERM	PRI	RES. TRIANGLE PK, NC
BIOTROL, INC.	PRI	CHASKA, MN
BIOTX (BIOSCIENCES CORP. OF TEXAS)	PRI	HOUSTON, TX
CIBA-CORNING DIAGNOSTIC CORP.	SUB	MEDFIELD, MA
CYGNUS RESEARCH CORP.	PRI	REDWOOD CITY, CA
EMBREX, INC.	PRI	RES. TRIANGLE PK, NC
ENDOGEN, INC.	PRI	BOSTON, MA
IMMUNOVISION, INC.	PRI	SPRINGDALE, AR
MARTEK	PRI	COLUMBIA, MD
NYGENE CORP.	PRI	YONKERS, NY
RESEARCH&DIAGNOSTIC ANTIBODIES	PRI	BERKELEY, CA
SYNTHETIC GENETICS, INC.	PRI	SAN DIEGO, CA
VIROSTAT, INC.	PRI	PORTLAND, OR
1986		
AQUASYNERGY, LTD.	PRI	KINSTON, NC
BINAX, INC.	PRI	SOUTH PORTLAND, ME
CELLMARK DIAGNOSTICS	SUB	GERMANTOWN, MD
DIAMEDIX, INC.	PRI	MIAMI, FL
ESCAGENETICS CORP.	PUB	SAN CARLOS, CA
FERMENTA ANIMAL HEALTH	SUB	KANSAS CITY, MO
FORGENE, INC.	PRI	RHINELANDER, WI
GENETIC THERAPY, INC.	PRI	GAITHERSBURG, MD
IDEC PHARMACEUTICAL	PRI	MOUNTAIN VIEW, CA

Company	Type	Location
1986 (Cont.)		
IMCERA BIOPRODUCTS, INC.	SUB	NORTHBROOK, IL
MOLECULAR BIOLOGY RESOURCES	PRI	MILWAUKEE, WI
MULTIPLE PEPTIDE SYSTEMS, INC.	PRI	SAN DIEGO, CA
NEUREX CORP.	PRI	MENLO PARK, CA
NORTH COAST BIOTECHNOLOGY, INC.	PRI	CLEVELAND HTS, OH
OMNI BIOCHEM, INC.	PRI	NATIONAL CITY, CA
PHARMAGENE	PRI	CHARLOTTE, NC
QUEST BIOTECHNOLOGY, INC.	PUB	DETROIT, MI
RHOMED, INC.	PRI	ALBUQUERQUE, NM
RICERCA, INC.	SUB	PAINESVILLE, OH
UPSTATE BIOTECHNOLOGY, INC.	PRI	LAKE PLACID, NY
WASHINGTON BIOLAB	PRI	AMES, IA
1987		
BACHEM BIOSCIENCE, INC.	PRI	PHILADELPHIA, PA
BIODESIGN, INC.	PRI	KENNEBUNKPORT, ME
BIOTECHNICA AGRICULTURE	SUB	OVERLAND PARK, KS
CARBOHYDRATES INT'L, INC.	SUB	CHICAGO, IL
HUNTER BIOSCIENCES	PRI	RALEIGH, NC
IBF BIOTECHNICS, INC.	SUB	SAVAGE, MD
MEDAREX, INC.	PRI	CLIFTON, NJ
PAMBEC LABORATORIES	PRI	GURNEE, IL
PROBIOLOGICS INT'L, INC.	PRI	RES. TRIANGLE PK, NC
PROGENX	PRI	SAN DIEGO, CA
SPHINX BIOTECHNOLOGIES	PRI	RES. TRIANGLE PK, NC
TEKTAGEN	PRI	MALVERN, PA
VIAGENE, INC.	PRI	SAN DIEGO, CA
1988		
TRANSGENIC SCIENCES, INC.	PRI	WORCESTER, MA
WELGEN MANUFACTURING, INC.	PRI	WARWICK, RI

Notes:
Company foundings are through February 1988. Type indicates financing type as PRI=Private; PUB=Public; SUB=Subsidiary.

Figure 4-1 Companies Founded per Year

FIGURE 4-1 COMPANIES FOUNDED PER YEAR

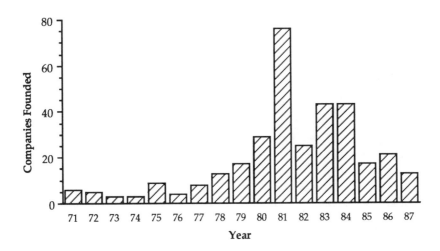

This figure shows the number of U.S. biotechnology companies founded each year from 1971 through 1987. Of the 360 biotechnology companies listed in Section 2, 93 percent were founded during this period of time.

FIGURE 4-2 THE FOUNDERS OF BIOTECHNOLOGY COMPANIES

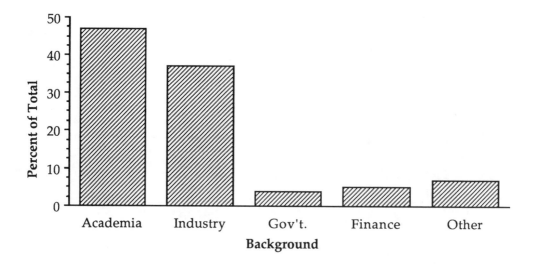

Previous positions of 292 founders of 121 U.S. biotechnology companies were assessed from data collected in a questionnaire-based study in mid-1977. The percentages of the total number of founders coming from the indicated backgrounds are given. The classification "Other" represents miscellaneous backgrounds including law and medicine.

Figure 4-3 Company Founders Then and Now

FIGURE 4-3 COMPANY FOUNDERS THEN AND NOW

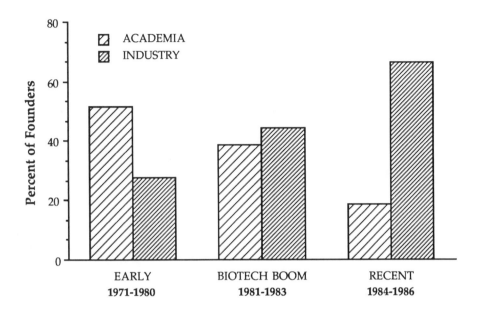

Company founders from the early companies (started 1971-1980), biotech boom year companies (1981-1983) and recently founded companies (1984-1986) were examined for backgrounds of the founders. A total of 230 founders from 121 of the biotechnology firms were categorized as working in either academia or industry prior to founding their companies. This figure demonstrates a shift of founders' backgrounds from mostly academic backgrounds in the early years, to an almost equal mix during the boom years, to a more than three-to-one majority of founders coming from industry in the recently founded companies.

SECTION 5 COMPANY AREAS OF INTEREST

Section 5 Areas of Interest

INTRODUCTION

The classification codes in Table 5-1 define the coding used by our Information Program to define primary and secondary interests of biotechnology companies for our databases. A section of companies with primary or secondary interest in each of these classifications follows in Listings 5-1 through 5-29. Please note that there are no sections for Consultants or Venture Capital (codes U and Z, respectively) as these companies fell beyond the definition we used to identify biotechnology companies for this Guide -- companies using the new technologies for research and product creation. Each listing has a subsection of companies with primary focus in the specified area followed by a subsection of companies with secondary focus in the area. Products, employee number and the primary (1°) and secondary (2°) industry classification codes are listed for each company to facilitate selection of companies of interest.

Following the listings by classifications are three tables that analyze the foci of the U.S. biotechnology companies. Table 5-2 analyzes the number and percent of all companies with primary or secondary interest in each classification, as well as the number and percent of companies with primary and secondary focus in the classification combined. Table 5-3 ranks the areas of concentration by number of companies working in that area with a primary focus, and Table 5-4 has a similar ranking with both primary and secondary foci combined. These tables allow an overview of areas of concentration in the U.S. biotechnology industry.

Table 5-1 Classification Codes

TABLE 5-1 CLASSIFICATION CODES FOR COMPANIES

Code	Industry Classification
A	Agriculture, Animal
B	Agriculture, Plant
C	Biomass Conversion
D	Biosensors/Bioelectronics
E	Bioseparations
F	Biotechnology Equipment
G	Biotechnology Reagents
H	Cell Culture, General
I	Chemicals, Commodity
J	Chemicals, Specialty (includes proteins and enzymes)
K	Diagnostics, Clinical Human
L	Energy
M	Food Production/Processing
N	Mining
O	Production/Fermentation
P	Therapeutics
Q	Vaccines
R	Waste Disposal/Treatment
S	Aquaculture
T	Marine Natural Products (includes algae)
U	Consulting
V	Veterinary (all animal health care)
W	Research
X	Immunological Products (non-pharmaceutical)
Y	Toxicology
Z	Venture Capital/Financing
1	Biomaterials
2	Fungi
3	Drug Delivery Systems
4	Medical Devices
5	Testing/Analytical Services

LISTING 5-1 COMPANIES IN ANIMAL AGRICULTURE (A)

Primary Focus

BIO TECHNIQUES LABS, INC.   Year: 1982 Empl: 25   Code–1°:A
   FEED ADDITIVES

BIOKYOWA, INC.   Year: 1983 Empl: 70   Code–1°:A 2°:P
   L-LYSINE FEED ADDITIVE FOR
   POULTRY & SWINE; RX

EMBREX, INC.   Year: 1985 Empl: 20   Code–1°:A
   BIOLOGICALS FOR POULTRY IN-OVO APPLICATIONS

GENETIC ENGINEERING, INC.   Year: 1980 Empl: 23   Code–1°:A 2°:V
   ANIMAL SEX SELECTION, BULL SEMEN LABS

GRANADA GENETICS CORP.   Year: 1979 Empl: 3   Code–1°:A 2°:M
   BULL GENETICS, SHRIMP, FOOD

IMMUCELL CORP.   Year: 1982 Empl: 17   Code–1°:A 2°:GHQ
   HUMAN DX; REAGENTS; ANIMAL DX; VX

MOLECULAR GENETICS, INC.   Year: 1979 Empl: 63   Code–1°:A 2°:BKV
   DX; AG DX; BGH; VET DX; SCOURS VX

SYNTRO CORP.   Year: 1981 Empl: 84   Code–1°:A 2°:V
   ANIMAL VX & DX; ANIMAL HEALTH PRODUCTS

Secondary Focus

AGRACETUS   Year: 1981 Empl: 65   Code–1°:B 2°:AV
   PESTICIDES, VET RX, ANIMAL VX,
   MICROBIAL INOCULATIONS

AMGEN   Year: 1980 Empl: 340   Code–1°:P 2°:AJKG
   RESEARCH REAGENTS, EGF, IGF, IF,
   EPO, IL-2, CSF, BGH, HGH

BETHYL LABS, INC.   Year: 1977 Empl: 8   Code–1°:X 2°:KA
   ANTISERA & CUSTOM ANTISERA SERVICE

BIO HUMA NETICS   Year: 1984 Empl: 15   Code–1°:B 2°:AR
   SOIL CONDITIONERS; FEED SUPPLEMENTS;
   WASTE TREATMENT

BIOMED RESEARCH LABS, INC.   Year: 1974 Empl: 18   Code–1°:V 2°:A
   FISH BIOLOGICS, VET PRODUCTS

Listing 5-1 Animal Agriculture

Secondary Focus (Cont.)

CALIFORNIA BIOTECHNOLOGY   Year: 1981 Empl: 155   Code–1°:P 2°:AKV3
   RX; ANF, EPO, SURFACTANT, HGH, RENIN,
   PTH; VET RX, NASAL DRUG DELIVERY

CENTRAL BIOLOGICS, INC.   Year: 1982 Empl: 7   Code–1°:Q 2°:AHX
   VET VX; CUSTOM SERA; CONTRACT RESEARCH

CETUS CORP.   Year: 1971 Empl: 711 Code–1°:P 2°:BKAV
   RX; IF; IL-2; MABS; TNF; CSF; INSULIN;
   ANTIBIOTICS; VET VX

DNAX RESEARCH INSTITUTE   Year: 1980 Empl: 105 Code–1°:P 2°:XWA
   RESEARCH; MABS; RX

FERMENTA ANIMAL HEALTH   Year: 1986 Empl: 359 Code–1°:V 2°:A
   VET DX & VX; PESTICIDES

GENETIC DIAGNOSTICS CORP.   Year: 1981 Empl: 17   Code–1°:K 2°:AQ
   DX TEST KITS FOR ABUSE DRUGS, DISEASES

IDETEK, INC.   Year: 1983 Empl: 14   Code–1°:Y 2°:MAB
   DX & TOX TEST KITS FOR FOOD
   & AGRICULTURE

IMMUNEX CORP.   Year: 1981 Empl: 153 Code–1°:P 2°:AG
   RX; IL-2; GM-CSF; IL-1; IL-4; LYMPHOKINES; MAF

IMMUNO MODULATORS LABS   Year: 1981 Empl: 25   Code–1°:P 2°:AGKH
   RX; IF; AIDS RX; HCG; COSMETICS

INTEGRATED GENETICS, INC.   Year: 1981 Empl: 200 Code–1°:P 2°:KAY
   DX; RX; AIDS DX; FOOD TOX DX; CANCER DX;
   TPA; HCG; EPO; HEP B VX

KIRKEGAARD & PERRY LABS   Year: 1979 Empl: 40   Code–1°:V 2°:AGX
   POULTRY DX; VET DX;
   IMMUNOLOGICAL REAGENTS

LUCKY BIOTECH CORP.   Year: 1984 Empl: 4   Code–1°:P 2°:QJKA
   RX; IF; IF-A; IF-B; IF-G; IL-2; HEP B DX;
   PORCINE GROWTH HORMONE

MYCOGEN CORP.   Year: 1982 Empl: 50   Code–1°:B 2°:A1
   PLANT HERBICIDES/PESTICIDES
   BASED ON NATURAL PATHOGENS

Secondary Focus (Cont.)

NEOGEN CORP.   Year: 1981 Empl: 40   Code–1°:B 2°:AVFY
    FOOD, PLANT HEALTH, ANIMAL HEALTH, DX

O.C.S. LABORATORIES, INC.   Year: 1983 Empl: 12   Code–1°:J 2°:DAC
    CUSTOM MADE SYNTHETIC GENES & PEPTIDES

PHARMACIA LKB BIOTECH.   Year: 1960   Code–1°:E 2°:GVAJ
    CHROMATOGRAPHY SEPARATION CHEMICALS &
    INSTRUMENTS; ENZYMES; REAGENTS; BIOCHEMICALS

PROBIOLOGICS INT'L, INC.   Year: 1987 Empl: 5   Code–1°:P 2°:AVW
    RX, PROBIOTICS, ANTIBIOTICS FOR
    HUMAN & ANIMAL USE

REPLIGEN CORP.   Year: 1981 Empl: 95   Code–1°:J 2°:QAB
    ENZYMES, AIDS VX, PROTEINS, PROTEIN A,
    HTLVIII, PESTICIDES

RIBI IMMUNOCHEM RESEARCH   Year: 1981 Empl: 35   Code–1°:V 2°:AG
    VET ANTI-TUMOR PRODUCTS; VET DX; VET VX

RICERCA, INC.   Year: 1986 Empl: 160   Code–1°:Y 2°:BAO
    FERMENTATION AND RESEARCH

SYNGENE PRODUCTS   Year: 1980 Empl: 20   Code–1°:P 2°:VA
    PRODUCTS FOR ANIMAL AND
    HUMAN HEALTH

UNIGENE LABORATORIES, INC.   Year: 1980 Empl: 30   Code–1°:P 2°:JA
    PEPTIDES, HORMONES, ENZYMES,
    CALCITONIN, GROWTH FACTORS

UNIVERSITY GENETICS CO.   Year: 1981 Empl: 59   Code–1°:B 2°:AV
    CLONED OR CULTURED PLANTS;
    MONOCLONAL CELLS & EMBRYOS

Listing 5-2 Plant Agriculture

LISTING 5-2 COMPANIES IN PLANT AGRICULTURE (B)

Primary Focus

ADVANCED GENETIC SCIENCES   Year: 1979 Empl: 130   Code–1°:B 2°:J
    PLANT & AG RESEARCH, CHEMICALS,
    SEEDS, FROSTBAN, SNOMAX

AGDIA, INC.   Year: 1981 Empl: 8   Code–1°:B 2°:KY
    PLANT AND FOOD TOX DX FOR VIRUSES

AGRACETUS   Year: 1981 Empl: 65   Code–1°:B 2°:AV
    TERMITICIDE, VET RX, ANIMAL VX,
    MICROBIAL INOCULATIONS, PESTICIDES

AGRI-DIAGNOSTICS ASSOC.   Year: 1983 Empl: 22   Code–1°:B 2°:Y
    TURF & CROP DISEASE DX KITS,
    FOOD PATHOGEN DX

AGRIGENETICS CORP.   Year: 1975 Empl: 750   Code–1°:B
    SEED AND PLANT GENETICS, CORN

BENTECH LABORATORIES   Year: 1984 Empl: 7   Code–1°:B 2°:JO
    CROP YIELD ENHANCING AGENT;
    ANTIFUNGAL; WOUND HEALING

BIO HUMA NETICS   Year: 1984 Empl: 15   Code–1°:B 2°:AR
    SOIL CONDITIONERS; FEED SUPPLMTS;
    WASTE TREATMENT

BIOTECHNICA AGRICULTURE   Year: 1987 Empl: 50   Code–1°:B
    DISEASE & PEST RESISTANT CORN, SOYBEAN, HAY,
    WHEAT; ENHANCED NUTRITIONAL VALUES OF CROPS

CALGENE, INC.   Year: 1980 Empl: 165   Code–1°:B 2°:MJ
    TOMATOES, COTTON, CORN, SUNFLOWER,
    RAPESEED, FOODS; PESTICIDES, PLANTS, TREES

CROP GENETICS INTERNATIONAL Year: 1981 Empl: 75   Code–1°:B
    DISEASE-FREE SUGARCANE SEED, BIOINSECTICIDES

DEKALB-PFIZER GENETICS   Year: 1982 Empl: 235   Code–1°:B
    HYBRID SEEDS, CORN, SOY, ALFALFA, CROP PLANTS

DNA PLANT TECHNOLOGY CORP.   Year: 1981 Empl: 80   Code–1°:B 2°:M
    FOODS AND PLANTS, AG PRODS., TOMATOES,
    RICE, COFFEE, SNACKS, OILS, CORN

Primary Focus (Cont.)

ECOGEN, INC.   Year: 1983 Empl: 75   Code–1°:B
  BIOPESTICIDES; MICROBIAL AG PRODUCTS; INSECTICIDES

EMTECH RESEARCH   Year: 1983 Empl: 30   Code–1°:B 2°:W
  BIOTECH R&D FOR COSMETICS AND AGRICULTURE

ESCAGENETICS CORP.   Year: 1986 Empl: 50   Code–1°:B 2°:M
  TISSUE CULTURE FLAVORS; POTATO SEEDS;
  TOMATOES; MICROPROPAGATED DATE PALMS

FORGENE, INC.   Year: 1986 Empl: 4   Code–1°:B
  FORESTRY APPLICATIONS-NEW GENETIC
  TECHNOLOGIES FOR TREE CROPS

HAWAII BIOTECHNOLOGY GROUP Year: 1982 Empl: 10   Code–1°:B
  IMMUNOTOXINS, PESTICIDE DETECTION KIT

INT'L PLANT RESEARCH INST.   Year: 1978 Empl: 50   Code–1°:B 2°:MFG
  DATE PALMS; OIL PALMS; CEREALS;
  FOOD PLANTS; GENETIC ENGINEERING REAGENTS

MAIZE GENETIC RESOURCES   Year: 1984 Empl: 2   Code–1°:B 2°:W
  PROPRIETARY TECHNOL. FOR CORN BREEDING
  STRESS-RESISTANT VARIETIES

MYCOGEN CORP.   Year: 1982 Empl: 50   Code–1°:B 2°:A1
  PLANT HERBICIDES/PESTICIDES BASED
  ON NATURAL PATHOGENS

NATIVE PLANTS, INC.   Year: 1973 Empl: 625   Code–1°:B 2°:HJM
  POTATO, FLOWER, VEGETABLE SEEDS;
  SOIL INOCULANTS; PLANTS

NEOGEN CORP.   Year: 1981 Empl: 40   Code–1°:B 2°:AVFY
  FOOD, PLANT HEALTH, ANIMAL HEALTH, DX

NITRAGIN CO.   Year: 1898 Empl: 80   Code–1°:B 2°:J
  NITROGEN FIXING BACTERIA INOCULANTS

PHYTOGEN   Year: 1980 Empl: 20   Code–1°:B
  IMPROVED CROP PLANTS; MUNG
  BEANS, COTTON, SOY

PLANT GENETICS, INC.   Year: 1981 Empl: 63   Code–1°:B 2°:M
  POTATOES; HYBRID VEGS; ALFALFA;
  CORN; SEEDS

Listing 5-2 Plant Agriculture

Primary Focus (Cont.)

R & A PLANT/SOIL, INC.   Year: 1978 Empl: 6   Code–1°:B 2°:TO
   ALGAL SOIL CONDITIONERS, FERTILIZERS

SUNGENE TECHNOLOGIES CORP.   Year: 1981 Empl: 28   Code–1°:B
   PLANT BREEDING, CORN, SUNFLOWER, SEEDS

UNITED AGRISEEDS, INC.   Year: 1981 Empl: 192   Code–1°:B
   PLANTS, SEEDS

UNIVERSITY GENETICS CO.   Year: 1981 Empl: 59   Code–1°:B 2°:AV
   CLONED OR CULTURED PLANTS;
   MONOCLONAL CELLS & EMBRYOS

ZOECON CORP.   Year: 1968 Empl: 130   Code–1°:B 2°:IJ
   PLANT IMPROVEMENT; SEEDS; AG CHEMICALS

Secondary Focus

ABC RESEARCH CORP.   Year: 1967 Empl: 82   Code–1°:W 2°:JBHR
   CONTRACT RESEARCH; CUSTOM STARTER
   CULTURES; BACTERIAL PROTEIN; ANTIOXIDANTS

AMERICAN BIOGENETICS CORP.   Year: 1984 Empl: 10   Code–1°:I 2°:JBLR
   PROCESS DESIGN MICROORGANISMS; CHEMICALS

ASTRE CORPORATE GROUP   Year: 1967 Empl: 148   Code–1°:R 2°:B
   AMMONIA DEGRADER; BIOPESTICIDES;
   LIPASE-PRODUCING MICROBE

BIOTECHNICA INTERNATIONAL   Year: 1981 Empl: 145   Code–1°:J 2°:KQMB
   INOCULANT, VX, DENTAL DX; ENZYMES,
   INDUSTRIAL YEAST, LITE BEER, FOODS

BOEHRINGER MANNHEIM DIAGN. Year: 1975 Empl: 100   Code–1°:K 2°:PGBH
   DX; RX; REAGENTS; DIGOXIN; ANIMAL CELL
   TISSUE CULTURE, PLANT DX

CETUS CORP.   Year: 1971 Empl: 711   Code–1°:P 2°:BKAV
   RX; IF; IL-2; MABS; TNF; CSF; INSULIN;
   ANTIBIOTICS; VET VX

CYANOTECH CORP.   Year: 1983 Empl: 25   Code–1°:M 2°:BHJ
   PHYCOBILIPROTEIN; SPIRULINA; BETA-CAROTENE

GENETICS INSTITUTE, INC.   Year: 1981 Empl: 302   Code–1°:P 2°:BK
   RX; DX; GM-CSF; EPO; TPA; HSA; AIDS RX;
   IL-3; IF; FACTOR VIII; SEEDS; CORN

Secondary Focus (Cont.)

IDETEK, INC.   Year: 1983 Empl: 14   Code–1°:Y 2°:MAB
   DX & TOX TEST KITS FOR FOOD & AGRICULTURE

INGENE, INC.   Year: 1980 Empl: 65   Code–1°:P 2°:JB
   RX; BONE & CARTILAGE RX;
   CANCER RX; SWEETENERS; PESTICIDE

INTERFERON SCIENCES, INC.   Year: 1981 Empl: 70   Code–1°:P 2°:GHB
   IF-A; IF-G; MABS

LIPOSOME TECHNOLOGY, INC.   Year: 1981 Empl: 65   Code–1°:3 2°:VBPJ
   LIPOSOMES FOR RX, DRUG DELIVERY,
   VETERINARY USE, AGRICULTURE

MICROGENESYS, INC.   Year: 1983 Empl: 32   Code–1°:Q 2°:KB
   AIDS DX; AIDS VX; VIRAL INSECTICIDE;
   DEVELOPING HUMAN INFECTIOUS DISEASE VX

MOLECULAR GENETICS, INC.   Year: 1979 Empl: 63   Code–1°:A 2°:BKV
   DX; AG DX; BGH; VET DX; SCOURS VX

PROVESTA CORP.   Year: 1975 Empl: 50   Code–1°:J 2°:LBMP
   FERMENTATION-BASED FLAVORS; ENZYMES;
   YEAST FEEDS; PHEROMONES FOR PEST CONTROL; TPA

REPLIGEN CORP.   Year: 1981 Empl: 95   Code–1°:J 2°:QAB
   ENZYMES, AIDS VX, PROTEINS,
   PROTEIN A, HTLVIII, PESTICIDES

RICERCA, INC.   Year: 1986 Empl: 160   Code–1°:Y 2°:BAO
   FERMENTATION AND RESEARCH

WESTBRIDGE RESEARCH GROUP   Year: 1982 Empl: 15   Code–1°:J 2°:B
   PLANT GROWTH REGULATOR; WATER TREATMENT

Listing 5-3 Biomass Conversion

LISTING 5-3 COMPANIES IN BIOMASS CONVERSION (C)

Primary Focus
REPAP TECHNOLOGIES ✓   Year: 1981 Empl: 16   Code–1°:C
   BIOMASS CONVERSION FOR
   ALCOHOLS, SUGARS

Secondary Focus
MYCOSEARCH, INC.✓   Year: 1979 Empl: 6   Code–1°:2 2°:OCHR
   CULTURES OF RARE & NEW FUNGI;
   LIVE CULTURES; SPECIALTY CHEMICALS

NEUSHUL MARICULTURE ✓   Year: 1978 Empl: 6   Code–1°:S 2°:PMHC
   ANTI-VIRAL; ENZYMES; MOLLUSCAN SEED & CULTURE

O.C.S. LABORATORIES, INC. ✓   Year: 1983 Empl: 12   Code–1°:J 2°:DAC
   CUSTOM MADE SYNTHETIC GENES & PEPTIDES

LISTING 5-4 COMPANIES IN BIOSENSORS/BIOELECTRONICS (D)

Primary Focus

BIOCHEM TECHNOLOGY, INC. Year: 1977 Empl: 12 Code–1°:D 2°:I
    SENSORS; COMPUTER PRODUCTS &
    SOFTWARE; NADH DETECTOR

BIOMATERIALS INTERNATIONAL Year: 1981 Empl: 30 Code–1°:D 2°:K
    RAMAN TECHNOL. FOR RESPIRATORY
    APPLICATIONS; GAS MONITOR

GENTRONIX LABORATORIES, INC. Year: 1972 Empl: 6 Code–1°:D
    BIOCHIPS, BIOELECTRONICS

INTEGRATED CHEMICAL SENSORS Year: 1984 Empl: 8 Code–1°:D
    INSTRUMENTS AND BIOSENSORS

Secondary Focus

BEND RESEARCH, INC. Year: 1975 Empl: 60 Code–1°:E 2°:D
    SEPARATION; BIOSENSORS;
    IMMOBILIZED ENZYME

BIOCONTROL SYSTEMS Year: 1985 Empl: 29 Code–1°:R 2°:YD
    TESTS FOR FOOD, WATER; BIOSENSORS

BIOSCIENCE MANAGEMENT, INC. Year: 1984 Empl: 8 Code–1°:J 2°:RDF
    BACTERIAL CULTURES; INOCULANTS;
    INSTRUMENTS

BIOSPHERICS, INC. Year: 1967 Empl: 250 Code–1°:R 2°:YIDF
    WASTEWATER PROCESS &
    MONITORS & PILOT PLANT

BIOTHERM . Year: 1985 Empl: 10 Code–1°:K 2°:FDW
    DX; CARDIOVASCULAR DX

CYGNUS RESEARCH CORP. Year: 1985 Empl: 18 Code–1°:3 2°:DKP
    DRUG DELIVERY SYSTEM; CONTRACTS TO RX COMPANIES

LIFECODES CORP. Year: 1982 Empl: 53 Code–1°:K 2°:D
    REFERENCE LAB DX USING DNA PROBES

O.C.S. LABORATORIES, INC. Year: 1983 Empl: 12 Code–1°:J 2°:DAC
    CUSTOM MADE SYNTHETIC
    GENES & PEPTIDES

Listing 5-4 Biosensors/Bioelectronics

Secondary Focus (Cont.)
ORGANON TEKNIKA CORP.   Year: 1978 Empl: 270   Code–1°:K 2°:DFWX
   HEALTH CARE/MEDICAL DX

QUEUE SYSTEMS, INC.   Year: 1980 Empl: 85   Code–1°:H 2°:DEF
   CRYO PRESERVATION; CELL CULTURE;
   ENVIRONMENTAL SYSTEMS

LISTING 5-5 COMPANIES IN BIOSEPARATIONS (E)

Primary Focus

A/G TECHNOLOGY CORP.   Year: 1981 Empl: 8   Code–1°:E 2°:F
   GAS SEPARATION; ULTRA & MICRO FILTRATION

BEND RESEARCH, INC.   Year: 1975 Empl: 60   Code–1°:E 2°:D
   SEPARATION; BIOSENSORS; IMMOBILIZED ENZYME

BIOCONSEP, INC.   Year: 1983 Empl: 6   Code–1°:E 2°:F
   BIOLOGICAL SEPARATIONS R&D AND EQUIPMENT

BIOPROBE INTERNATIONAL, INC.   Year: 1983 Empl: 15   Code–1°:E 2°:KF
   SPECIALIZED GELS; SEPARATION DEVICE

IMRE CORP.   Year: 1981 Empl: 25   Code–1°:E 2°:P
   IMMUNO BLOOD PURIFIER SYSTEMS AND PROTEIN A

PHARMACIA LKB BIOTECH.   Year: 1960   Code–1°:E 2°:GVAJ
   CHROMATOGRAPHY SEPARATION CHEMICALS
   & INSTRUMENTS; ENZYMES; REAGENTS; BIOCHEMICALS

SEPRACOR, INC.   Year: 1984 Empl: 40   Code–1°:E 2°:FJ
   MEMBRANE COLUMNS; ANTI-
   FACTOR VII & ANTI-TPA; SEPARATIONS

VIAGENE, INC.   Year: 1987 Empl: 9   Code–1°:E 2°:P
   GENE TRANSFER TECHNOL TO TREAT INFECTIONS,
   MALIGNANCIES, GENE RELATED DISEASES

Secondary Focus

ADVANCED BIOTECHNOLOGIES   Year: 1982 Empl: 17   Code–1°:G 2°:EHW
   MEDIA, SERA, GROWTH FACTORS, DNA,
   RESEARCH SERVICES

ADVANCED MAGNETICS, INC.   Year: 1981 Empl: 40   Code–1°:K 2°:E
   DX RIA KITS, MAGNETIC SEPARATION REAGENTS

ADVANCED MINERAL TECHNOL.   Year: 1982 Empl: 17   Code–1°:R 2°:NE
   WASTE TREATMENT & METAL RECOVERY

AMERICAN QUALEX INT'L   Year: 1981 Empl: 10   Code–1°:G 2°:FEK
   CUSTOM MABS; ENZYMES;
   DNA PROBES; INSTRUMENTS

Listing 5-5 Bioseparations

Secondary Focus (Cont.)

BIO-RECOVERY, INC.   Year: 1983 Empl: 49   Code–1°:F 2°:E
 FILTRATION MEMBRANE SYSTEMS,
 LARGE SCALE PRODUCTION

BIOGROWTH, INC.   Year: 1985 Empl: 16   Code–1°:P 2°:FE
 WOUND/BONE RX; SEPARATION; PROTEIN
 CHARACTERIZATION; HPLC/FPLC EQUIPMENT

COAL BIOTECH CORP.   Year: 1984 Empl: 10   Code–1°:N 2°:E
 MICROBIAL COAL DESULFURIZATION

ELECTRO-NUCLEONICS, INC. Year: 1960 Empl: 720 Code–1°:K 2°:EFGH
 RX; DX; AIDS DX; IL-2; HTLV-III
 ANTIBODY TEST; TOXIN DX

ENZYME CENTER, INC.   Year: 1978 Empl: 30   Code–1°:O 2°:E
 LARGE SCALE FERMENTATION; ENZYME
 SEPARATION & PURIFICATION

FLOW LABORATORIES, INC.   Year: 1961 Empl: 200 Code–1°:G 2°:EF
 MEDIA, FERMENTATION
 EQUIPMENT, REAGENTS

GENZYME CORP.   Year: 1981 Empl: 185 Code–1°:P 2°:KJGE
 RX, DX FOR HUMAN HEALTH CARE; MABS;
 EYE PRODUCTS, SKIN PRODUCTS; SURFACTANTS

HYCLONE LABORATORIES, INC. Year: 1975 Empl: 107 Code–1°:H 2°:EO
 IMMUNOCHEMICALS; ANIMAL
 SERA; HYDBRIDOMAS

MELOY LABORATORIES, INC. Year: 1970 Empl: 246 Code–1°:W 2°:EP
 SERVICES; REAGENTS; R&D; IF

NORTH COAST BIOTECHNOLOGY Year: 1986 Empl: 2   Code–1°:J 2°:PEFK
 ENZYMES FOR RX MANUFACTURING

QUEUE SYSTEMS, INC.   Year: 1980 Empl: 85   Code–1°:H 2°:DEF
 CRYO PRESERVATION; CELL CULTURE;
 ENVIRONMENTAL SYSTEMS

LISTING 5-6 COMPANIES IN BIOTECHNOLOGY EQUIPMENT (F)

Primary Focus

AN-CON GENETICS   Year: 1982 Empl: 3   Code–1°:F
   EQUIPMENT; INSTRUMENTS

APPLIED BIOSYSTEMS, INC.   Year: 1981 Empl: 350   Code–1°:F 2°:G
   DNA SYNTHESIZER; PEPTIDE
   SEQUENCER; REAGENTS

APPLIED PROTEIN TECHNOLOGIES Year: 1984 Empl: 5   Code–1°:F 2°:JG
   PEPTIDE & PROTEIN SYNTHESIZERS,
   CHEMICALS, REAGENTS

BIO-RECOVERY, INC.   Year: 1983 Empl: 49   Code–1°:F 2°:E
   FILTRATION MEMBRANE SYSTEMS,
   LARGE SCALE PRODUCTION

BIOTECHNOLOGY DEVEL. CORP.   Year: 1982 Empl: 35   Code–1°:F 2°:PJQ3
   AIDS VX; EQUIPMENT; REAGENTS;
   DRUG DELIVERY SYSTEMS

ENDOTRONICS, INC.   Year: 1981 Empl: 45   Code–1°:F 2°:HPQ
   RX; DX; MABS; APS SERIES; ACUSYST-P;
   VX; BIOLOGICALS; HEP-B VX

INTELLIGENETICS, INC.   Year: 1980 Empl: 52   Code–1°:F 2°:5
   GENETIC ENGINEERING SOFTWARE;
   DNA & PROTEIN ANALYSIS

MULTIPLE PEPTIDE SYSTEMS, INC.   Year: 1986 Empl: 6   Code–1°:F 2°:G
   HYDROGEN FLUORIDE CLEAVAGE
   EQUIPMENT; PEPTIDE SYNTHESIS

NYGENE CORP.   Year: 1985 Empl: 18   Code–1°:F 2°:H
   BIOAFFINITY SEPARATION SYSTEMS;
   HYBRIDOMA PRODUCTION

VEGA BIOTECHNOLOGIES, INC.   Year: 1979 Empl: 60   Code–1°:F 2°:G
   PEPTIDE SYNTHESIZERS;
   REAGENTS; ENZYMES

Secondary Focus

A/G TECHNOLOGY CORP.   Year: 1981 Empl: 8   Code–1°:E 2°:F
   GAS SEPARATION; ULTRA &
   MICRO FILTRATION

Listing 5-6 Biotechnology Equipment

Secondary Focus (Cont.)

ABN   Year: 1981 Empl: 45   Code–1°:K 2°:GF
   DX, REAGENTS, BT INSTRUMENTS,
   DNA PROBES, BLOTTING SYSTEMS

AMERICAN QUALEX INT'L, INC.   Year: 1981 Empl: 10   Code–1°:G 2°:FEK
   CUSTOM MABS; ENZYMES;
   DNA PROBES; INSTRUMENTS

AQUASYNERGY, LTD.   Year: 1986 Empl: 5   Code–1°:T 2°:SWF
   CO2 GENERATOR FOR AQUACULTURE
   PLANT PRODUCTION; AQUATIC PLANT/ANIMAL CROPS

BIOCONSEP, INC.   Year: 1983 Empl: 6   Code–1°:E 2°:F
   BIOLOGICAL SEPARATIONS R&D AND EQUIPMENT

BIOGROWTH, INC.   Year: 1985 Empl: 16   Code–1°:P 2°:FE
   WOUND/BONE RX; SEPARATION; PROTEIN
   CHARACTERIZATION; HPLC/FPLC EQUIPMENT

BIOPROBE INTERNATIONAL, INC. Year: 1983 Empl: 15   Code–1°:E 2°:KF
   SPECIALIZED GELS; SEPARATION DEVICE

BIOSCIENCE MANAGEMENT, INC. Year: 1984 Empl: 8   Code–1°:J 2°:RDF
   BACTERIAL CULTURES; INOCULANTS;
   INSTRUMENTS

BIOSEARCH, INC.   Year: 1977 Empl: 100   Code–1°:G 2°:F
   REAGENT CHEMICALS AND
   GENE/PEPTIDE INSTRUMENTS

BIOSPHERICS, INC.   Year: 1967 Empl: 250   Code–1°:R 2°:YIDF
   WASTEWATER PROCESS & MONITORS
   & PILOT PLANT

BIOSTAR MEDICAL PRODUCTS   Year: 1983 Empl: 30   Code–1°:K 2°:WF
   DX & TEST KITS FOR AUTOIMMUNE
   DISEASES, INSTRUMENTATION

BIOTHERM   Year: 1985 Empl: 10   Code–1°:K 2°:FDW
   DX; CARDIOVASCULAR DX

BIOTROL, INC.   Year: 1985 Empl: 30   Code–1°:R 2°:F
   PROPRIETARY BACTERIA; BIOREACTORS FOR
   ENVIRONMENTAL SVCS; DEVELOPING DEGRADERS

CHARLES RIVER BIOTECH. SERV.   Year: 1983 Empl: 55   Code–1°:X 2°:HF
   MABS, CELL CULTURE SYSTEMS

Secondary Focus (Cont.)

DIAGNOSTIC TECHNOLOGY, INC.   Year: 1980 Empl: 60   Code–1°:K 2°:GF
  DX, BLOOD TEST KITS, REAGENTS

ELECTRO-NUCLEONICS, INC.   Year: 1960 Empl: 720   Code–1°:K 2°:EFGH
  RX; DX; AIDS DX; IL-2; HTLV-III
  ANTIBODY TEST; TOXIN DX

ENVIRONMENTAL DIAGN.   Year: 1983 Empl: 55   Code–1°:Y 2°:KF
  ENVIRONMENTAL AND TOXIN TESTING
  IN FOODS, ANIMALS & HUMANS

FLOW LABORATORIES, INC.   Year: 1961 Empl: 200   Code–1°:G 2°:EF
  MEDIA, FERMENTATION
  EQUIPMENT, REAGENTS

GEN-PROBE, INC.   Year: 1984 Empl: 127   Code–1°:K 2°:FG
  DNA PROBE DX TECH FOR RESEARCH & CLINIC

HAZLETON BIOTECHNOLOGIES   Year: 1983 Empl: 500   Code–1°:G 2°:FX
  LAB REAGENTS AND EQUIPMENT, MEDIA, MABS

IBF BIOTECHNICS, INC.   Year: 1987 Empl: 5   Code–1°:G 2°:F
  TISSUE CULTURE REAGENTS; CHROMATOGRAPHY
  COLUMNS, MONITORS; SERUM SUBSTITUTES

INT'L BIOTECHNOLOGIES, INC.   Year: 1982 Empl: 85   Code–1°:G 2°:FKW
  NUCLEIC ACIDS, CLONING, SEQUENCING,
  GENOME ANALYSES & ELECTROPHORESIS SYSTEMS

INT'L PLANT RESEARCH INST.   Year: 1978 Empl: 50   Code–1°:B 2°:MFG
  DATE PALMS; OIL PALMS; CEREALS; FOOD
  PLANTS; GENETIC ENGINEERING REAGENTS

LABSYSTEMS, INC.   Year: 1983 Empl: 10   Code–1°:K 2°:HJMF
  MICROBIOLOGY ASSAYS & EQUIPMENT; MAB'S

MOLECULAR BIOLOGY RESOURCES Year: 1986 Empl: 10   Code–1°:G 2°:JOF
  REAGENTS; DNA & RNA MODIFYING ENZYMES
  & NUCLEIC ACIDS; HYBRIDIZATION APPARATUS

NEOGEN CORP.   Year: 1981 Empl: 40   Code–1°:B 2°:AVFY
  FOOD, PLANT HEALTH, ANIMAL HEALTH, DX

NORTH COAST BIOTECHNOLOGY   Year: 1986 Empl: 2   Code–1°:J 2°:PEFK
  ENZYMES FOR RX MANUFACTURING

Listing 5-6 Biotechnology Equipment

Secondary Focus (Cont.)

ORGANON TEKNIKA CORP.   Year: 1978 Empl: 270   Code–1°:K 2°:DFWX
   HEALTH CARE/MEDICAL DX

PROMEGA CORP.   Year: 1978 Empl: 85   Code–1°:G 2°:FJ
   REAGENTS FOR MOLECULAR
   BIOLOGY; ENZYMES

QUEUE SYSTEMS, INC.   Year: 1980 Empl: 85   Code–1°:H 2°:DEF
   CRYO PRESERVATION; CELL CULTURE;
   ENVIRONMENTAL SYSTEMS

SEPRACOR, INC.   Year: 1984 Empl: 40   Code–1°:E 2°:FJ
   MEMBRANE COLUMNS-ANTI-FACTOR
   VII & ANTI-TPA; SEPARATIONS

TECHNICLONE INT'L, INC.   Year: 1981 Empl: 15   Code–1°:K 2°:PFH
   CANCER DX, TISSUE CULTURES, MAB,
   MAB CULTURING EQUIPMENT

UNITED STATES BIOCHEMICAL   Year: 1973   Code–1°:G 2°:F
   REAGENTS AND EQUIPMENT

VECTOR LABORATORIES, INC.   Year: 1976 Empl: 50   Code–1°:G 2°:F
   MOLECULAR BIOLOGICAL
   REAGENTS AND EQUIPMENT

LISTING 5-7 COMPANIES IN BIOTECHNOLOGY REAGENTS (G)

Primary Focus

ADVANCED BIOTECHNOLOGIES   Year: 1982 Empl: 17   Code–1°:G 2°:EHW
    MEDIA, SERA, GROWTH FACTORS,
    DNA, RESEARCH SERVICES

ALFACELL CORP.   Year: 1981 Empl: 22   Code–1°:G 2°:PX
    TUMOR TOXINS, REAGENTS, SERA

AMERICAN BIOCLINICAL   Year: 1977 Empl: 24   Code–1°:G 2°:KP
    REAGENTS; HORMONE DX KITS; RX

AMERICAN QUALEX INT'L, INC.   Year: 1981 Empl: 10   Code–1°:G 2°:FEK
    CUSTOM MABS, MABS; ENZYMES; DNA
    PROBES; INSTRUMENTS

BIODESIGN, INC.   Year: 1987 Empl: 3   Code–1°:G 2°:PY
    IMMUNOLOGICAL REAGENTS; BIODESIGN
    ANTISERA, PURIFIED BIOLOGICAL COMPOUNDS

BIOSEARCH, INC.   Year: 1977 Empl: 100   Code–1°:G 2°:F
    REAGENT CHEMICALS AND GENE/
    PEPTIDE INSTRUMENTS

BIOTEST DIAGNOSTICS CORP.   Year: 1946 Empl: 10   Code–1°:G 2°:X
    REAGENTS AND MAB MARKERS

CHEMGENES   Year: 1981 Empl: 8   Code–1°:G
    INTERMEDIATES FOR DNA SYNTHESIS

CHEMICAL DYNAMICS CORP.   Year: 1972 Empl: 45   Code–1°:G
    DNA TECHNOL REAGENTS, RE'S, ETC.

CHEMICON INTERNATIONAL,   Year: 1981 Empl: 8   Code–1°:G 2°:K
    MABS; GROWTH FACTORS; IMMUNOLOGICAL
    CELL GROWTH ASSAY KIT

CLINICAL SCIENCES, INC.   Year: 1971 Empl: 50   Code–1°:G 2°:KV
    BIOCHEMICALS FOR DX AND VET USES

COLLABORATIVE RESEARCH, INC. Year: 1961 Empl: 150   Code–1°:G 2°:WK5
    DX SERVICES (DNA PROBE); IF; RENNIN;
    UROKINASE; IL-2; ENZYMES; TPA; REAGENTS

CONSOLIDATED BIOTECHNOLOGY Year: 1983 Empl: 15   Code–1°:G 2°:M
    ENZYMES, FOOD RELATED RESEARCH

124

Listing 5-7 Biotechnology Reagents

Primary Focus (Cont.)

DIAGNOSTIC PRODUCTS, CORP.   Year: 1972 Empl: 270   Code–1°:G 2°:K
   DX KITS, MABS; REAGENTS

EPITOPE, INC.   Year: 1979 Empl: 60   Code–1°:G 2°:K
   DX; AIDS DX; VIRAL DETECTION KITS;
   WESTERN BLOT, MAB TO HIV

FLOW LABORATORIES, INC.   Year: 1961 Empl: 200   Code–1°:G 2°:EF
   MEDIA, FERMENTATION EQUIPMENT, REAGENTS

HAZLETON BIOTECHNOLOGIES   Year: 1983 Empl: 500   Code–1°:G 2°:FX
   LAB REAGENTS AND EQUIPMENT, MEDIA, MABS

IBF BIOTECHNICS, INC.   Year: 1987 Empl: 5   Code–1°:G 2°:F
   TISSUE CULTURE REAGENTS; CHROMATOGRAPHY
   COLUMNS, MONITORS; SERUM SUBSTITUTES

IMMUNOSYSTEMS, INC.   Year: 1981 Empl: 4   Code–1°:G 2°:X
   ENZYME IMMUNOASSAYS

INTER-AMERICAN RES. ASSOC.   Year: 1979 Empl: 10   Code–1°:G 2°:XWK
   PURIFIED ANTIGENS, ANTISERA,
   BLOOD PRODUCTS

INT'L BIOTECHNOLOGIES, INC.   Year: 1982 Empl: 85   Code–1°:G 2°:FKW
   NUCLEIC ACIDS, CLONING, SEQUENCING,
   GENOME ANALYSES & ELECTROPHORESIS SYSTEMS

INTERNATIONAL ENZYMES   Year: 1983 Empl: 4   Code–1°:G 2°:JX
   ENZYMES; ANTISERA; SERUM-FREE
   MEDIA SUPPLIES

JACKSON IMMUNORESEARCH   Year: 1982 Empl: 16   Code–1°:G 2°:J
   AFFINITY PURIFICATION PROTEINS;
   PURE PROTEINS

MARCOR DEVELOPMENT CO.   Year: 1977 Empl: 6   Code–1°:G 2°:J
   REAGENTS FOR FERMENTATION, ENZYMES

MOLECULAR BIOLOGY RESOURCES Year: 1986 Empl: 10   Code–1°:G 2°:JOF
   REAGENTS; DNA & RNA MODIFYING ENZYMES
   & NUCLEIC ACIDS; HYBRIDIZATION APPARATUS

MOLECULAR GENETIC RESOURCES Year: 1983 Empl: 5   Code–1°:G 2°:K
   ENZYMES; REVERSE TRANSCRIPTASE;
   TERMINAL TRANSFERASE; MRNA ANALYSIS KIT

Primary Focus (Cont.)

MONOCLONETICS INT'L   Year: 1984 Empl: 5   Code–1°:G 2°:HKP
    HUMAN SCREENING TESTS & DX, EPITOPE SCREENS

PEL-FREEZ BIOLOGICALS, INC.   Year: 1911 Empl: 175   Code–1°:G 2°:JX
    VARIOUS SPECIALTY REAGENTS, MAB SERA

PROMEGA CORP.   Year: 1978 Empl: 85   Code–1°:G 2°:FJ
    REAGENTS FOR MOLECULAR BIOLOGY; ENZYMES

SCOTT LABORATORIES   Year: 1981 Empl: 375   Code–1°:G 2°:J
    REAGENTS AND SPECIALTY CHEMICALS

STRATAGENE, INC.   Year: 1984 Empl: 77   Code–1°:G 2°:W
    REAGENTS; MOLECULAR BIOLOGICALS;
    GENE DX; CUSTOM RESEARCH

SUMMA MEDICAL CORP.   Year: 1978 Empl: 33   Code–1°:G 2°:KX
    HISTOCHEMICAL MARKERS AND DX KITS, CANCER DX

UNITED STATES BIOCHEMICAL   Year: 1973   Code–1°:G 2°:F
    REAGENTS AND EQUIPMENT

UNIV. MICRO REFERENCE LAB   Year: 1980 Empl: 6   Code–1°:G
    BACTERIAL REFERENCE STRAINS; BATCH
    PROCESSING OF MICROBIAL STRAINS

UPSTATE BIOTECHNOLOGY, INC.   Year: 1986 Empl: 4   Code–1°:G
    GROWTH FACTORS, EGF, BRAIN EXTRACTS
    FROM PITUITARY; SERUM SUBSTITUTE

VECTOR LABORATORIES, INC.   Year: 1976 Empl: 50   Code–1°:G 2°:F
    MOLECULAR BIOLOGICAL REAGENTS AND EQUIPMENT

ZYMED LABORATORIES, INC.   Year: 1980 Empl: 20   Code–1°:G 2°:KJ
    IMMUNOCHEMICALS, MABS, PROTEINS, ENZYMES

Secondary Focus

ABN   Year: 1981 Empl: 45   Code–1°:K 2°:GF
    DX, REAGENTS, BT INSTRUMENTS,
    DNA PROBES, BLOTTING SYSTEMS

ALLELIC BIOSYSTEMS   Year: 1984 Empl: 7   Code–1°:V 2°:JG
    VET TOX TESTS, PROBES, OLIGONUCLEOTIDES, REAGENTS

AMERICAN BIOTECHNOLOGY CO.   Year: 1984 Empl: 5   Code–1°:X 2°:PGHJ
    GROWTH FACTORS, IMMUNOMODULATORS

Listing 5-7 Biotechnology Reagents

Secondary Focus (Cont.)

AMERICAN DIAGNOSTICA, INC.   Year: 1983 Empl: 9   Code–1°:K 2°:G
   DX, TPA ELISA, TPA ACTIVE ASSAY KITS

AMGEN   Year: 1980 Empl: 215   Code–1°:P 2°:AJKG
   RESEARCH REAGENTS; RX, EGF, IGF, IF,
   EPO, IL-2, CSF, BGH, HGH

ANTIVIRALS, INC.   Year: 1980 Empl: 5   Code–1°:K 2°:PG
   NEU-GENES (TM) FOR DX & RX USE

APPLIED BIOSYSTEMS, INC.   Year: 1981 Empl: 350   Code–1°:F 2°:G
   DNA SYNTHESIZER; PEPTIDE SEQUENCER; REAGENTS

APPLIED PROTEIN TECHNOL.   Year: 1984 Empl: 5   Code–1°:F 2°:JG
   PEPTIDE & PROTEIN SYNTHESIZERS,
   CHEMICALS, REAGENTS

ATLANTIC ANTIBODIES   Year: 1972 Empl: 60   Code–1°:K 2°:G
   DX TEST KITS; DX; MABS; REAGENTS

BACHEM, INC.   Year: 1971 Empl: 45   Code–1°:J 2°:G
   SYNTHETIC BIOACTIVE PEPTIDES;
   GROWTH FACTORS, CRF

BEHRING DIAGNOSTICS   Year: 1952 Empl: 100   Code–1°:K 2°:G
   AB'S; DX; REAGENTS (CALBIOCHEM)

BERKELEY ANTIBODY CO., INC.   Year: 1983 Empl: 27   Code–1°:X 2°:GHP
   CUSTOM ANTIBODIES, MAB & POLYCLONALS

BINAX, INC.   Year: 1986 Empl: 17   Code–1°:K 2°:VGX
   MABS, DX TEST KITS, VET DX

BIOGEN   Year: 1979 Empl: 250   Code–1°:P 2°:QGJ
   RX; DX; TNF, IL-2, CSF, TPA, HSA,
   FACTOR VIII, IF-G, HEP B ANTIGEN

BIOGENEX LABORATORIES   Year: 1981 Empl: 42   Code–1°:K 2°:PG
   RIA DX TEST; IMMUNOHISTO-
   CHEMICALS; KITS; REAGENTS

BIOMEDICAL TECHNOLOGIES   Year: 1981 Empl: 10   Code–1°:J 2°:HGK
   CELL GROWTH FACTORS; CELL CULTURE PRODUCTS;
   ASSAY RIA KITS; MAB; POLYCLONALS; PROTEINS

BIOMERICA, INC.   Year: 1971 Empl: 35   Code–1°:K 2°:GV
   PREGNANCY, ALLERGY, BLOOD DX; VET DX & RX

Secondary Focus (Cont.)

BIONETICS RESEARCH   Year: 1985 Empl: 200   Code–1°:K 2°:GP
   AIDS DX, SALMONELLA EIA, RADIOLABELLED
   COMPOUNDS, LIPOPROTEINS, MELATONIN

BIOPRODUCTS FOR SCIENCE, INC.   Year: 1985 Empl: 19   Code–1°:X 2°:G
   ANTIBODIES; ANTIGEN MARKERS; REAGENTS

BOEHRINGER MANNHEIM DIAGN. Year: 1975 Empl: 1000   Code–1°:K 2°:PGBH
   DX; RX; REAGENTS; DIGOXIN; ANIMAL
   CELL TISSUE CULTURE, PLANT DX

CENTOCOR   Year: 1979 Empl: 246   Code–1°:K 2°:PJGQ
   MAB RX; CANCER, HEP B DX; MABS; AIDS VX

CREATIVE BIOMOLECULES   Year: 1981 Empl: 70   Code–1°:J 2°:PGW
   PEPTIDES, CONTRACT R AND D, TPA, EGF

CYTOTECH, INC.   Year: 1982 Empl: 43   Code–1°:K 2°:G
   AIDS DX, REAGENTS

DIAGNOSTIC TECHNOLOGY, INC.   Year: 1980 Empl: 60   Code–1°:K 2°:GF
   DX, BLOOD TEST KITS, REAGENTS

DIAMEDIX, INC.   Year: 1986 Empl: 35   Code–1°:K 2°:G
   DX KITS FOR INFECTIOUS &
   AUTOIMMUNE DISEASES

ELCATECH   Year: 1984 Empl: 4   Code–1°:W 2°:PQXG
   RESEARCH; COAGULATION ASSAY TECHNOLOGY

ELECTRO-NUCLEONICS, INC.   Year: 1960 Empl: 720   Code–1°:K 2°:EFGH
   RX; DX; AIDS DX; IL-2; HTLV-III
   ANTIBODY TEST; TOXIN DX

ENZO-BIOCHEM, INC.   Year: 1976 Empl: 110   Code–1°:K 2°:GP
   ENZYMES; LECTINS; FILTERS; NUCLEOTIDES;
   NON-RADIOACTIVE DNA PROBES; MAB; DX

GEN-PROBE, INC.   Year: 1984 Empl: 127   Code–1°:K 2°:FG
   DNA PROBE DX TECH FOR RESEARCH & CLINIC

GENZYME CORP.   Year: 1981 Empl: 185   Code–1°:P 2°:KJGE
   RX, DX FOR HUMAN HEALTH CARE; MABS;
   EYE PRODUCTS, SKIN PRODUCTS; SURFACTANTS

IMMUCELL CORP.   Year: 1982 Empl: 17   Code–1°:A 2°:GHQ
   HUMAN DX; REAGENTS; ANIMAL DX; VX

Listing 5-7 Biotechnology Reagents

Secondary Focus (Cont.)

IMMUNEX CORP.   Year: 1981 Empl: 153   Code–1°:P 2°:AG
   RX; IL-2; GM-CSF; IL-1; IL-4; LYMPHOKINES; MAF

IMMUNEX, INC.   Year: 1981 Empl: 5   Code–1°:K 2°:PG
   DX; MABS, POLYCLONAL AB'S TO HAPTENS;
   RADIOIMMUNOASSAY KITS; REAGENTS

IMMUNO MODULATORS LABS   Year: 1981 Empl: 25   Code–1°:P 2°:AGKH
   RX; IF; AIDS RX; HCG; COSMETICS

IMMUNOMEDICS, INC.   Year: 1983 Empl: 55   Code–1°:P 2°:KGL
   DX; RX; TEST KITS; CANCER DX; AUTOIMMUNE
   DISEASE ASSAY; ALLERGY PRODUCTS

INCELL CORP.   Year: 1982 Empl: 23   Code–1°:J 2°:G
   PEPTIDE HORMONES; AMINO ACIDS;
   REAGENTS; CHEMICALS

INTER-CELL TECHNOLOGIES, INC.   Year: 1982 Empl: 9   Code–1°:X 2°:GKP
   BIOLOGICALS; IMMUNOLOGICAL
   REAGENTS; LYMPHOKINES

INTERFERON SCIENCES, INC.   Year: 1981 Empl: 70   Code–1°:P 2°:GHB
   IF-A; IF-G; MABS

INT'L PLANT RESEARCH INST.   Year: 1978 Empl: 50   Code–1°:B 2°:MFG
   DATE PALMS; OIL PALMS; CEREALS;
   FOOD PLANTS; GENETIC ENGINEERING REAGENTS

KIRKEGAARD & PERRY LABS, INC.   Year: 1979 Empl: 40   Code–1°:V 2°:AGX
   POULTRY DX; VET DX;
   IMMUNOLOGICAL REAGENTS

LIFE TECHNOLOGIES, INC.   Year: 1983 Empl: 1300   Code–1°:K 2°:GP
   IF; PROBE ASSAY DX; REAGENTS

MARTEK   Year: 1985 Empl: 16   Code–1°:J 2°:KG
   STABILIZATOPE-LABELLED BIOCHEMICALS, C-13 &
   DEUTRIUM-LABELLED LUBRICANTS, AMINO ACIDS

MOLECULAR BIOSYSTEMS, INC.   Year: 1981 Empl: 55   Code–1°:K 2°:JKG
   DX; DNA & RNA PROBES, ALBUMEX, EXTRACTOR KITS

MULTIPLE PEPTIDE SYSTEMS, INC. Year: 1986 Empl: 6   Code–1°:F 2°:G
   HYDROGEN FLUORIDE CLEAVAGE
   EQUIPMENT; PEPTIDE SYNTHESIS

Secondary Focus (Cont.)

MUREX CORP.   Year: 1984 Empl: 180   Code–1°:K 2°:PGQ
    MAB REAGENTS & ANTIBODY KITS; DX

OMNI BIOCHEM, INC.   Year: 1986 Empl: 5   Code–1°:J 2°:G
    PEPTIDES; ENZYMES; AMINO ACIDS; REAGENTS

ONCOR, INC.   Year: 1983 Empl: 30   Code–1°:K 2°:G
    AIDS DX, DX REAGENTS, DNA PROBES FOR CANCER,
    HTLV-III RNA PROBE, B CELL & T CELL PROBES

PENINSULA LABORATORIES, INC.   Year: 1971 Empl: 75   Code–1°:J 2°:G
    PEPTIDES, ENZYMES, SPECIALTY REAGENTS

PHARMACIA LKB BIOTECH.   Year: 1960   Code–1°:E 2°:GVAJ
    CHROMATOGRAPHY SEPARATION CHEMICALS &
    INSTRUMENTS; ENZYMES; REAGENTS; BIOCHEMICALS

RES. & DIAGNOS. ANTIBODIES   Year: 1985   Code–1°:K 2°:XG
    DX KITS, MAB REAGENTS

RIBI IMMUNOCHEM RESEARCH   Year: 1981 Empl: 35   Code–1°:V 2°:AG
    VET ANTI-TUMOR PRODUCTS; VET DX; VET VX

SCRIPPS LABORATORIES   Year: 1980 Empl: 60   Code–1°:P 2°:JGK
    BULK HUMAN PROTEINS; HORMONES; ENZYMES; DX

SERAGEN, INC.   Year: 1979 Empl: 250   Code–1°:X 2°:PG
    CANCER RX; MABS; REAGENTS; ANTISERA

SYNTHETIC GENETICS, INC.   Year: 1985 Empl: 6   Code–1°:J 2°:G
    CUSTOM & SYNTHETIC DNA & PEPTIDES

T CELL SCIENCES, INC.   Year: 1984 Empl: 65   Code–1°:P 2°:KG
    RX; MABS; CANCER DX; T CELL
    PRODUCTS FOR ARTHRITIS; IL-2

TRITON BIOSCIENCES, INC.   Year: 1983 Empl: 180   Code–1°:P 2°:KGW
    RX, IF-B, DX RESEARCH REAGENTS, MAB, RESEARCH

VEGA BIOTECHNOLOGIES, INC.   Year: 1979 Empl: 60   Code–1°:F 2°:G
    PEPTIDE SYNTHESIZERS; REAGENTS; ENZYMES

WASHINGTON BIOLAB   Year: 1986 Empl: 10   Code–1°:J 2°:GW
    AIDS RX; ENZYMES, RAW MATERIALS
    FOR BIOTECHNOLOGY

Listing 5-8 Cell Culture

LISTING 5-8 COMPANIES IN CELL CULTURE (H)

Primary Focus

BIONIQUE LABS, INC.   Year: 1983 Empl: 11   Code–1°:H 2°:K
CELL CULTURE, CULTURE CHAMBER; MYCOPLASMA
DETECTION SYSTEM; CELL PROTECTIVE DEVICE; DX

HYCLONE LABORATORIES, INC.   Year: 1975 Empl: 107   Code–1°:H 2°:EO
IMMUNOCHEMICALS; ANIMAL SERA; HYDBRIDOMAS

IMCERA BIOPRODUCTS, INC.   Year: 1986 Empl: 5   Code–1°:H 2°:P
IGF, RDNA MURINE EGF FOR LARGE -SCALE CELL
CULTURE; RDNA & EXTRACTED ADHESION PEPTIDES

KARYON TECHNOLOGY, LTD.   Year: 1984 Empl: 25   Code–1°:H
TISSUE PROCESS CULTURE TECHNOLOGY

QUEUE SYSTEMS, INC.   Year: 1980 Empl: 85   Code–1°:H 2°:DEF
CRYO PRESERVATION; CELL CULTURE;
ENVIRONMENTAL SYSTEMS

Secondary Focus

ABC RESEARCH CORP.   Year: 1967 Empl: 82   Code–1°:W 2°:JBHR
CONTRACT RESEARCH; CUSTOM STARTER
CULTURES; BACTERIAL PROTEIN; ANTIOXIDANTS

ADVANCED BIOTECHNOLOGIES   Year: 1982 Empl: 17   Code–1°:G 2°:EHW
MEDIA, SERA, GROWTH FACTORS, DNA, RESEARCH SERVICES

AMERICAN BIOTECHNOLOGY CO. Year: 1984 Empl: 5   Code–1°:X 2°:PGHJ
GROWTH FACTORS, IMMUNOMODULATORS

APPLIED MICROBIOLOGY, INC.   Year: 1983 Empl: 20   Code–1°:V 2°:KH
RX FOR BOVINE MASTITIS, TOXIC SHOCK SYNDROME
DX, B. SUBTILIS PRODUCTION, STAPHYLOCIDE

BERKELEY ANTIBODY CO. INC.   Year: 1983 Empl: 27   Code–1°:X 2°:GHP
CUSTOM ANTIBODIES, MAB & POLYCLONALS

BIO-RESPONSE, INC.   Year: 1972 Empl: 45   Code–1°:P 2°:HYK
NON-RDNA TPA; MASS CELL CULTURE; DX KITS; MABS

BIO-TECHNICAL RESOURCES   Year: 1962 Empl: 25   Code–1°:M 2°:IJH
YEAST/BACTERIAL PRODS IN FOOD, BREWING, CHEMICALS

Secondary Focus (Cont.)

BIOMEDICAL TECHNOLOGIES   Year: 1981 Empl: 7   Code–1°:J 2°:HGK
   CELL GROWTH FACTORS; CELL CULTURE PRODUCTS;
   ASSAY RIA KITS; MAB; POLYCLONALS; PROTEINS

BIOPOLYMERS, INC.   Year: 1984 Empl: 30   Code–1°:1 2°:HJ
   CELL & TISSUE, OPHTHALMIC PRODUCTS;
   MOLLUSK ADHESIVES

BOEHRINGER MANNHEIM DIAGN. Year: 1975 Empl: 1000 Code–1°:K 2°:PGBH
   DX; RX; REAGENTS; DIGOXIN; ANIMAL
   CELL TISSUE CULTURE, PLANT DX

CENTRAL BIOLOGICS, INC.   Year: 1982 Empl: 7   Code–1°:Q 2°:AHXW
   VET VX; CUSTOM SERA; CONTRACT RESEARCH

CHARLES RIVER BIOTECH. SERV.   Year: 1983 Empl: 55   Code–1°:X 2°:HF
   MABS, CELL CULTURE SYSTEMS

COLLAGEN CORP.   Year: 1975 Empl: 225   Code–1°:4 2°:HP1
   MEDICAL DEVICES; COLLAGEN
   IMPLANTS; TISSUE CULTURES

CYANOTECH CORP.   Year: 1983 Empl: 25   Code–1°:M 2°:BHJ
   PHYCOBILIPROTEIN; SPIRULINA; BETA-CAROTENE

EARL-CLAY LABORATORIES, INC.   Year: 1984 Empl: 7   Code–1°:X 2°:H
   MABS; ULTRACLONE GROWTH CHAMBERS IN
   TISSUE CULTURE; LYMPHOCYTE STIMULATING ASSAY

ELECTRO-NUCLEONICS, INC.   Year: 1960 Empl: 720   Code–1°:K 2°:EFGH
   RX; DX; AIDS DX; IL-2; HTLV-III
   ANTIBODY TEST; TOXIN DX

ENDOTRONICS, INC.   Year: 1981 Empl: 45   Code–1°:F 2°:HPQ
   RX; DX; MABS; APS SERIES; ACUSYST-P; VX;
   BIOLOGICALS; HEP B VX

GENENTECH   Year: 1976 Empl: 1500 Code–1°:P 2°:HKOQ
   RX; DX; VX; INSULIN, IF, TNF, RENIN,
   FACTOR VIII, PROTROPIN HGH, ACTIVASE TPA

HANA BIOLOGICS, INC.   Year: 1979 Empl: 130   Code–1°:P 2°:H
   CELL IMPLANTS TO TREAT DISEASE

IMMUCELL CORP.   Year: 1982 Empl: 17   Code–1°:A 2°:GHQ
   HUMAN DX; REAGENTS; ANIMAL DX; VX

Listing 5-8 Cell Culture

Secondary Focus (Cont.)

IMMUNO MODULATORS LABS Year: 1981 Empl: 25 Code–1°:P 2°:AGKH
  RX; IF; AIDS RX; HCG; COSMETICS

IMMUNOMED CORP. Year: 1979 Empl: 17 Code–1°:Q 2°:HP
  VX; IMMUNOTHERAPEUTICS; RABIES VX

INTERFERON SCIENCES, INC. Year: 1981 Empl: 70 Code–1°:P 2°:GHB
  IF-A; IF-G; MABS

LABSYSTEMS, INC. Year: 1983 Empl: 10 Code–1°:K 2°:HJMF
  MICROBIOLOGY ASSAYS & EQUIP; MAB'S

LEE BIOMOLECULAR RES. LABS Year: 1980 Empl: 16 Code–1°:P 2°:HKQ
  IF; ANTI-IF SERA

MONOCLONETICS INT'L, INC. Year: 1984 Empl: 5 Code–1°:G 2°:HKP
  HUMAN SCREENING TESTS & DX, EPITOPE SCREENS

MYCOSEARCH, INC. Year: 1979 Empl: 6 Code–1°:2 2°:OCHR
  CULTURES OF RARE & NEW FUNGI; LIVE
  CULTURES; SPECIALTY CHEMICALS

NATIVE PLANTS, INC. Year: 1973 Empl: 625 Code–1°:B 2°:HJM
  POTATO, FLOWER, VEGETABLE SEEDS; SOIL
  INOCULANTS; PLANTS

NEUSHUL MARICULTURE Year: 1978 Empl: 6 Code–1°:S 2°:PMHC
  ANTI-VIRAL; ENZYMES; MOLLUSCAN
  SEEDING & CULTURE

NYGENE CORP. Year: 1985 Empl: 18 Code–1°:F 2°:H
  BIOAFFINITY SEPARATION SYSTEMS;
  HYBRIDOMA PRODUCTION

PHARMAGENE Year: 1986 Empl: 3 Code–1°:P 2°:HIJK
  RX; CHEMICALS; SPECIFIC SITE
  RECOMBINATION PROCESSES

SEAPHARM, INC. Year: 1983 Empl: 50 Code–1°:T 2°:PH
  MARINE RX PRODUCTS; TISSUE CULTURE

TECHNICLONE INT'L, INC. Year: 1981 Empl: 15 Code–1°:K 2°:PFH
  CANCER DX, TISSUE CULTURES,
  MAB'S, MAB CULTURING EQUIPMENT

LISTING 5-9 COMPANIES IN COMMODITY CHEMICALS (I)

Primary Focus

AMERICAN BIOGENETICS CORP.   Year: 1984 Empl: 10   Code–1°:I 2°:JBLR
    PROCESS DESIGN MICROORGANISMS; CHEMICALS

GENEX, INC.   Year: 1977 Empl: 80   Code–1°:I 2°:JO
    COMMODITY & SPECIALTY CHEMICALS; ENZYMES,
    PROTEINS, DRAIN CLEANER; VITAMIN B-12

Secondary Focus

BIO-TECHNICAL RESOURCES, INC. Year: 1962 Empl: 25   Code–1°:M 2°:IJH
    YEAST/BACTERIAL PRODS IN FOOD,
    BREWING, CHEMICALS

BIOCHEM TECHNOLOGY, INC.   Year: 1977 Empl: 12   Code–1°:D 2°:I
    SENSORS; COMPUTER PRODS &
    SOFTWARE; NADH DETECTOR

BIOSPHERICS, INC.   Year: 1967 Empl: 250   Code–1°:R 2°:YIDF
    WASTEWATER PROCESS &
    MONITORS & PILOT PLANT

PHARMAGENE   Year: 1986 Empl: 3   Code–1°:P 2°:HIJK
    RX; CHEMICALS; SPECIFIC SITE
    RECOMBINATION PROCESSES

ZOECON CORP.   Year: 1968 Empl: 130   Code–1°:B 2°:IJ
    PLANT IMPROVEMENT; SEEDS;
    AG CHEMICALS

Listing 5-10 Specialty Chemicals

LISTING 5-10 COMPANIES IN SPECIALTY CHEMICALS (J)

Primary Focus

BACHEM BIOSCIENCE, INC.   Year: 1987   Empl: 15   Code–1°:J   2°:PQ
    AMINO ACIDS & DERIVATIVES; BIOACTIVE
    PEPTIDES; ENZYME SUBSTRATES; VX EPITOPES

BACHEM, INC.   Year: 1971   Empl: 45   Code–1°:J   2°:G
    SYNTHETIC BIOACTIVE PEPTIDES; GROWTH FACTORS, CRF

BIOMEDICAL TECHNOLOGIES   Year: 1981   Empl: 10   Code–1°:J   2°:HGK
    CELL GROWTH FACTORS; CELL CULTURE PROD.;
    ASSAY RIA KITS; MAB; POLYCLONALS; PROTEINS

BIOSCIENCE MANAGEMENT, INC.   Year: 1984   Empl: 8   Code–1°:J   2°:RDF
    BACTERIAL CULTURES; INOCULANTS; INSTRUMENTS

BIOTECHNICA INTERNATIONAL   Year: 1981   Empl: 145   Code–1°:J   2°:KQMB
    INOCULANT, VX, DENTAL DX; ENZYMES,
    INDUSTRIAL YEAST, LITE BEER, FOODS

CALZYME LABORATORIES, INC.   Year: 1983   Empl: 10   Code–1°:J
    ENZYMES, BIOCHEMICALS, PROTEINS

CAMBRIDGE RESEARCH BIOCHEM.   Year: 1980   Empl: 75   Code–1°:J   2°:K
    MABS; PEPTIDES, ONCO PROTEIN PRODUCTS;
    DEVELOPING EPITOPE SCANNING PRODUCTS

CARBOHYDRATES INT'L, INC.   Year: 1987   Empl: 25   Code–1°:J 2°:K
    INFECTIOUS DISEASE, BLOOD TYPE DX;
    SUGAR STRUCTURES TO RAISE MABS

CREATIVE BIOMOLECULES   Year: 1981   Empl: 70   Code–1°:J   2°:PGW
    PEPTIDES, CONTRACT R&D, TPA, EGF

DELTOWN CHEMURGIC CORP.   Year: 1968   Empl: 90   Code–1°:J
    PEPTONES AND HYDROLASES FOR
    RX; CHEMICALS

ENZYME BIO-SYSTEMS LTD.   Year: 1983   Empl: 101   Code–1°:J   2°:M
    INDUSTRIAL ENZYME PRODUCTS USED
    FOR STARCH, BREWING & BAKING

GENE-TRAK SYSTEMS   Year: 1981   Empl: 196   Code–1°:J   2°:PK
    RX PROTEINS, DNA PROBE-BASED DX'S; DEVELOPING
    CV PROTEINS, HEMOPOIETIC GROWTH FACTOR

Primary Focus (Cont.)

GENENCOR, INC.   Year: 1982 Empl: 130   Code–1°:J 2°:M
    ENZYMES IN FOOD/BEVERAGE PROCESSING

√ INCELL CORP.   Year: 1982 Empl: 23   Code–1°:J 2°:G
    PEPTIDE HORMONES; AMINO ACIDS; REAGENTS; CHEMICALS

√ INFERGENE CO.   Year: 1984 Empl: 25   Code–1°:J 2°:KM
    DX FOR HERPES, OTHERS; ENZYMES;
     BAKING & BREWING STRAINS, GLUCOAMYLASE STRAIN

√ LIFECORE BIOMEDICAL, INC.   Year: 1975 Empl: 41   Code–1°:J 2°:4
    HYALURONIC ACID, MEDICAL DEVICES

√ MARTEK   Year: 1985 Empl: 16   Code–1°:J 2°:KG
    STABILIZATOPE-LABELLED BIOCHEMICALS,
    C-13 & DEUTERIUM-LABELLED LUBRICANTS, AMINO ACIDS

√ MICROBIO RESOURCES   Year: 1981 Empl: 30   Code–1°:J 2°:MRPO
    FOOD ADDITIVES & COLORANTS; SPECIAL CHEMICALS

√ NORTH COAST BIOTECHNOLOGY   Year: 1986 Empl: 2   Code–1°:J 2°:PEFK
    ENZYMES FOR RX MANUFACTURING

√ O.C.S. LABORATORIES, INC.   Year: 1983 Empl: 12   Code–1°:J 2°:DAC
    CUSTOM MADE SYNTHETIC GENES & PEPTIDES

√ OMNI BIOCHEM, INC.   Year: 1986 Empl: 5   Code–1°:J 2°:G
    PEPTIDES; ENZYMES; AMINO ACIDS; REAGENTS

√ PAMBEC LABORATORIES   Year: 1987 Empl: 6   Code–1°:J 2°:KP
    CUSTOM/CONTRACT WORK ON GENETIC ENGINEERING,
    PROKARYOTES, EUKARYOTES, DNA PROBES, AIDS RX

√ PENINSULA LABORATORIES, INC.   Year: 1971 Empl: 75   Code–1°:J 2°:G
    PEPTIDES, ENZYMES, SPECIALTY REAGENTS

√ PROVESTA CORP.   Year: 1975 Empl: 50   Code–1°:J 2°:LBMP
     FERMENTATION-BASED FLAVORS; ENZYMES;
    YEAST FEEDS; PHEROMONES FOR PEST CONTROL; TPA

REPLIGEN CORP.   Year: 1981 Empl: 95   Code–1°:J 2°:QAB
    ENZYMES, AIDS VX, PROTEINS, PROTEIN A, HTLVIII, PESTICIDES

√ SOUTHERN BIOTECH. ASSOC.   Year: 1982 Empl: 12   Code–1°:J 2°:X
    COLLAGEN; COLLAGEN AB'S; CUSTOM
    CONJUGATION AB DEVELOPMENT

Listing 5-10 Specialty Chemicals

Primary Focus (Cont.)

SYNTHETIC GENETICS, INC.   Year: 1985 Empl: 6   Code–1°:J 2°:G
    CUSTOM & SYNTHETIC DNA & PEPTIDES

VIROSTAT, INC.   Year: 1985 Empl: 1   Code–1°:J 2°:K
    IMMUNOCHEMICALS FOR INFECTIOUS DISEASE RESEARCH

WASHINGTON BIOLAB   Year: 1986 Empl: 10   Code–1°:J 2°:GW
    AIDS RX; ENZYMES, RAW MATERIALS
    FOR BIOTECHNOLOGY

WESTBRIDGE RESEARCH GROUP   Year: 1982 Empl: 15   Code–1°:J 2°:B
    PLANT GROWTH REGULATOR; WATER TREATMENT

Secondary Focus

ABC RESEARCH CORP.   Year: 1967 Empl: 82   Code–1°:W 2°:JBHR
    CONTRACT RESEARCH; CUSTOM STARTER
    CULTURES; BACTERIAL PROTEIN; ANTIOXIDANTS

ADVANCED GENETIC SCIENCES   Year: 1979 Empl: 130   Code–1°:B 2°:J
    PLANT & AG RESEARCH, CHEMICALS, SEEDS,
    FROSTBAN, SNOMAX

ALLELIC BIOSYSTEMS   Year: 1984 Empl: 7   Code–1°:V 2°:JG
    VET TOX TESTS, PROBES, OLIGOS, REAGENTS

AMERICAN BIOGENETICS CORP.   Year: 1984 Empl: 10   Code–1°:I 2°:JBLR
    PROCESS DESIGN MICROORGANISMS; CHEMICALS

AMERICAN BIOTECHNOLOGY CO.   Year: 1984 Empl: 5   Code–1°:X 2°:PGHJ
    GROWTH FACTORS, IMMUNOMODULATORS

AMGEN   Year: 1980 Empl: 340   Code–1°:P 2°:AJKG
    RESEARCH REAGENTS, EGF, IGF, IF,
    EPO, IL-2, CSF, BGH, HGH

APPLIED PROTEIN TECHNOLOGIES   Year: 1984 Empl: 5   Code–1°:F 2°:JG
    PEPTIDE & PROTEIN SYNTHESIZERS,
    CHEMICALS, REAGENTS

BENTECH LABORATORIES   Year: 1984 Empl: 7   Code–1°:B 2°:JO
    CROP YIELD ENHANCING AGENT;
    ANTIFUNGAL; WOUND HEALING

BIO-TECHNICAL RESOURCES, INC.   Year: 1962 Empl: 25   Code–1°:M 2°:IJH
    YEAST/BACTERIAL PRODS IN FOOD,
    BREWING, CHEMICALS.

Secondary Focus (Cont.)

BIOGEN   Year: 1979 Empl: 250   Code–1°:P 2°:QGJ
  RX; DX; TNF, IL-2, CSF, TPA, HSA,
  FACTOR VIII, IF-G, HEP B ANTIGEN

BIOPOLYMERS, INC.   Year: 1984 Empl: 30   Code–1°:1 2°:HJ
  CELL & TISSUE, OPHTHALMIC PRODUCTS;
  MOLLUSK ADHESIVES

BIOPURE   Year: 1984 Empl: 31   Code–1°:O 2°:JPX
  CUSTOM FERMENTATION OF RX, MAB, CHEMICALS

BIOTECHNOLOGY DEVEL. CORP.   Year: 1982 Empl: 35   Code–1°:F 2°:PJQ3
  AIDS VX; EQUIPMENT; REAGENTS;
  DRUG DELIVERY SYSTEMS

CALGENE, INC.   Year: 1980 Empl: 165   Code–1°:B 2°:MJ
  TOMATOES, COTTON, CORN, SUNFLOWER,
  RAPESEED, FOODS; PESTICIDES, PLANTS, TREES

CENTOCOR   Year: 1979 Empl: 246   Code–1°:K 2°:PJGQ
  MAB RX; CANCER, HEP B DX; MABS; AIDS VX

CHIRON CORP.   Year: 1981 Empl: 300   Code–1°:P 2°:QKJV
  AIDS DX; DNA PROBE DX; WOUND HEALING;
  HEP B VX; VET VX, SOD; ENZYMES; IF

COORS BIOTECH PRODUCTS CO.   Year: 1984 Empl: 185   Code–1°:M 2°:OJ
  INDUSTRIAL DEGREASER; FOOD PRODUCTS;
  SPECIALTY CHEMICALS

CYANOTECH CORP.   Year: 1983 Empl: 25   Code–1°:M 2°:BHJ
  PHYCOBILIPROTEIN; SPIRULINA; BETA-CAROTENE

E-Y LABORATORIES, INC.   Year: 1978 Empl: 20   Code–1°:X 2°:KPJ
  ANTISERA; LECTINS; REAGENT; MABS; DX

GENEX, INC.   Year: 1977 Empl: 80   Code–1°:I 2°:JO
  COMMODITY & SPECIALTY CHEMICALS;
  ENZYMES, PROTEINS, DRAIN CLEANER; VITAMIN B-12

GENZYME CORP.   Year: 1981 Empl: 185   Code–1°:P 2°:KJGE
  RX, DX FOR HUMAN HEALTH CARE; MABS;
  EYE PRODUCTS, SKIN PRODUCTS; SURFACTANTS

IMMUNOTECH CORP.   Year: 1980 Empl: 15   Code–1°:K 2°:J
  IMMUNODIAGNOSTIC TEST KITS, DX

Listing 5-10 Specialty Chemicals

Secondary Focus (Cont.)

INGENE, INC.   Year: 1980 Empl: 65   Code–1°:P 2°:JB
  RX; BONE & CARTILAGE RX; CANCER RX;
  SWEETENERS; PESTICIDE

INTERNATIONAL ENZYMES, INC.   Year: 1983 Empl: 4   Code–1°:G 2°:JX
  ENZYMES; ANTISERA; SERUM-FREE MEDIA SUPPLIES

JACKSON IMMUNORESEARCH   Year: 1982 Empl: 16   Code–1°:G 2°:J
  AFFINITY PURIFICATION PROTEINS; PURE PROTEINS

LABSYSTEMS, INC.   Year: 1983 Empl: 10   Code–1°:K 2°:HJMF
  MICROBIOLOGY ASSAYS & EQUIP; MABS

LIPOSOME COMPANY, INC.   Year: 1981 Empl: 67   Code–1°:3 2°:PQJ
  LIPOSOMES; DRUG DELIVERY; CANCER RX; VX

LIPOSOME TECHNOLOGY, INC.   Year: 1981 Empl: 65   Code–1°:3 2°:VBPJ
  LIPOSOMES FOR RX, DRUG DELIVERY,
  VETERINARY, AGRICULTURE

LUCKY BIOTECH CORP.   Year: 1984 Empl: 4   Code–1°:P 2°:QJKA
  RX; IF; IF-A; IF-B; IF-G; IL-2; HEP B DX;
  PORCINE GROWTH HORMONE

MARCOR DEVELOPMENT CO.   Year: 1977 Empl: 6   Code–1°:G 2°:J
  REAGENTS FOR FERMENTATION, ENZYMES

MARINE BIOLOGICALS, INC.   Year: 1981 Empl: 2   Code–1°:T 2°:J
  MARINE BYPRODUCTS; LIMULUS AMEBOCYTE LYSATE

MOLECULAR BIOLOGY RESOURCES Year: 1986 Empl: 10   Code–1°:G 2°:JOF
  REAGENTS; DNA & RNA MODIFYING
  ENZYMES & NUCLEIC ACIDS; HYBRIDIZATION APPARATUS

MOLECULAR BIOSYSTEMS, INC.   Year: 1981 Empl: 55   Code–1°:K 2°:JKG
  DX; DNA & RNA PROBES, ALBUMEX, EXTRACTOR KITS

NATIVE PLANTS, INC.   Year: 1973 Empl: 625   Code–1°:B 2°:HJM
  POTATO, FLOWER, VEGETABLE SEEDS;
  SOIL INOCULANTS; PLANTS

NITRAGIN CO.   Year: 1898 Empl: 80   Code–1°:B 2°:J
  NITROGEN FIXING BACTERIA INOCULANTS

OCEAN GENETICS   Year: 1981 Empl: 35   Code–1°:T 2°:JP
  SPECIAL CHEMICALS & RX OF MARINE ALGAE

Secondary Focus (Cont.)

PEL-FREEZ BIOLOGICALS, INC.   Year: 1911 Empl: 175   Code–1°:G 2°:JX
   VARIOUS SPECIALTY REAGENTS, MAB SERA

PETROFERM, INC.   Year: 1977 Empl: 68   Code–1°:L 2°:J
   MICROBIALS FOR IMPROVED FUELS

PHARMACIA LKB BIOTECH.   Year: 1960   Code–1°:E 2°:GVAJ
   CHROMATOGRAPHY SEPARATION CHEMICALS
   & INSTRUMENTS; ENZYMES; REAGENTS; BIOCHEMICALS

PHARMAGENE   Year: 1986 Empl: 3   Code–1°:P 2°:HIJK
   RX; CHEMICALS; SPECIFIC SITE RECOMBINATION PROCESSES

PROMEGA CORP.   Year: 1978 Empl: 85   Code–1°:G 2°:FJ
   REAGENTS FOR MOLECULAR BIOLOGY; ENZYMES

PROTATEK INTERNATIONAL, INC. Year: 1984 Empl: 37   Code–1°:K 2°:YJPV
   VIRUS, DX TEST KITS; RX; SPECIALIZED
   MICROBES; VET VX, RX, DX

SCOTT LABORATORIES   Year: 1981 Empl: 375   Code–1°:G 2°:J
   REAGENTS AND SPECIALTY CHEMICALS

SCRIPPS LABORATORIES   Year: 1980 Empl: 60   Code–1°:P 2°:JGK
   BULK HUMAN PROTEINS; HORMONES; ENZYMES; DX

SEPRACOR, INC.   Year: 1984 Empl: 40   Code–1°:E 2°:FJ
   MEMBRANE COLUMNS-ANTI-FACTOR VII &
   ANTI-TPA; SEPARATIONS

UNIGENE LABORATORIES, INC.   Year: 1980 Empl: 30   Code–1°:P 2°:JA
   PEPTIDES, HORMONES, ENZYMES,
   CALCITONIN, GROWTH FACTORS

UNITED BIOMEDICAL, INC.   Year: 1983 Empl: 10   Code–1°:K 2°:JXQ
   DX, MABS, SPECIALTY PROTEINS, AIDS VX

XENOGEN   Year: 1981   Code–1°:K 2°:JW
   RDNA RESEARCH FOR DX AND CHEMICALS;
   DX FOR BREAST CANCER, GONORRHEA

ZOECON CORP.   Year: 1968 Empl: 130   Code–1°:B 2°:IJ
   PLANT IMPROVEMENT; SEEDS; AG CHEMICALS

ZYMED LABORATORIES, INC.   Year: 1980 Empl: 20   Code–1°:G 2°:KJ
   IMMUNOCHEMICALS, MABS, PROTEINS, ENZYMES

Listing 5-11 Clinical Diagnostics

LISTING 5-11 COMPANIES IN CLINICAL DIAGNOSTICS (K)

Primary Focus

ABN   Year: 1981 Empl: 45   Code–1°:K 2°:GF
  DX, REAGENTS, BT INSTRUMENTS,
  DNA PROBES, BLOTTING SYSTEMS

ADVANCED MAGNETICS, INC.   Year: 1981 Empl: 40   Code–1°:K 2°:E
  DX RIA KITS, MAGNETIC SEPARATION REAGENTS

AMERICAN DIAGNOSTICA, INC.   Year: 1983 Empl: 9   Code–1°:K 2°:G
  DX, TPA ELISA, TPA ACTIVE ASSAY KITS

ANTIVIRALS, INC.   Year: 1980 Empl: 5   Code–1°:K 2°:PG
  NEU-GENES (TM) FOR DX & RX USE

ATLANTIC ANTIBODIES   Year: 1972 Empl: 60   Code–1°:K 2°:G
  DX TEST KITS; DX; MABS; REAGENTS

BEHRING DIAGNOSTICS   Year: 1952 Empl: 100   Code–1°:K 2°:G
  AB'S; DX; REAGENTS (CALBIOCHEM)

BINAX, INC.   Year: 1986 Empl: 17   Code–1°:K 2°:VGX
  MABS, DX TEST KITS, VET DX

BIOGENEX LABORATORIES   Year: 1981 Empl: 42   Code–1°:K 2°:PG
  RIA DX TEST; IMMUNOHISTOCHEMICALS;
  KITS; REAGENTS

BIOMERICA, INC.   Year: 1971 Empl: 35   Code–1°:K 2°:GV
  PREGNANCY, ALLERGY, OCCULT BLOOD DX;
  LAB PRODUCTS; VET DX & RX

BIONETICS RESEARCH   Year: 1985 Empl: 200   Code–1°:K 2°:GP
  AIDS DX, SALMONELLA EIA, RADIOLABELLED
  COMPOUNDS, LIPOPROTEINS, MELATONIN

BIOSTAR MEDICAL PRODUCTS   Year: 1983 Empl: 30   Code–1°:K 2°:WF
  DX & TEST KITS FOR AUTOIMMUNE
  DISEASES, INSTRUMENTATION

BIOTECH RESEARCH LABS, INC.   Year: 1973 Empl: 167   Code–1°:K 2°:PQ
  MABS FOR DX, AIDS DX, CANCER DX; AIDS VX, IL-2

BIOTHERM   Year: 1985 Empl: 10   Code–1°:K 2°:FDW
  DX, CARDIOVASCULAR DX

Primary Focus (Cont.)

BIOTX   Year: 1985 Empl: 20   Code–1°:K 2°:P
   DX CANCER NUCLEAR ANTIGEN DX KIT,
   ABS TO ONCOGENES & GROWTH FACTORS

BMI, INC.   Year: 1984 Empl: 10   Code–1°:K 2°:V
   DX; RX; OTC PREG. TEST, ENZYME IMMUNOASSAYS,
   ALLERGY TESTS, BOVINE PROGESTERONE RX

BOEHRINGER MANNHEIM DIAGN. Year: 1975 Empl: 1000 Code–1°:K 2°:PGBH
   DX; RX; REAGENTS; DIGOXIN; ANIMAL
   CELL TISSUE CULTURE, PLANT DX

BRAIN RESEARCH, INC.   Year: 1968   Code–1°:K
   CANCER DX

CALIFORNIA INTEGRATED DIAGN. Year: 1981 Empl: 9   Code–1°:K
   DX KITS FOR SEXUALLY TRANSMITTED
   DISEASES; MABS

CAMBRIDGE BIOSCIENCE CORP.   Year: 1981 Empl: 97   Code–1°:K 2°:QPV
   DX KITS FOR AIDS, ADENOVIRUS,
   ROTAVIRUS; VET VX, DX

CELLMARK DIAGNOSTICS   Year: 1986 Empl: 30   Code–1°:K
   GENE PROBING, GENETIC TESTING; DNA FINGERPRINTING.

CELLULAR PRODUCTS   Year: 1982 Empl: 101   Code–1°:K 2°:P
   MABS, DIPSTIK (TM) DX TESTS KITS; AIDS DX; IF

CENTOCOR   Year: 1979 Empl: 246   Code–1°:K 2°:PJGQ
   MAB RX; CANCER, HEP B DX; MABS; AIDS VX

CIBA-CORNING DIAGN. CORP.   Year: 1985 Empl: 750   Code–1°:K
   HUMAN CLINICAL DX

CLINETICS CORP.   Year: 1979 Empl: 10   Code–1°:K
   PREGNANCY ASSAY KIT, DX

CLONTECH LABORATORIES, INC.   Year: 1984 Empl: 30   Code–1°:K
   CLONING, TPA, BLOOD CLOTTING FACTORS

COVALENT TECHNOLOGY CORP.   Year: 1981 Empl: 16   Code–1°:K 2°:Y
   DX FOR HORMONE, DISEASE; BIOWAR DX;
   TROPICAL DISEASE DX

CYTOGEN CORP.   Year: 1980 Empl: 140   Code–1°:K 2°:PX
   RX; MABS; DX; CANCER RX

Listing 5-11 Clinical Diagnostics

Primary Focus (Cont.)

CYTOTECH, INC.   Year: 1982 Empl: 43   Code–1°:K 2°:G
  AIDS DX, REAGENTS

DIAGNON CORP.   Year: 1981 Empl: 70   Code–1°:K 2°:Y
  DX ASSAYS FOR FERRITIN, B12/FOLATE,
  HERPES, VIRAL TREATMENT, TOXIN ASSAYS

DIAGNOSTIC TECHNOLOGY, INC.   Year: 1980 Empl: 60   Code–1°:K 2°:GF
  DX, BLOOD TEST KITS, REAGENTS

DIAMEDIX, INC.   Year: 1986 Empl: 35   Code–1°:K 2°:G
  DX KITS FOR INFECTIOUS & AUTOIMMUNE DISEASES

DIGENE DIAGNOSTICS, INC.   Year: 1984 Empl: 20   Code–1°:K
  DNA PROBE DX, VIRAL DETECTION KITS

ELECTRO-NUCLEONICS, INC.   Year: 1960 Empl: 720   Code–1°:K 2°:EFGH
  RX; DX; AIDS DX; IL-2; HTLV-III
  ANTIBODY TEST; TOXIN DX

ENDOGEN, INC.   Year: 1985 Empl: 7   Code–1°:K 2°:X
  DX ASSAYS (QUANTITATIVE); DEVELOPING
  HUMAN MABS.

ENZO-BIOCHEM, INC.   Year: 1976 Empl: 110   Code–1°:K 2°:GP
  ENZYMES; LECTINS; FILTERS; NUCLEOTIDES;
  NON-RADIOACTIVE DNA PROBES; MAB; DX

GAMMA BIOLOGICALS, INC.   Year: 1969 Empl: 166   Code–1°:K
  REAGENT-BASED DX TESTS FOR HUMAN BLOOD SERA

GEN-PROBE, INC.   Year: 1984 Empl: 127   Code–1°:K 2°:FG
  DNA PROBE DX TECH FOR RESEARCH & CLINIC

GENELABS, INC.   Year: 1984 Empl: 90   Code–1°:K 2°:P
  AIDS RX, CYTOKINES, NON-A, NON-B HEPATITIS,
  SOLID PHASE BLOOD TESTING

GENESIS LABS, INC.   Year: 1984 Empl: 29   Code–1°:K
  DX; BLOOD GLUCOSE TEST; DRUG MONITOR

GENETIC DIAGNOSTICS CORP.   Year: 1981 Empl: 17   Code–1°:K 2°:AQ
  DX TEST KITS FOR ABUSE DRUGS, DISEASES

GENETIC SYSTEMS CORP.   Year: 1980 Empl: 400   Code–1°:K 2°:P
  MAB BASED HUMAN DX AND RX; AIDS DX;
  HERPES DX; LEGIONNAIRES DX

Primary Focus (Cont.)

HYGEIA SCIENCES   Year: 1980 Empl: 70   Code–1°:K
   DX OVULATION PREDICTOR; PREGNANCY TEST KITS

ICL SCIENTIFIC, INC.   Year: 1981 Empl: 70   Code–1°:K
   DX

IMCLONE SYSTEMS, INC.   Year: 1984 Empl: 50   Code–1°:K 2°:PQ
   DX; VX; RX; AIDS DX; HEP B DX; DNA PROBES,
   IMMUNODIAGNOSTICS FOR VIRAL DISEASES

IMMUNEX, INC.   Year: 1981 Empl: 5   Code–1°:K 2°:PG
   DX; MABS, POLYCLONAL AB'S TO HAPTENS;
   RADIOIMMUNOASSAY KITS; REAGENTS

IMMUNO CONCEPTS, INC.   Year: 1981 Empl: 20   Code–1°:K
   DX; ANTINUCLEAR AB TESTS; FLUORESCENT
   DNA TEST; AUTO ANTIBODY TEST SYSTEMS

IMMUNOTECH CORP.   Year: 1980 Empl: 15   Code–1°:K 2°:J
   IMMUNODIAGNOSTIC TEST KITS, DX

KALLESTAD DIAGNOSTICS   Year: 1967 Empl: 330   Code–1°:K
   DX PRODUCTS

LABSYSTEMS, INC.   Year: 1983 Empl: 10   Code–1°:K 2°:HJMF
   MICROBIOLOGY ASSAYS & EQUIP; MABS

LIFE TECHNOLOGIES, INC.   Year: 1983 Empl: 1300   Code–1°:K 2°:GP
   IF; PROBE ASSAY DX; REAGENTS

LIFECODES CORP.   Year: 1982 Empl: 53   Code–1°:K 2°:D
   REFERENCE LAB DX USING DNA PROBES

MAST IMMUNOSYSTEMS, INC.   Year: 1979 Empl: 45   Code–1°:K
   ALLERGY DX'S; ADAPTING PATENTED
   TECHNOLOGY TO INFECTIOUS DISEASES

MICROGENICS   Year: 1981 Empl: 70   Code–1°:K
   DX ASSAYS; DIGOXIN ASSAY

MOLECULAR BIOSYSTEMS, INC.   Year: 1981 Empl: 55   Code–1°:K 2°:JKG
   DX; DNA & RNA PROBES, ALBUMEX,
   EXTRACTOR KITS

MOLECULAR DIAGNOSTICS, INC.   Year: 1981 Empl: 30   Code–1°:K 2°:P
   RX FOR DIABETES, CANCER, INFECTIOUS
   DISEASES, HUMAN GENETIC DISEASES

Listing 5-11 Clinical Diagnostics

Primary Focus (Cont.)

MONOCLONAL ANTIBODIES, INC.   Year: 1979 Empl: 110   Code–1°:K 2°:V
   DX TEST KITS-PREGNANCY, REPRODUCTION,
   INFERTILITY; VET DX

MUREX CORP.   Year: 1984 Empl: 180   Code–1°:K 2°:PGQ
   MAB REAGENTS & ANTIBODY KITS; DX

ONCOGENE SCIENCE, INC.   Year: 1983 Empl: 65   Code–1°:K 2°:PX
   ONCOGENE PROBES FOR DX AND RX, MABS

ONCOR, INC.   Year: 1983 Empl: 30   Code–1°:K 2°:G
   AIDS DX, DX REAGENTS, DNA PROBES FOR
   CANCER, HTLV-III RNA PROBE, B CELL & T CELL PROBES

ORGANON TEKNIKA CORP.   Year: 1978 Empl: 270   Code–1°:K 2°:DFWX
   HEALTH CARE/MEDICAL DX

POLYCELL, INC.   Year: 1983 Empl: 6   Code–1°:K
   DX; MABS; IMAGING FOR HEART DISEASE,
   OVARIAN CANCER DX

PROGENX   Year: 1987 Empl: 20   Code–1°:K 2°:P
   CANCER DX MARKERS; CANCER RX

PROTATEK INT'L, INC.   Year: 1984 Empl: 37   Code–1°:K 2°:YJPV
   VIRUS, DX TEST KITS; RX; SPECIALIZED
   MICROBES; VET VX, RX, DX

PROTEINS INT'L, INC.   Year: 1983 Empl: 6   Code–1°:K 2°:X
   IMMUNOLOGICAL DX; CUSTOM MABS

QUEST BIOTECHNOLOGY, INC.   Year: 1986 Empl: 9   Code–1°:K 2°:P
   BISPECIFIC MAB; IN VITRO DX; CANCER RX & IMAGING

QUIDEL   Year: 1981 Empl: 325   Code–1°:K
   DX ASSAYS-PREGNANCY, FERTILITY,
   STREP-A, ALLERGY, INFECTIOUS DISEASE

RECOMTEX CORP.   Year: 1983 Empl: 12   Code–1°:K
   GENETIC MEDICAL DX

RES. & DIAGNOS. ANTIBODIES   Year: 1985   Code–1°:K 2°:XG
   DX KITS, MAB REAGENTS

SENETEK PLC   Year: 1983 Empl: 14   Code–1°:K 2°:P
   MAB-BASED DX FOR ALZHEIMER'S DISEASE;
   SKIN AGING REVERSAL RX

Primary Focus (Cont.)

SYVA CO.   Year: 1966 Empl: 1000   Code–1°:K 2°:K
    DX PRODUCTS; THERAPEUTIC DRUG
    MONITORING; DRUG ABUSE/TOXINS

TECHNICLONE INT'L, INC.   Year: 1981 Empl: 15   Code–1°:K 2°:PFH
    CANCER DX, TISSUE CULTURES, MAB,
    MAB CULTURING EQUIP.

TECHNOGENETICS, INC.   Year: 1983 Empl: 30   Code–1°:K 2°:P
    MABS FOR IMMUNOCYTOLOGY, HUMAN DX KITS

THREE-M (3M) DIAGNOSTIC SYS.   Year: 1981 Empl: 130   Code–1°:K
    DX

UNITED BIOMEDICAL, INC.   Year: 1983 Empl: 10   Code–1°:K 2°:JXQ
    DX, MABS, SPECIALTY PROTEINS, AIDS VX

XENOGEN   Year: 1981   Code–1°:K 2°:JW
    RDNA RESEARCH FOR DX AND CHEMICALS;
    DX FOR BREAST CANCER, GONORRHEA

Secondary Focus

AGDIA, INC.   Year: 1981 Empl: 8   Code–1°:B 2°:KY
    PLANT AND FOOD TOX DX FOR VIRUSES

AGRITECH SYSTEMS, INC.   Year: 1984 Empl: 90   Code–1°:V 2°:K
    IMMUNOASSAYS, DX; VET DX; TESTS
    FOR FEED/FOOD CONTAMINANTS

ALPHA I BIOMEDICALS, INC.   Year: 1982 Empl: 11   Code–1°:P 2°:QKV
    PEPTIDES & VX FOR DX/RX; AIDS RX, VX;
    DEVELOPING THYMOSIN

AMERICAN BIOCLINICAL   Year: 1977 Empl: 24   Code–1°:G 2°:KP
    REAGENTS; HORMONE DX KITS; RX

AMERICAN LABORATORIES, INC.   Year: 1962 Empl: 26   Code–1°:P 2°:KMO
    RAW MATERIALS OF BIOTECHNOLOGY PRODUCTS

AMERICAN QUALEX INT'L, INC.   Year: 1981 Empl: 10   Code–1°:G 2°:FEK
    CUSTOM MABS; ENZYMES;
    DNA PROBES; INSTRUMENTS

AMGEN   Year: 1980 Empl: 340   Code–1°:P 2°:AJKG
    RESEARCH REAGENTS, EGF, IGF, IF,
    EPO, IL-2, CSF, BGH, HGH

Listing 5-11 Clinical Diagnostics

Secondary Focus (Cont.)

ANGENICS   Year: 1980   Empl: 50   Code–1°:Y 2°:K
    SCREENING TESTS FOR DRUG AND MILK ABUSE

ANTIBODIES, INC.   Year: 1961   Empl: 35   Code–1°:X 2°:VK
    ANTISERA, MABS, VET DX, HUMAN DX

APPLIED BIOTECHNOLOGY   Year: 1982   Empl: 45   Code–1°:V 2°:QKP
    VET VX; AIDS, TB VX; CANCER DX, RX

APPLIED GENETICS LABS, INC.   Year: 1984   Empl: 11   Code–1°:Y 2°:K
    GENETIC TOX TEST, DNA PROBES

APPLIED MICROBIOLOGY, INC.   Year: 1983   Empl: 20   Code–1°:V 2°:KH
    RX FOR BOVINE MASTITIS, TOXIC SHOCK
    SYNDROME DX, B. SUBTILIS PROD'N., STAPHYLOCIDE

BETHYL LABS, INC.   Year: 1977   Empl: 8   Code–1°:X 2°:KA
    ANTISERA & CUSTOM ANTISERA SERVICE

BIO-RESPONSE, INC.   Year: 1972   Empl: 45   Code–1°:P 2°:HYK
    NON-RDNA TPA; MASS CELL CULTURE; DX KITS; MABS

BIOMATERIALS INT'L, INC.   Year: 1981   Empl: 30   Code–1°:D 2°:K
    RAMAN TECHNOL. FOR RESPIRATORY
    APPLICATIONS; GAS MONITOR

BIOMEDICAL TECHNOLOGIES   Year: 1981   Empl: 7   Code–1°:J 2°:HGK
    CELL GROWTH FACTORS; CELL CULTURE
    PRODUCTS; ASSAY RIA KITS; MAB; POLYCLONALS;

BIONIQUE LABS, INC.   Year: 1983   Empl: 11   Code–1°:H 2°:K
    CELL CULTURE,CULTURE CHAMBER; MYCOPLASMA
    DETECTION SYSTEM; CELL PROTECTIVE DEVICE; DX

BIOPROBE INT'L, INC.   Year: 1983   Empl: 15   Code–1°:E 2°:KF
    SPECIALIZED GELS; SEPARATION DEVICE

BIOTECHNICA INT'L   Year: 1981   Empl: 145   Code–1°:J 2°:KQMB
    INOCULANT, VX, DENTAL DX; ENZYMES,
    INDUSTRIAL YEAST, LITE BEER, FOODS

BIOTECHNOLOGY GENERAL CORP. Year: 1980   Empl: 130   Code–1°:P 2°:KV
    SOD, BGH, HGH, RX, SURFACTANTS, FUNGICIDE

BIOTHERAPEUTICS, INC.   Year: 1984   Empl: 120   Code–1°:P 2°:KY
    CANCER RX & DX; BIOLOGICAL RESPONSE MODIFIERS

Secondary Focus (Cont.)

CALIFORNIA BIOTECHNOLOGY Year: 1981 Empl: 155 Code–1°:P 2°:AKV3
   RX; ANF, EPO, SURFACTANT, HGH, RENIN,
   PTH; VET RX, NASAL DRUG DELIVERY

CAMBRIDGE RESEARCH BIOCHEM. Year: 1980 Empl: 75 Code–1°:J 2°:K
   MABS; PEPTIDES, ONCO PROTEIN PRODUCTS;
   DEVELOPING EPITOPE SCANNING PRODUCTS

CARBOHYDRATES INT'L, INC. Year: 1987 Empl: 25 Code–1°:J 2°:K
   INFECTIOUS DISEASE, BLOOD TYPE DX;
   SUGAR STRUCTURES TO RAISE MABS

CEL-SCI CORP. Year: 1983 Empl: 2 Code–1°:P 2°:KQX
   RX; LYMPHOKINES; AIDS VX; IL-1, IL-2

CETUS CORP. Year: 1971 Empl: 711 Code–1°:P 2°:BKAV
   RX; IF; IL-2; MABS; TNF; CSF; INSULIN;
   ANTIBIOTICS; VET VX

CHEMICON INT'L, INC. Year: 1981 Empl: 8 Code–1°:G 2°:K
   MABS; GROWTH FACTORS; IMMUNOLOGICAL
   CELL GROWTH ASSAY KIT

CHIRON CORP. Year: 1981 Empl: 300 Code–1°:P 2°:QKJV
   AIDS DX; DNA PROBE DX; WOUND HEALING;
   HEP B VX; VET VX, SOD; ENZYMES; IF

CISTRON BIOTECHNOLOGY, INC. Year: 1984 Empl: 30 Code–1°:P 2°:XK
   IL-1, IL-2, DX

CLINICAL SCIENCES, INC. Year: 1971 Empl: 50 Code–1°:G 2°:KV
   BIOCHEMICALS FOR DX AND VET USES

CODON CORP. Year: 1980 Empl: 120 Code–1°:V 2°:KP
   VET VX; HUMAN DX

COLLABORATIVE RESEARCH Year: 1961 Empl: 150 Code–1°:G 2°:WK5
   DX SERVICES (DNA PROBE); IF; RENNIN;
   UROKINASE; IL-2; ENZYMES; TPA; REAGENTS

COOPER DEVELOPMENT CO. Year: 1980 Empl: 900 Code–1°:P 2°:K
   ALPHA-1 ANTITRYPSIN, RX, MABS

CYGNUS RESEARCH CORP. Year: 1985 Empl: 18 Code–1°:3 2°:DKP
   DRUG DELIVERY SYSTEM; CONTRACTS
   TO RX COMPANIES

Listing 5-11 Clinical Diagnostics

Secondary Focus (Cont.)

DAKO CORP.   Year: 1979 Empl: 50   Code–1°:X 2°:K
  POLYCLONAL AB'S & MABS,
  CLINICAL TEST KITS, PROTEINS

DIAGNOSTIC PRODUCTS, CORP.   Year: 1972 Empl: 270   Code–1°:G 2°:K
  DX KITS, MABS; REAGENTS

E-Y LABORATORIES, INC.   Year: 1978 Empl: 20   Code–1°:X 2°:KPJ
  ANTISERA; LECTINS; REAGENT; MABS; DX

ENVIRONMENTAL DIAGNOSTICS Year: 1983 Empl: 55   Code–1°:Y 2°:KF
  ENVIRONMENTAL AND TOXIN TESTING
  IN FOODS, ANIMALS & HUMANS

ENZYME TECHNOLOGY CORP.   Year: 1981 Empl: 26   Code–1°:O 2°:MK
  GLUCOAMYLASE, ALPHA-AMYLASE ENZYMES

EPITOPE, INC.   Year: 1979 Empl: 60   Code–1°:G 2°:K
  DX; AIDS DX; VIRAL DETECTION KITS;
  WESTERN BLOT, MAB TO HIV

GAMETRICS, LTD.   Year: 1974 Empl: 20   Code–1°:V 2°:K
  VET DX; VET & HUMAN SEX SELECTION, ISOLATE X/Y SPERM

GENE-TRAK SYSTEMS   Year: 1981 Empl: 196   Code–1°:J 2°:PK
  RX PROTEINS, DNA PROBE-BASED DX'S; DEVELOPING
  CV PROTEINS/HEMOPOIETIC GROWTH FACTOR

GENENTECH   Year: 1976 Empl: 1500   Code–1°:P 2°:HKOQ
  RX; DX; VX; INSULIN, IF, TNF, RENIN,
  FACTOR VIII, PROTROPIN HGH, ACTIVASE TPA

GENETICS INSTITUTE, INC.   Year: 1981 Empl: 302   Code–1°:P 2°:BK
  RX; DX; GM-CSF; EPO; TPA; HSA; AIDS RX;
  IL-3; IF; FACTOR VIII; SEEDS; CORN

GENZYME CORP.   Year: 1981 Empl: 185   Code–1°:P 2°:KJGE
  RX, DX FOR HUMAN HEALTH CARE; MABS;
  EYE PRODUCTS, SKIN PRODUCTS; SURFACTANTS

HYBRITECH, INC.   Year: 1978 Empl: 708   Code–1°:P 2°:K
  RX; DX; MABS; INSULIN; HEMATROPE
  GROWTH HORMONE; CANCER DX

IGEN   Year: 1982 Empl: 60   Code–1°:X 2°:KW
  MABS FOR HUMAN PATHOGENS;
  CONTRACT RESEARCH

Secondary Focus (Cont.)

IMMUNO MODULATORS LABS   Year: 1981 Empl: 25   Code–1°:P 2°:AGKH
 RX; IF; AIDS RX; HCG; COSMETICS

IMMUNOMEDICS, INC.   Year: 1983 Empl: 55   Code–1°:P 2°:KGL
 DX; RX; TEST KITS; CANCER DX; AUTOIMMUNE
 DISEASE ASSAY; ALLERGY PRODUCTS

IMMUNOVISION, INC.   Year: 1985 Empl: 6   Code–1°:P 2°:KV
 AB'S; PURIFIED PROTEINS FOR AUTOIMMUNE
 DISEASE; BLOOD COMPLEMENTS, ANIMAL TISSUE; DX

IMREG, INC.   Year: 1981 Empl: 50   Code–1°:P 2°:KX
 IMMUNOREGULATORS; CANCER DX;
 AIDS, ARTHRITIS RX

INFERGENE CO.   Year: 1984 Empl: 25   Code–1°:J 2°:KM
 DX FOR HERPES, OTHERS; ENZYMES; BAKING
 & BREWING STRAINS, GLUCOAMYLASE STRAIN

INTEGRATED GENETICS, INC.   Year: 1981 Empl: 200   Code–1°:P 2°:KAY
 DX; RX; AIDS DX; FOOD TOXIN DX;
 CANCER DX; TPA; HCG; EPO; HEP B VX

INTER-AMERICAN RES. ASSOC.   Year: 1979 Empl: 10   Code–1°:G 2°:XWK
 PURIFIED ANTIGENS, ANTISERA, BLOOD PRODS

INTER-CELL TECHNOLOGIES, INC. Year: 1982 Empl: 9   Code–1°:X 2°:GKP
 BIOLOGICALS; IMMUNOLOGICAL
 REAGENTS; LYMPHOKINES

INT'L BIOTECHNOLOGIES, INC. Year: 1982 Empl: 85   Code–1°:G 2°:FKW
 NUCLEIC ACIDS, CLONING, SEQUENCING,
 GENOME ANALYSES & ELECTROPHORESIS SYSTEMS

INVITRON CORP.   Year: 1984 Empl: 165   Code–1°:P 2°:K
 DX; RX; TPA, HGH, FACTOR VIII; CELL PRODUCTS

LEE BIOMOLECULAR RES. LABS   Year: 1980 Empl: 16   Code–1°:P 2°:HKQ
 IF; ANTI-IF SERA

LUCKY BIOTECH CORP.   Year: 1984 Empl: 4   Code–1°:P 2°:QJKA
 RX; IF; IF-A; IF-B; IF-G; IL-2; HEP B DX;
 PORCINE GROWTH HORMONE

MARTEK   Year: 1985 Empl: 16   Code–1°:J 2°:KG
 STABILIZATOPE-LABELLED BIOCHEMICALS, C-13 &
 DEUTERIUM-LABELLED LUBRICANTS, AMINO ACIDS

Listing 5-11 Clinical Diagnostics

Secondary Focus (Cont.)

MEDAREX, INC.   Year: 1987 Empl: 3   Code–1°:P 2°:K
   MAB RX & DX

MICROGENESYS, INC.   Year: 1983 Empl: 32   Code–1°:Q 2°:KB
   AIDS DX; AIDS VX; VIRAL INSECTICIDE;
   DEVELOPING HUMAN INFECTIOUS DISEASE VX

MOLECULAR BIOSYSTEMS, INC.   Year: 1981 Empl: 55   Code–1°:K 2°:JKG
   DX; DNA & RNA PROBES, ALBUMEX, EXTRACTOR KITS

MOLECULAR GENETIC RESOURCES Year: 1983 Empl: 5   Code–1°:G 2°:K
   ENZYMES; REVERSE TRANSCRIPTASE; TERMINAL
   TRANSFERASE; MRNA ANALYSIS KIT

MOLECULAR GENETICS, INC.   Year: 1979 Empl: 63   Code–1°:A 2°:BKV
   DX; AG DX; BGH; VET DX; SCOURS VX

MONOCLONETICS INT'L, INC.   Year: 1984 Empl: 5   Code–1°:G 2°:HKP
   HUMAN SCREENING TESTS & DX, EPITOPE SCREENS

NEORX CORP.   Year: 1984 Empl: 195   Code–1°:P 2°:K
   MELANOMA IMAGING AGENT WITH TECHNETIUM

NORTH COAST BIOTECH., INC.   Year: 1986 Empl: 2   Code–1°:J 2°:PEFK
   ENZYMES FOR RX MANUFACTURING

PAMBEC LABORATORIES   Year: 1987 Empl: 6   Code–1°:J 2°:KP
   CUSTOM/CONTRACT WORK ON GENETIC ENGINEERING
   PROKARYOTES, EUKARYOTES, DNA PROBES, AIDS RX

PHARMAGENE   Year: 1986 Empl: 3   Code–1°:P 2°:HIJK
   RX; CHEMICALS; SPECIFIC SITE
   RECOMBINATION PROCESSES

PRAXIS BIOLOGICS   Year: 1983 Empl: 180   Code–1°:Q 2°:XPK
   VX; TEST KITS; RX; DX

RHOMED, INC.   Year: 1986 Empl: 8   Code–1°:3 2°:PK
   DX, AB DELIVERY SYSTEM FOR CANCER
   DX & RX; IMMUNOASSAYS

SCRIPPS LABORATORIES   Year: 1980 Empl: 60   Code–1°:P 2°:JGK
   BULK HUMAN PROTEINS; HORMONES; ENZYMES; DX

SERONO LABS   Year: 1971 Empl: 198   Code–1°:P 2°:KX
   RX AND DX; FERTILITY RX, GROWTH RX,
   HGH, IMMUNOLOGICAL RX, VIROLOGICAL RX

Secondary Focus (Cont.)

SUMMA MEDICAL CORP.   Year: 1978 Empl: 33   Code–1°:G 2°:KX
   HISTOCHEMICAL MARKERS AND
   DX KITS, CANCER DX

SYNBIOTICS CORP.   Year: 1982 Empl: 89   Code–1°:V 2°:KQ
   VET DX; HUMAN DX; VX

SYVA CO.   Year: 1966 Empl: 1000 Code–1°:K 2°:K
   DX PRODUCTS; THERAPEUTIC DRUG
   MONITORING; DRUG ABUSE/TOXINS

T CELL SCIENCES, INC.   Year: 1984 Empl: 65   Code–1°:P 2°:KG
   RX; MABS; CANCER DX; T CELL
   PRODUCTS FOR ARTHRITIS; IL-2

TRITON BIOSCIENCES, INC.   Year: 1983 Empl: 180   Code–1°:P 2°:KGW
   RX, IF-B, DX RESEARCH REAGENTS,
   MABS, RESEARCH

VIROSTAT, INC.   Year: 1985 Empl: 1   Code–1°:J 2°:K
   IMMUNOCHEMICALS FOR INFECTIOUS
   DISEASE RESEARCH

VIVIGEN, INC.   Year: 1981 Empl: 70   Code–1°:5 2°:K
   GENETIC TESTING LAB, AMNIOCENTESIS
   CHROMOSOME ANALYSIS

ZYMED LABORATORIES, INC.   Year: 1980 Empl: 20   Code–1°:G 2°:KJ
   IMMUNOCHEMICALS, MABS,
   PROTEINS, ENZYMES

Listing 5-12 Energy

LISTING 5-12 COMPANIES IN ENERGY (L)

Primary Focus

PETROFERM, INC.   Year: 1977 Empl: 68   Code–1°:L 2°:J
   MICROBIALS FOR IMPROVED FUELS

PETROGEN, INC.   Year: 1980 Empl: 6   Code–1°:L
   MICROBES FOR IMPROVED OIL RECOVERY

Secondary Focus

AMERICAN BIOGENETICS CORP.   Year: 1984 Empl: 10   Code–1°:I 2°:JBLR
   PROCESS DESIGN MICROORGANISMS; CHEMICALS

IMMUNOMEDICS, INC.   Year: 1983 Empl: 55   Code–1°:P 2°:KGL
   DX; RX; TEST KITS; CANCER DX; AUTOIMMUNE
   DISEASE ASSAY; ALLERGY PRODUCTS

PROVESTA CORP.   Year: 1975 Empl: 50   Code–1°:J 2°:LBMP
   FERMENTATION-BASED FLAVORS; ENZYMES;
   YEAST FEEDS; PHEROMONES FOR PEST CONTROL; TPA

LISTING 5-13 COMPANIES IN FOOD PRODUCTION/PROCESSING (M)

Primary Focus

BIO-TECHNICAL RESOURCES, INC. Year: 1962 Empl: 25 Code–1°:M 2°:IJH
 YEAST/BACTERIAL PRODS IN FOOD,
 BREWING, CHEMICALS

BIOTICS RESEARCH CORP. Year: 1975 Empl: 22 Code–1°:M
 NUTRITIONAL SUPPLEMENTS

COORS BIOTECH PRODUCTS CO. Year: 1984 Empl: 185 Code–1°:M 2°:OJ
 INDUSTRIAL DEGREASER; FOOD
 PRODUCTS; SPECIALTY CHEMICALS

CYANOTECH CORP. Year: 1983 Empl: 25 Code–1°:M 2°:BHJ
 PHYCOBILIPROTEIN; SPIRULINA; BETA-CAROTENE

IGENE BIOTECHNOLOGY, INC. Year: 1981 Empl: 30 Code–1°:M 2°:OS
 MILK/EGG WHITE REPLACER, CALCIUM & MILK
 MINERALS FOOD SUPPLEMENT, FISH AQUACULTURE

Secondary Focus

AMERICAN LABORATORIES, INC. Year: 1962 Empl: 26 Code–1°:P 2°:KMO
 RAW MATERIALS OF BIOTECHNOLOGY PRODUCTS

APPLIED DNA SYSTEMS, INC. Year: 1982 Empl: 7 Code–1°:P 2°:MR
 HUMAN COLLAGENASE; MABS; OXYRASE;
 IN VITRO CHEMOSENSITIVITY ASSAY

BIOTECHNICA INTERNATIONAL Year: 1981 Empl: 145 Code–1°:J 2°:KQMB
 INOCULANT, VX, DENTAL DX; ENZYMES,
 INDUSTRIAL YEAST, LITE BEER, FOODS

CALGENE, INC. Year: 1980 Empl: 165 Code–1°:B 2°:MJ
 TOMATOES, COTTON, CORN, SUNFLOWER,
 RAPESEED, FOODS; PESTICIDES, PLANTS, TREES

CONSOLIDATED BIOTECHNOLOGY Year: 1983 Empl: 15 Code–1°:G 2°:M
 ENZYMES, FOOD-RELATED RESEARCH

DNA PLANT TECHNOLOGY CORP. Year: 1981 Empl: 80 Code–1°:B 2°:M
 FOODS AND PLANTS, AG PRODUCTS, TOMATOES,
 RICE, COFFEE, SNACKS, OILS, CORN

Listing 5-13 Food Production/Processing

Secondary Focus (Cont.)

ENZYME BIO-SYSTEMS, LTD.   Year: 1983 Empl: 101   Code–1°:J 2°:M
   INDUSTRIAL ENZYME PRODUCTS USED
   FOR STARCH, BREWING & BAKING.

ENZYME TECHNOLOGY CORP.   Year: 1981 Empl: 26   Code–1°:O 2°:MK
   GLUCOAMYLASE ENZYME; ALPHA-AMYLASE ENZYME

ESCAGENETICS CORP.   Year: 1986 Empl: 50   Code–1°:B 2°:M
   TISSUE CULTURE FLAVORS; POTATO SEEDS;
   TOMATOES; MICROPROPAGATED DATE PALMS

GENENCOR, INC.   Year: 1982 Empl: 130   Code–1°:J 2°:M
   ENZYMES IN FOOD/BEVERAGE PROCESSING

GRANADA GENETICS CORP.   Year: 1979 Empl: 3   Code–1°:A 2°:M
   BULL GENETICS, SHRIMP, FOOD

IDETEK, INC.   Year: 1983 Empl: 14   Code–1°:Y 2°:MAB
   DX & TOX TEST KITS FOR FOOD & AGRICULTURE

INFERGENE CO.   Year: 1984 Empl: 25   Code–1°:J 2°:KM
   DX FOR HERPES, OTHERS; ENZYMES; BAKING
   & BREWING STRAINS, GLUCOAMYLASE STRAIN

INT'L PLANT RESEARCH INST.   Year: 1978 Empl: 50   Code–1°:B 2°:MFG
   DATE PALMS; OIL PALMS; CEREALS;
   FOOD PLANTS; GENETIC ENGINEERING REAGENTS

LABSYSTEMS, INC.   Year: 1983 Empl: 10   Code–1°:K 2°:HJMF
   MICROBIOLOGY ASSAYS & EQUIPMENT; MAB'S

MICROBIO RESOURCES   Year: 1981 Empl: 30   Code–1°:J 2°:MRPO
   FOOD ADDITIVES & COLORANTS; SPECIAL CHEMICALS

NATIVE PLANTS, INC.   Year: 1973 Empl: 625   Code–1°:B 2°:HJM
   POTATO, FLOWER, VEGETABLE SEEDS; SOIL INOCULANTS; PLANTS

NEUSHUL MARICULTURE   Year: 1978 Empl: 6   Code–1°:S 2°:PMHC
   ANTI-VIRAL; ENZYMES; MOLLUSCAN
   SEED & CULTURE

PLANT GENETICS, INC.   Year: 1981 Empl: 63   Code–1°:B 2°:M
   POTATOES; HYBRID VEGS; ALFALFA; CORN; SEEDS

PROVESTA CORP.   Year: 1975 Empl: 50   Code–1°:J 2°:LBMP
   FERMENTATION-BASED FLAVORS; ENZYMES;
   YEAST FEEDS; PHEROMONES FOR PEST CONTROL; TPA

LISTING 5-14 COMPANIES IN MINING (N)

Primary Focus
COAL BIOTECH CORP.   Year: 1984 Empl: 10   Code–1°:N 2°:E
  MICROBIAL COAL DESULFURIZATION

Secondary Focus
ADVANCED MINERAL TECHNOL.   Year: 1982 Empl: 17   Code–1°:R 2°:NE
  WASTE TREATMENT & METAL RECOVERY

Listing 5-15   Production/Fermentation

LISTING 5-15 COMPANIES IN PRODUCTION/FERMENTATION (O)

Primary Focus

BIOPURE   Year: 1984 Empl: 31   Code–1°:O 2°:JPX
   CUSTOM FERMENTATION OF RX, MAB, CHEMICALS

ENZYME CENTER, INC.   Year: 1978 Empl: 30   Code–1°:O 2°:E
   LARGE SCALE FERMENTATION; ENZYME
   SEPARATION & PURIFICATION

ENZYME TECHNOLOGY CORP.   Year: 1981 Empl: 26   Code–1°:O 2°:MK
   GLUCOAMYLASE, ALPHA-AMYLASE ENZYMES

VERAX CORP.   Year: 1978 Empl: 80   Code–1°:O
   LARGE SCALE CELL CULTURE PRODUCTION,
   MAMMALIAN CELL GROWTH

Secondary Focus

AMERICAN LABORATORIES, INC.   Year: 1962 Empl: 26   Code–1°:P 2°:KMO
   RAW MATERIALS OF BIOTECHNOLOGY PRODUCTS

BENTECH LABORATORIES   Year: 1984 Empl: 7   Code–1°:B 2°:JO
   CROP YIELD ENHANCING AGENT;
   ANTIFUNGAL; WOUND HEALING

COORS BIOTECH PRODUCTS CO.   Year: 1984 Empl: 185   Code–1°:M 2°:OJ
   INDUSTRIAL DEGREASER; FOOD
   PRODUCTS; SPECIALTY CHEMICALS

GENENTECH   Year: 1976 Empl: 1500 Code–1°:P 2°:HKOQ
   RX; DX; VX; INSULIN, IF, TNF, RENIN,
   FACTOR VIII, PROTROPIN HGH, ACTIVASE TPA

GENEX, INC.   Year: 1977 Empl: 80   Code–1°:I 2°:JO
   COMMODITY & SPECIALTY CHEMICALS;
   ENZYMES, PROTEINS, DRAIN CLEANER; VITAMIN B-12

HYCLONE LABORATORIES, INC.   Year: 1975 Empl: 107   Code–1°:H 2°:EO
   IMMUNOCHEMICALS; ANIMAL SERA; HYDBRIDOMAS

IGENE BIOTECHNOLOGY, INC.   Year: 1981 Empl: 30   Code–1°:M 2°:OS
   MILK/EGG WHITE REPLACER, CALCIUM &
   MILK MINERALS FOOD SUPPLEMENT, FISH AQUACULTURE

MICROBIO RESOURCES   Year: 1981 Empl: 30   Code–1°:J 2°:MRPO
   FOOD ADDITIVES & COLORANTS; SPECIAL CHEMICALS

Secondary Focus (Cont.)

MOLECULAR BIOLOGY RESOURCES Year: 1986 Empl: 10   Code–1°:G 2°:JOF
   REAGENTS; DNA & RNA MODIFYING ENZYMES
   & NUCLEIC ACIDS; HYBRIDIZATION APPARATUS

MYCOSEARCH, INC.   Year: 1979 Empl: 6   Code–1°:2 2°:OCHR
   CULTURES OF RARE & NEW FUNGI; LIVE
   CULTURES; SPECIALTY CHEMICALS

R & A PLANT/SOIL, INC.   Year: 1978 Empl: 6   Code–1°:B 2°:TO
   ALGAL SOIL CONDITIONERS, FERTILIZERS, ETC.

RICERCA, INC.   Year: 1986 Empl: 160   Code–1°:Y 2°:BAO
   FERMENTATION AND RESEARCH

Listing 5-16 Therapeutics

LISTING 5-16 COMPANIES IN THERAPEUTICS (P)

Primary Focus

ALPHA I BIOMEDICALS, INC.   Year: 1982 Empl: 11   Code–1°:P 2°:QKV
  PEPTIDES & VX FOR DX/RX; AIDS RX, VX;
  DEVELOPING THYMOSIN

AMERICAN LABORATORIES, INC.   Year: 1962 Empl: 26   Code–1°:P 2°:KMO
  RAW MATERIALS OF BIOTECHNOLOGY PRODUCTS

AMGEN   Year: 1980 Empl: 340   Code–1°:P 2°:AJKG
  RESEARCH REAGENTS, EGF, IGF, IF,
  EPO, IL-2, CSF, BGH, HGH

APPLIED DNA SYSTEMS, INC.   Year: 1982 Empl: 7   Code–1°:P 2°:MR
  HUMAN COLLAGENASE; MABS; OXYRASE;
  IN VITRO CHEMOSENSITIVITY ASSAY

AUTOMEDIX SCIENCES, INC.   Year: 1978 Empl: 7   Code–1°:P
  RX PRODUCTS IN DEVELOPMENT

BIO-RESPONSE, INC.   Year: 1972 Empl: 45   Code–1°:P 2°:HYK
  NON-RDNA TPA; MASS CELL CULTURE; DX KITS; MABS

BIOGEN   Year: 1979 Empl: 250   Code–1°:P 2°:QGJ
  RX; DX; TNF, IL-2, CSF, TPA, HSA,
  FACTOR VIII, IF-G, HEP B ANTIGEN

BIOGROWTH, INC.   Year: 1985 Empl: 16   Code–1°:P 2°:FE
  WOUND/BONE RX; SEPARATION; PROTEIN
  CHARACTERIZATION; HPLC/FPLC EQUIPMENT

BIOTECHNOLOGY GENERAL CORP. Year: 1980 Empl: 130   Code–1°:P 2°:KV
  SOD, BGH, HGH, RX, SURFACTANTS, FUNGICIDE

BIOTHERAPEUTICS, INC.   Year: 1984 Empl: 120   Code–1°:P 2°:KY
  CANCER RX & DX; BIOLOGICAL RESPONSE MODIFIERS

BIOTHERAPY SYSTEMS, INC.   Year: 1984 Empl: 8   Code–1°:P
  MABS FOR CANCER THERAPY

CALIFORNIA BIOTECHNOLOGY   Year: 1981 Empl: 155   Code–1°:P 2°:AKV3
  RX; ANF, EPO, SURFACTANT, HGH, RENIN,
  PTH; VET RX, NASAL DRUG DELIVERY

CEL-SCI CORP.   Year: 1983 Empl: 2   Code–1°:P 2°:KQX
  RX; LYMPHOKINES; AIDS VX; IL-1, IL-2

Primary Focus (Cont.)

CETUS CORP.   Year: 1971 Empl: 711   Code–1°:P 2°:BKAV
   RX; IF; IL-2; MABS; TNF; CSF; INSULIN;
   ANTIBIOTICS; VET VX

CHIRON CORP.   Year: 1981 Empl: 300   Code–1°:P 2°:QKJV
   AIDS DX; DNA PROBE DX; WOUND HEALING;
   HEP B VX; VET VX, SOD; ENZYMES; IF

CISTRON BIOTECHNOLOGY, INC.   Year: 1982 Empl: 30   Code–1°:P 2°:XK
   IL-1, IL-2, DX

COOPER DEVELOPMENT CO.   Year: 1980 Empl: 900   Code–1°:P 2°:K
   ALPHA-1 ANTITRYPSIN, RX, MABS

DAMON BIOTECH, INC.   Year: 1978 Empl: 101   Code–1°:P 2°:X3
   MABS, DRUG DELIV., CELL ENCAPSULATION
   SYSTEM; AIDS DX; TPA

DNAX RESEARCH INSTITUTE   Year: 1980 Empl: 105   Code–1°:P 2°:XWA
   RESEARCH; MABS; RX

ELANEX PHARMACEUTICALS, INC. Year: 1984 Empl: 10   Code–1°:P 2°:
   EPO, BLOOD-RELATED PRODUCTS

ENZON   Year: 1965 Empl: 22   Code–1°:P 2°:V
   RX; SOD FOR HUMAN AND VET USE

ENZON, INC.   Year: 1981 Empl: 31   Code–1°:P
   HUMAN RX; ENZYMES

EXOVIR, INC.   Year: 1981 Empl: 5   Code–1°:P
   RX, TOPICAL FOR HSV-II, OTHER SKIN RX

GENENTECH   Year: 1976 Empl: 1500 Code–1°:P 2°:HKOQ
   RX; DX; VX; INSULIN, IF, TNF, RENIN,
   FACTOR VIII, PROTROPIN HGH, ACTIVASE TPA

GENETIC THERAPY, INC.   Year: 1986 Empl: 4   Code–1°:P 2°:V3
   ANIMAL/HUMAN GENE THERAPY; RX; RETROVIRAL
   VECTORS;TISSUE TARGETED DELIVERY SYSTEMS

GENETICS INSTITUTE, INC.   Year: 1981 Empl: 302   Code–1°:P 2°:BK
   RX; DX; GM-CSF; EPO; TPA; HSA; AIDS RX;
   IL-3; IF; FACTOR VIII; SEEDS; CORN

Listing 5-16 Therapeutics

Primary Focus (Cont.)

GENZYME CORP.   Year: 1981 Empl: 185   Code–1°:P 2°:KJGE
    RX, DX FOR HUMAN HEALTH CARE; MABS;
    EYE PRODUCTS, SKIN PRODUCTS; SURFACTANTS

HANA BIOLOGICS, INC.   Year: 1979 Empl: 130   Code–1°:P 2°:H
    CELL IMPLANTS TO TREAT DISEASE

HOUSTON BIOTECHNOLOGY, INC.   Year: 1984 Empl: 30   Code–1°:P
    OPHTHAMOLOGY AND NEUROLOGY PRODUCTS

HYBRITECH, INC.   Year: 1978 Empl: 708   Code–1°:P 2°:K
    RX; DX; MABS; INSULIN; HEMATROPE
    GROWTH HORMONE; CANCER DX

IDEC PHARMACEUTICAL   Year: 1986 Empl: 45   Code–1°:P
    MAB (ANTI-IDIOTYPE)

IMMUNEX CORP.   Year: 1981 Empl: 153   Code–1°:P 2°:AG
    RX; IL-2; GM-CSF; IL-1; IL-4; LYMPHOKINES; MAF

IMMUNO MODULATORS LABS   Year: 1981 Empl: 25   Code–1°:P 2°:AGKH
    RX; IF; AIDS RX; HCG; COSMETICS

IMMUNOMEDICS, INC.   Year: 1983 Empl: 55   Code–1°:P 2°:KGL
    DX; RX; TEST KITS; CANCER DX; AUTOIMMUNE
    DISEASE ASSAY; ALLERGY PRODUCTS

IMMUNOVISION, INC.   Year: 1985 Empl: 6   Code–1°:P 2°:KV
    AB'S; PURIFIED PROTEINS FOR AUTOIMMUNE
    DISEASE; BLOOD COMPLEMENTS; DX

IMREG, INC.   Year: 1981 Empl: 50   Code–1°:P 2°:KX
    IMMUNOREGULATORS; CANCER DX; AIDS, ARTHRITIS RX

INGENE, INC.   Year: 1980 Empl: 65   Code–1°:P 2°:JB
    RX; BONE & CARTILAGE RX; CANCER
    RX; SWEETENERS; PESTICIDE

INTEGRATED GENETICS, INC.   Year: 1981 Empl: 200   Code–1°:P 2°:KAY
    DX; RX; AIDS DX; FOOD TOXIN DX; CANCER DX;
    TPA; HCG; EPO; HEP B VX

INTERFERON SCIENCES, INC.   Year: 1981 Empl: 70   Code–1°:P 2°:GHB
    IF-A; IF-G; MABS

INVITRON CORP.   Year: 1984 Empl: 165   Code–1°:P 2°:K
    DX; RX; TPA, HGH, FACTOR VIII; CELL PRODUCTS

Primary Focus (Cont.)

KIRIN-AMGEN   Year: 1984 Empl: 5   Code–1°:P
   EPO

LEE BIOMOLECULAR RES. LABS   Year: 1980 Empl: 16   Code–1°:P 2°:HKQ
   IF; ANTI-IF SERA

LUCKY BIOTECH CORP.   Year: 1984 Empl: 4   Code–1°:P 2°:QJKA
   RX; IF; IF-A; IF-B; IF-G; IL-2; HEP B DX;
   PORCINE GROWTH HORMONE

LYPHOMED, INC.   Year: 1981 Empl: 1200   Code–1°:P
   RX, VITAMIN SUPPLEMENTS FOR
   INTRAVENOUS USE; ANTIBIOTICS

MEDAREX, INC.   Year: 1987 Empl: 3   Code–1°:P 2°:K
   MAB RX & DX

NATIONAL GENO SCIENCES   Year: 1980 Empl: 4   Code–1°:P
   LEUCOCYTE IF

NEORX CORP.   Year: 1984 Empl: 194   Code–1°:P 2°:K
   MELANOMA IMAGING AGENT WITH TECHNETIUM

NEUREX CORP.   Year: 1986 Empl: 22   Code–1°:P
   RX, NEUROPEPTIDES, NEUROHORMONES

PHARMAGENE   Year: 1986 Empl: 3   Code–1°:P 2°:HIJK
   RX; CHEMICALS; SPECIFIC SITE
   RECOMBINATION PROCESSES

PROBIOLOGICS INT'L, INC.   Year: 1987 Empl: 5   Code–1°:P 2°:AVW
   RX, PROBIOTICS, ANTIBIOTICS FOR
   HUMAN & ANIMAL USE

SCRIPPS LABORATORIES   Year: 1980 Empl: 60   Code–1°:P 2°:JGK
   BULK HUMAN PROTEINS; HORMONES; ENZYMES; DX

SERONO LABS   Year: 1971 Empl: 198   Code–1°:P 2°:KX
   RX AND DX; FERTILITY RX, GROWTH RX,
   HGH, IMMUNOLOGICAL RX, VIROLOGICAL RX

SPHINX BIOTECHNOLOGIES   Year: 1987 Empl: 3   Code–1°:P 2°:W
   CELL REGULATION THERAPEUTICS

SYNERGEN, INC.   Year: 1981 Empl: 101   Code–1°:P 2°:VQW
   RX; RESEARCH LUNG DISEASE; FGF; VET VX

Listing 5-16 Therapeutics

Primary Focus (Cont.)

SYNGENE PRODUCTS   Year: 1980 Empl: 20   Code–1°:P 2°:VA
   PRODUCTS FOR ANIMALS AND HUMAN HEALTH

T CELL SCIENCES, INC.   Year: 1984 Empl: 65   Code–1°:P 2°:KG
   RX; MABS; CANCER DX; T CELL PRODUCTS
   FOR ARTHRITIS; IL-2

TRITON BIOSCIENCES, INC.   Year: 1983 Empl: 180   Code–1°:P 2°:KGW
   RX, IF-B, DX RESEARCH REAGENTS, MAB, RESEARCH

UNIGENE LABORATORIES, INC.   Year: 1980 Empl: 30   Code–1°:P 2°:JA
   PEPTIDES, HORMONES, ENZYMES,
   CALCITONIN, GROWTH FACTORS

VIRAGEN   Year: 1980 Empl: 15   Code–1°:P
   IF, PEPTIDES, RX

WELGEN MANUFACTURING, INC.   Year: 1988 Empl: 200   Code–1°:P
   RX, TPA, IF-A

XOMA CORP.   Year: 1981 Empl: 149   Code–1°:P 2°:X
   MAB BASED ANTICANCER RX

ZYMOGENETICS, INC.   Year: 1981 Empl: 95   Code–1°:P 2°:W
   RX, WOUND HEALING, COAGULATION,
   PROTEINS, ANTI-INFECTIVES

Secondary Focus

ALFACELL CORP.   Year: 1981 Empl: 22   Code–1°:G 2°:PX
   TUMOR TOXINS, REAGENTS, SERA

AMERICAN BIOCLINICAL   Year: 1977 Empl: 24   Code–1°:G 2°:KP
   REAGENTS; HORMONE DX KITS; RX

AMERICAN BIOTECHNOLOGY CO.   Year: 1984 Empl: 5   Code–1°:X 2°:PGHJ
   GROWTH FACTORS, IMMUNOMODULATORS

ANTIVIRALS, INC.   Year: 1980 Empl: 5   Code–1°:K 2°:PG
   NEU-GENES (TM) FOR DX & RX USE

APPLIED BIOTECHNOLOGY   Year: 1982 Empl: 45   Code–1°:V 2°:QKP
   VET VX; AIDS, TB VX; CANCER DX, RX

BACHEM BIOSCIENCE, INC.   Year: 1987 Empl: 15   Code–1°:J 2°:PQ
   AMINO ACIDS & DERIVATIVES; BIOACTIVE
   PEPTIDES; ENZYME SUBSTRATES; VX EPITOPES

Secondary Focus (Cont.)

BERKELEY ANTIBODY CO., INC.   Year: 1983 Empl: 27   Code–1°:X 2°:GHP
  CUSTOM ANTIBODIES, MAB & POLYCLONALS

BIODESIGN, INC.   Year: 1987 Empl: 3   Code–1°:G 2°:PY
  IMMUNOLOGICAL REAGENTS; BIODESIGN ANTISERA,
  PURIFIED BIOLOG. COMPDS; HAZARDOUS CHEMICALS

BIOGENEX LABORATORIES   Year: 1981 Empl: 42   Code–1°:K 2°:PG
  RIA DX TEST; IMMUNOHISTOCHEMICALS;
  KITS; REAGENTS

BIOKYOWA, INC.   Year: 1983 Empl: 70   Code–1°:A 2°:P
  L-LYSINE FEED ADDITIVE FOR POULTRY/
  SWINE, DEVELOPING RX

BIONETICS RESEARCH   Year: 1985 Empl: 200   Code–1°:K 2°:GP
  AIDS DX, SALMONELLA EIA, RADIOLABELLED
  COMPOUNDS, LIPOPROTEINS, MELATONIN

BIOPURE   Year: 1984 Empl: 31   Code–1°:O 2°:JPX
  CUSTOM FERMENTATION OF RX, MAB, CHEMICALS

BIOTECH RESEARCH LABS, INC.   Year: 1973 Empl: 167   Code–1°:K 2°:PQ
  MABS FOR DX, AIDS DX, CANCER DX; AIDS VX, IL-2

BIOTECHNOLOGY DEVEL. CORP.   Year: 1982 Empl: 35   Code–1°:F 2°:PJQ3
  AIDS VX; EQUIPMENT; REAGENTS;
  DRUG DELIVERY SYSTEMS

BIOTX   Year: 1985 Empl: 20   Code–1°:K 2°:P
  DX CANCER NUCLEAR ANTIGEN DX KIT, ABS
  TO ONCOGENES & GROWTH FACTORS

BOEHRINGER MANNHEIM DIAGN. Year: 1975 Empl: 1000 Code–1°:K 2°:PGBH
  DX; RX; REAGENTS; DIGOXIN; ANIMAL
  CELL TISSUE CULTURE, PLANT DX

CAMBRIDGE BIOSCIENCE CORP.   Year: 1981 Empl: 97   Code–1°:K 2°:QPV
  DX KITS FOR AIDS, ADENOVIRUS,
  ROTAVIRUS; VET VX, DX

CELLULAR PRODUCTS   Year: 1982 Empl: 101   Code–1°:K 2°:P
  MABS, DIPSTIK (TM) DX TESTS KITS; AIDS DX; IF

CENTOCOR   Year: 1979 Empl: 246   Code–1°:K 2°:PJGQ
  MAB RX; CANCER, HEP B DX; MABS; AIDS VX

Listing 5-16 Therapeutics

Secondary Focus (Cont.)

CODON CORP.   Year: 1980 Empl: 120   Code–1°:V 2°:KP
    VET VX; HUMAN DX

COLLAGEN CORP.   Year: 1975 Empl: 225   Code–1°:4 2°:HP1
    MEDICAL DEVICES; COLLAGEN IMPLANTS;
    TISSUE CULTURES

CREATIVE BIOMOLECULES   Year: 1981 Empl: 70   Code–1°:J 2°:PGW
    PEPTIDES, CONTRACT R&D, TPA, EGF

CYGNUS RESEARCH CORP.   Year: 1985 Empl: 18   Code–1°:3 2°:DKP
    DRUG DELIVERY SYSTEM; CONTRACTS
    TO DRUG COMPANIES

CYTOGEN CORP.   Year: 1980 Empl: 140   Code–1°:K 2°:PX
    RX; MABS; DX; CANCER RX

E-Y LABORATORIES, INC.   Year: 1978 Empl: 20   Code–1°:X 2°:KPJ
    ANTISERA; LECTINS; REAGENT; MABS; DX

ELCATECH   Year: 1984 Empl: 4   Code–1°:W 2°:PQXG
    RESEARCH; COAGULATION ASSAY TECHNOLOGY

ENDOTRONICS, INC.   Year: 1981 Empl: 45   Code–1°:F 2°:HPQ
    RX; DX; MABS; APS SERIES; ACUSYST-P;
    VX; BIOLOGICALS; HEP B VX

ENZO-BIOCHEM, INC.   Year: 1976 Empl: 110   Code–1°:K 2°:GP
    ENZYMES; LECTINS; FILTERS; NUCLEOTIDES;
    NON-RADIOACTIVE DNA PROBES; MAB; DX

GENE-TRAK SYSTEMS   Year: 1981 Empl: 196   Code–1°:J 2°:PK
    RX PROTEINS, DNA PROBE-BASED DX'S; DEVELOPING
    CV PROTEINS/HEMOPOIETIC GROWTH FACTOR

GENELABS, INC.   Year: 1984 Empl: 90   Code–1°:K 2°:P
    AIDS RX, CYTOKINES, NON-A, NON-B
    HEPATITIS, SOLID PHASE BLOOD TESTING

GENETIC SYSTEMS CORP.   Year: 1980 Empl: 400   Code–1°:K 2°:P
    MAB BASED HUMAN DX AND RX; AIDS DX;
    HERPES DX; LEGIONNAIRES DX

IMCERA BIOPRODUCTS, INC.   Year: 1986 Empl: 5   Code–1°:H 2°:P
    IGF, RDNA MURINE EGF FOR LARGE SCALE
    CELL CULTURE; RDNA & EXTRACTED ADHESION PEPTIDES

Secondary Focus (Cont.)

IMCLONE SYSTEMS, INC.   Year: 1984 Empl: 50   Code–1°:K 2°:PQ
   DX; VX; RX; AIDS DX; HEP B DX; DNA PROBES,
   IMMUNODIAGNOSTICS FOR VIRAL DISEASES

IMMUNETECH PHARMACEUT.   Year: 1981 Empl: 45   Code–1°:X 2°:P
   IMMUNE ACTIVE PEPTIDES AS RX

IMMUNEX, INC.   Year: 1981 Empl: 5   Code–1°:K 2°:PG
   DX; MABS, POLYCLONAL AB'S TO HAPTENS;
   RADIOIMMUNOASSAY KITS; REAGENTS

IMMUNOGEN, INC.   Year: 1981 Empl: 11   Code–1°:W 2°:P
   CONJUGATED ABS FOR TUMORS & DX

IMMUNOMED CORP.   Year: 1979 Empl: 17   Code–1°:Q 2°:HP
   VX; IMMUNOTHERAPEUTICS; RABIES VX

IMRE CORP.   Year: 1981 Empl: 25   Code–1°:E 2°:P
   IMMUNO BLOOD PURIFIER SYSTEMS AND PROTEIN A

INTER-CELL TECHNOLOGIES, INC.   Year: 1982 Empl: 9   Code–1°:X 2°:GKP
   BIOLOGICALS; IMMUNOLOGICAL REAGENTS; LYMPHOKINES

LIFE SCIENCES   Year: 1962 Empl: 20   Code–1°:W 2°:XP
   IF; RESEARCH ON VIRUSES, ANIMAL STRAINS

LIFE TECHNOLOGIES, INC.   Year: 1983 Empl: 1300   Code–1°:K 2°:GP
   IF; PROBE ASSAY DX; REAGENTS

LIPOSOME COMPANY, INC.   Year: 1981 Empl: 67   Code–1°:3 2°:PQJ
   LIPOSOMES; DRUG DELIVERY; CANCER RX; VX

LIPOSOME TECHNOLOGY, INC.   Year: 1981 Empl: 65   Code–1°:3 2°:VBPJ
   LIPOSOMES FOR RX, DRUG DELIVERY,
   VETERINARY, AGRICULTURE

MELOY LABORATORIES, INC.   Year: 1970 Empl: 246   Code–1°:W 2°:EP
   SERVICES; REAGENTS; RESEARCH
   & DEVELOPMENT; IF

MICROBIO RESOURCES   Year: 1981 Empl: 30   Code–1°:J 2°:MRPO
   FOOD ADDITIVES & COLORANTS; SPECIAL CHEMICALS

MOLECULAR DIAGNOSTICS, INC.   Year: 1981 Empl: 30   Code–1°:K 2°:P
   RX FOR DIABETES, CANCER, INFECTIOUS
   DISEASES, HUMAN GENETIC DISEASES

Listing 5-16 Therapeutics

Secondary Focus (Cont.)

MONOCLONETICS INT'L, INC.   Year: 1984   Empl: 5   Code–1°:G   2°:HKP
    HUMAN SCREENING TESTS & DX, EPITOPE SCREENS

MUREX CORP.   Year: 1984   Empl: 180   Code–1°:K   2°:PGQ
    MAB REAGENTS & ANTIBODY KITS; DX

NEUSHUL MARICULTURE   Year: 1978   Empl: 6   Code–1°:S   2°:PMHC
    ANTI-VIRALS; ENZYMES; MOLLUSCAN SEED & CULTURE

NORTH COAST BIOTECHNOLOGY   Year: 1986   Empl: 2   Code–1°:J   2°:PEFK
    ENZYMES FOR RX MANUFACTURING

OCEAN GENETICS   Year: 1981   Empl: 35   Code–1°:T   2°:JP
    SPECIAL CHEMICALS & RX OF MARINE ALGAE

ONCOGENE SCIENCE, INC.   Year: 1983   Empl: 65   Code–1°:K   2°:PX
    ONCOGENE PROBES FOR DX AND RX, MABS

PAMBEC LABORATORIES   Year: 1987   Empl: 6   Code–1°:J   2°:KP
    CUSTOM/CONTRACT WORK ON GENETIC ENGINEERING
    PROKARYOTES, EUKARYOTES, DNA PROBES, AIDS RX

PRAXIS BIOLOGICS   Year: 1983   Empl: 180   Code–1°:Q   2°:XPK
    VX; TEST KITS; RX; DX

PROGENX   Year: 1987   Empl: 20   Code–1°:K   2°:P
    CANCER DX MARKERS; CANCER RX

PROTATEK INTERNATIONAL, INC.   Year: 1984   Empl: 37   Code–1°:K   2°:YJPV
    VIRUS, DX TEST KITS; RX; SPECIALIZED
    MICROBES; VET VX, RX, DX

PROVESTA CORP.   Year: 1975   Empl: 50   Code–1°:J   2°:LBMP
    FERMENTATION-BASED FLAVORS; ENZYMES;
    YEAST FEEDS; PHEROMONES FOR PEST CONTROL; TPA

QUEST BIOTECHNOLOGY, INC.   Year: 1986   Empl: 9   Code–1°:K   2°:P
    BISPECIFIC MAB; IN VITRO DX; CANCER RX & IMAGING

RHOMED, INC.   Year: 1986   Empl: 8   Code–1°:3   2°:PK
    DX, AB DELIVERY SYSTEM FOR CANCER
    DX & RX; IMMUNOASSAYS

SEAPHARM, INC.   Year: 1983   Empl: 50   Code–1°:T   2°:PH
    MARINE RX PRODS; TISSUE CULTURE

Secondary Focus (Cont.)

SENETEK PLC   Year: 1983 Empl: 14   Code–1°:K 2°:P
   MAB-BASED DX FOR ALZHEIMER'S DISEASE;
   SKIN AGING REVERSAL RX

SERAGEN, INC.   Year: 1979 Empl: 250   Code–1°:X 2°:PG
   CANCER RX; MABS; REAGENTS; ANTISERA

TECHNICLONE INT'L, INC.   Year: 1981 Empl: 15   Code–1°:K 2°:PFH
   CANCER DX, TISSUE CULTURES, MAB,
   MAB CULTURING EQUIP.

TECHNOGENETICS, INC.   Year: 1983 Empl: 30   Code–1°:K 2°:P
   MABS FOR IMMUNOCYTOLOGY,
   HUMAN DX KITS

TRANSGENIC SCIENCES, INC.   Year: 1988 Empl: 3   Code–1°:V 2°:P
   DISEASE-RESISTANT POULTRY; USE OF TRANSGENIC
   ANIMALS FOR RX PRODUCTION; GENE THERAPY

VIAGENE, INC.   Year: 1987 Empl: 9   Code–1°:E 2°:P
   GENE TRANSFER TECHNOL. TO TREAT INFECTIONS,
   MALIGNANCIES, GENE RELATED DISEASES

Listing 5-17 Vaccines

LISTING 5-17 COMPANIES IN VACCINES (Q)

Primary Focus

CENTRAL BIOLOGICS, INC.　　　　Year: 1982 Empl: 7　　　Code–1°:Q 2°:AHXW
　　VET VX; CUSTOM SERA; CONTRACT RESEARCH

IMMUNOMED CORP.　　　　　　Year: 1979 Empl: 17　　Code–1°:Q 2°:HP
　　VX; IMMUNOTHERAPEUTICS; RABIES VX

MICROGENESYS, INC.　　　　　Year: 1983 Empl: 32　　Code–1°:Q 2°:KB
　　AIDS DX; AIDS VX; VIRAL INSECTICIDE;
　　DEVELOPING HUMAN INFECTIOUS DISEASE VX

NORDEN LABS　　　　　　　　Year: 1983 Empl: 145　Code–1°:Q 2°:V
　　VX; HEP B VX; VET VX & RX

PRAXIS BIOLOGICS　　　　　　Year: 1983 Empl: 180　Code–1°:Q 2°:XPK
　　VX; TEST KITS; RX; DX

Secondary Focus

ALPHA I BIOMEDICALS, INC.　　Year: 1982 Empl: 11　　Code–1°:P 2°:QKV
　　PEPTIDES & VX FOR DX/RX; AIDS RX,
　　VX; DEVELOPING THYMOSIN

APPLIED BIOTECHNOLOGY, INC.　Year: 1982 Empl: 45　Code–1°:V 2°:QKP
　　VET VX; AIDS, TB VX; CANCER DX, RX

BACHEM BIOSCIENCE, INC.　　Year: 1987 Empl: 15　Code–1°:J 2°:PQ
　　AMINO ACIDS & DERIVATIVES; BIOACTIVE
　　PEPTIDES; ENZYME SUBSTRATES; VX EPITOPES

BIOGEN　　　　　　　　　　Year: 1979 Empl: 250　Code–1°:P 2°:QGJ
　　RX; DX; TNF, IL-2, CSF, TPA, HSA, FACTOR VIII,
　　IF-G, HEP B ANTIGEN

BIOTECH RESEARCH LABS, INC.　Year: 1973 Empl: 167　Code–1°:K 2°:PQ
　　MABS FOR DX, AIDS DX, CANCER DX; AIDS VX, IL-2

BIOTECHNICA INTERNATIONAL　Year: 1981 Empl: 145　Code–1°:J 2°:KQMB
　　INOCULANT, VX, DENTAL DX; ENZYMES,
　　INDUSTRIAL YEAST, LITE BEER, FOODS

BIOTECHNOLOGY DEVEL. CORP.　Year: 1982 Empl: 35　Code–1°:F 2°:PJQ3
　　AIDS VX; EQUIPMENT; REAGENTS;
　　DRUG DELIVERY SYSTEMS

Secondary Focus (Cont.)

CAMBRIDGE BIOSCIENCE CORP.   Year: 1981 Empl: 97   Code–1°:K 2°:QPV
   DX KITS FOR AIDS, ADENOVIRUS,
   ROTAVIRUS; VET VX, DX

CEL-SCI CORP.   Year: 1983 Empl: 2   Code–1°:P 2°:KQX
   RX; LYMPHOKINES; AIDS VX; IL-1, IL-2

CENTOCOR   Year: 1979 Empl: 246   Code–1°:K 2°:PJGQ
   MAB RX; CANCER, HEP B DX; MABS; AIDS VX

CHIRON CORP.   Year: 1981 Empl: 300   Code–1°:P 2°:QKJV
   AIDS DX; DNA PROBE DX; WOUND HEALING;
   HEP B VX; VET VX, SOD; ENZYMES; IF

ELCATECH   Year: 1984 Empl: 4   Code–1°:W 2°:PQXG
   RESEARCH; COAGULATION ASSAY TECHNOLOGY

ENDOTRONICS, INC.   Year: 1981 Empl: 45   Code–1°:F 2°:HPQ
   RX; DX; MABS; APS SERIES; ACUSYST-P; VX;
   BIOLOGICALS; HEP B VX

GENENTECH   Year: 1976 Empl: 1500   Code–1°:P 2°:HKOQ
   RX; DX; VX; INSULIN, IF, TNF, RENIN,
   FACTOR VIII, PROTROPIN HGH, ACTIVASE TPA

GENETIC DIAGNOSTICS CORP.   Year: 1981 Empl: 17   Code–1°:K 2°:AQ
   DX TEST KITS FOR ABUSE DRUGS, DISEASES

IMCLONE SYSTEMS, INC.   Year: 1984 Empl: 50   Code–1°:K 2°:PQ
   DX; VX; RX; AIDS DX; HEP B DX; DNA PROBES,
   IMMUNODIAGNOSTICS FOR VIRAL DISEASES

IMMUCELL CORP.   Year: 1982 Empl: 17   Code–1°:A 2°:GHQ
   HUMAN DX; REAGENTS; ANIMAL DX; VX

LEE BIOMOLECULAR RES. LABS   Year: 1980 Empl: 16   Code–1°:P 2°:HKQ
   IF; ANTI-IF SERA

LIPOSOME CO. , INC.   Year: 1981 Empl: 67   Code–1°:3 2°:PQJ
   LIPOSOMES; DRUG DELIVERY; CANCER RX; VX

LUCKY BIOTECH CORP.   Year: 1984 Empl: 4   Code–1°:P 2°:QJKA
   RX; IF; IF-A; IF-B; IF-G; IL-2; HEP B DX;
   PORCINE GROWTH HORMONE

MUREX CORP.   Year: 1984 Empl: 180   Code–1°:K 2°:PGQ
   MAB REAGENTS & ANTIBODY KITS; DX

Listing 5-17 Vaccines

Secondary Focus (Cont.)

REPLIGEN CORP.   Year: 1981 Empl: 95   Code–1°:J 2°:QAB
   ENZYMES; AIDS VX; PROTEINS, PROTEIN A,
   HTLVIII; PESTICIDES

SYNBIOTICS CORP.   Year: 1982 Empl: 89   Code–1°:V 2°:KQ
   VET DX; HUMAN DX; VX

SYNERGEN, INC.   Year: 1981 Empl: 101   Code–1°:P 2°:VQW
   RX; RESEARCH LUNG DISEASE; FGF; VET VX

UNITED BIOMEDICAL, INC.   Year: 1983 Empl: 10   Code–1°:K 2°:JXQ
   DX, MABS, SPECIALTY PROTEINS, AIDS VX

LISTING 5-18 COMPANIES IN WASTE DISPOSAL/TREATMENT (R)

Primary Focus

ADVANCED MINERAL TECHNOL.   Year: 1982 Empl: 17   Code–1°:R 2°:NE
   WASTE TREATMENT & METAL RECOVERY

ASTRE CORPORATE GROUP   Year: 1967 Empl: 148   Code–1°:R 2°:B
   AMMONIA DEGRADER; BIOPESTICIDES;
   LIPASE-PRODUCING MICROBE

BIOCONTROL SYSTEMS   Year: 1985 Empl: 29   Code–1°:R 2°:YD
   TESTS FOR FOOD, WATER; BIOSENSORS

BIOSPHERICS, INC.   Year: 1967 Empl: 250   Code–1°:R 2°:YIDF
   WASTEWATER PROCESS & MONITORS & PILOT PLANT

BIOSYSTEMS, INC.   Year: 1976 Empl: 15   Code–1°:R
   MABS; GENERAL MICROBIAL R&D

BIOTROL, INC.   Year: 1985 Empl: 30   Code–1°:R 2°:F
   PROPRIETARY BACTERIA; BIOREACTORS
   FOR ENVIRONMENTAL SERVICES; DEGRADERS

HUNTER BIOSCIENCES   Year: 1987 Empl: 8   Code–1°:R
   BIORESTORATION OF CONTAMINATED WASTE SITES

MICROBE MASTERS   Year: 1982 Empl: 37   Code–1°:R
   INDUSTRIAL WASTEWATER TREATMENT
   USING STRESS-ACCLIMATED BACTERIAL PRODUCTS

Secondary Focus

ABC RESEARCH CORP.   Year: 1967 Empl: 82   Code–1°:W 2°:JBHR
   CONTRACT RESEARCH; CUSTOM STARTER
   CULTURES; BACTERIAL PROTEIN; ANTIOXIDANTS

AMERICAN BIOGENETICS CORP.   Year: 1984 Empl: 10   Code–1°:I 2°:JBLR
   PROCESS DESIGN MICROORGANISMS; CHEMICALS

APPLIED DNA SYSTEMS, INC.   Year: 1982 Empl: 7   Code–1°:P 2°:MR
   HUMAN COLLAGENASE; MABS; OXYRASE;
   IN VITRO CHEMOSENSITIVITY ASSAY

Listing 5-18 Waste Disposal/Treatment

Secondary Focus (Cont.)

BIO HUMA NETICS   Year: 1984 Empl: 15   Code–1°:B 2°:AR
  SOIL CONDITIONERS; FEED SUPPLMTS; WASTE TREATMENT

BIOSCIENCE MANAGEMENT, INC.   Year: 1984 Empl: 8   Code–1°:J 2°:RDF
  BACTERIAL CULTURES; INOCULANTS; INSTRUMENTS

MICROBIO RESOURCES   Year: 1981 Empl: 30   Code–1°:J 2°:MRPO
  FOOD ADDITIVES & COLORANTS; SPECIAL CHEMICALS

MYCOSEARCH, INC.   Year: 1979 Empl: 6   Code–1°:2 2°:OCHR
  CULTURES OF RARE & NEW FUNGI; LIVE
  CULTURES; SPECIALTY CHEMICALS

LISTING 5-19 COMPANIES IN AQUACULTURE (S)

Primary Focus

INCON CORP.   Year: 1983 Empl: 6   Code–1°:S 2°:WT
   R&D ON NEW PLANT/ANIMAL
   AQUACULTURE CROPS

NEUSHUL MARICULTURE   Year: 1978 Empl: 6   Code–1°:S 2°:PMHC
   ANTI-VIRALS; ENZYMES; MOLLUSCAN
   SEED & CULTURE

Secondary Focus

AQUASYNERGY, LTD.   Year: 1986 Empl: 5   Code–1°:T 2°:SWF
   CO_2 GENERATOR FOR AQUACULTURE
   PLANT PRODUCTION; AQUATIC PLANT/ANIMAL CROPS

IGENE BIOTECHNOLOGY, INC.   Year: 1981 Empl: 30   Code–1°:M 2°:OS
   MILK/EGG WHITE REPLACER, CALCIUM & MILK
   MINERALS FOOD SUPPLEMENT, FISH AQUACULTURE

Listing 5-20 Marine Natural Products

LISTING 5-20 COMPANIES IN MARINE NATURAL PRODUCTS (T)

Primary Focus

AQUASYNERGY, LTD.   Year: 1986 Empl: 5   Code–1°:T 2°:SWF
   CO2 GENERATOR FOR AQUACULTURE PLANT
   PRODUCTION; AQUATIC PLANT/ANIMAL CROPS

MARICULTURA, INC.   Year: 1984 Empl: 8   Code–1°:T
   MARINE NATURAL PRODUCTS & CHEMICALS

MARINE BIOLOGICALS, INC.   Year: 1981 Empl: 2   Code–1°:T 2°:J
   MARINE BYPRODUCTS; LIMULUS
   AMEBOCYTE LYSATE

OCEAN GENETICS   Year: 1981 Empl: 35   Code–1°:T 2°:JP
   SPECIAL CHEMICALS & RX OF MARINE ALGAE

SEAPHARM, INC.   Year: 1983 Empl: 50   Code–1°:T 2°:PH
   MARINE RX PRODS; TISSUE CULTURE

Secondary Focus

INCON CORP.   Year: 1983 Empl: 6   Code–1°:S 2°:WT
   R&D ON NEW PLANT/ANIMAL
   AQUACULTURE CROPS

R & A PLANT/SOIL, INC.   Year: 1978 Empl: 6   Code–1°:B 2°:TO
   ALGAL SOIL CONDITIONERS, FERTILIZERS, ETC.

LISTING 5-21 COMPANIES IN VETERINARY AREAS (V)

.

Primary Focus

A.M. BIOTECHNIQUES, INC.   Year: 1980 Empl: 5   Code–1°:V
   VET VX FOR PARVOVIRUS, CORONAVIRUS

AGRITECH SYSTEMS, INC.   Year: 1984 Empl: 90   Code–1°:V 2°:K
   IMMUNOASSAYS, DX; VET DX; TESTS
   FOR FEED/FOOD CONTAMINANTS

ALLELIC BIOSYSTEMS   Year: 1984 Empl: 7   Code–1°:V 2°:JG
   VET TOX TESTS, PROBES, OLIGOS, REAGENTS

AMBICO, INC.   Year: 1974 Empl: 25   Code–1°:V
   VET VX

AMTRON   Year: 1981 Empl: 16   Code–1°:V 2°:X
   ANIMAL DISEASE PREVENTATIVES,
   IMMUNOLOGICAL PRODUCTS

APPLIED BIOTECHNOLOGY, INC.   Year: 1982 Empl: 45   Code–1°:V 2°:QKP
   VET VX; AIDS, TB VX; CANCER DX, RX

APPLIED MICROBIOLOGY, INC.   Year: 1983 Empl: 20   Code–1°:V 2°:KH
   RX FOR BOVINE MASTITIS, TOXIC SHOCK
   SYNDROME DX, B. SUBTILIS PRODUCTION

BIOMED RESEARCH LABS, INC.   Year: 1974 Empl: 18   Code–1°:V 2°:A
   FISH BIOLOGICS, VET PRODUCTS

CODON CORP.   Year: 1980 Empl: 120   Code–1°:V 2°:KP
   VET VX; HUMAN DX

FERMENTA ANIMAL HEALTH   Year: 1986 Empl: 359   Code–1°:V 2°:A
   VET DX & VX; PESTICIDES

GAMETRICS, LTD.   Year: 1974 Empl: 20   Code–1°:V 2°:K
   VET DX; VET & HUMAN SEX
   SELECTION, ISOLATE SPERM

KIRKEGAARD & PERRY LABS   Year: 1979 Empl: 40   Code–1°:V 2°:AGX
   POULTRY DX; VET DX; IMMUNOLOGICAL
   REAGENTS

RIBI IMMUNOCHEM RESEARCH   Year: 1981 Empl: 35   Code–1°:V 2°:AG
   VET ANTI-TUMOR PRODUCTS; VET DX; VET VX

Listing 5-21 Veterinary Areas

Primary Focus (Cont.)
SYNBIOTICS CORP.   Year: 1982 Empl: 89   Code–1°:V 2°:KQ
   VET DX; HUMAN DX; VX

TRANSGENIC SCIENCES, INC.   Year: 1988 Empl: 3   Code–1°:V 2°:P
   DISEASE-RESISTANT POULTRY; TRANSGENIC
   ANIMALS FOR RX PRODUCTION; GENE THERAPY

Secondary Focus
AGRACETUS   Year: 1981 Empl: 65   Code–1°:B 2°:AV
   TERMITICIDE, VET RX, ANIMAL VX, MICROBIAL
   INOCULATIONS, PESTICIDES

ALPHA I BIOMEDICALS, INC.   Year: 1982 Empl: 11   Code–1°:P 2°:QKV
   PEPTIDES & VX FOR DX/RX; AIDS RX, VX;
   DEVELOPING THYMOSIN

ANTIBODIES, INC.   Year: 1961 Empl: 35   Code–1°:X 2°:VK
   ANTISERA, MABS, VET DX, HUMAN DX

BINAX, INC.   Year: 1986 Empl: 17   Code–1°:K 2°:VGX
   MABS, DX TEST KITS, VET DX

BIOMERICA, INC.   Year: 1971 Empl: 35   Code–1°:K 2°:GV
   PREGNANCY, ALLERGY, OCCULT BLOOD DX;
   LAB PRODUCTS; VET DX & RX

BIOTECHNOLOGY GENERAL CORP. Year: 1980 Empl: 130   Code–1°:P 2°:KV
   SOD, BGH, HGH, RX, SURFACTANTS, FUNGICIDE

BMI, INC.   Year: 1984 Empl: 10   Code–1°:K 2°:V
   DX; RX; OTC PREG. TEST, ENZYME IMMUNOASSAYS,
   ALLERGY TESTS, BOVINE PROGESTERONE RX

CALIFORNIA BIOTECHNOLOGY   Year: 1981 Empl: 155   Code–1°:P 2°:AKV3
   RX; ANF, EPO, SURFACTANT, HGH, RENIN,
   PTH; VET RX, NASAL DRUG DELIVERY

CAMBRIDGE BIOSCIENCE CORP.   Year: 1981 Empl: 97   Code–1°:K 2°:QPV
   DX KITS FOR AIDS, ADENOVIRUS,
   ROTAVIRUS; VET VX, DX

CETUS CORP.   Year: 1971 Empl: 711   Code–1°:P 2°:BKAV
   RX; IF; IL-2; MABS; TNF; CSF; INSULIN;
   ANTIBIOTICS; VET VX

Secondary Focus (Cont.)

CHIRON CORP.   Year: 1981 Empl: 300   Code–1°:P 2°:QKJV
   AIDS DX; DNA PROBE DX; WOUND HEALING;
   HEP B VX; VET VX, SOD; ENZYMES; IF

CLINICAL SCIENCES, INC.   Year: 1971 Empl: 50   Code–1°:G 2°:KV
   BIOCHEMICALS FOR DX AND VET USES

ENZON   Year: 1965 Empl: 22   Code–1°:P 2°:V
   RX; SOD FOR HUMAN AND VET USE

GENETIC ENGINEERING, INC.   Year: 1980 Empl: 23   Code–1°:A 2°:V
   ANIMAL SEX SELECTION, BULL SEMEN LABS

GENETIC THERAPY, INC.   Year: 1986 Empl: 4   Code–1°:P 2°:V3
   ANIMAL/HUMAN GENE THERAPY; RX; RETROVIRAL
   VECTORS;TISSUE TARGETED DELIVERY SYSTEMS

IMMUNOVISION, INC.   Year: 1985 Empl: 6   Code–1°:P 2°:KV
   AB'S; PURIFIED PROTEINS FOR AUTOIMMUNE
   DISEASE; BLOOD COMPLEMENTS, ANIMAL TISS; DX

LIPOSOME TECHNOLOGY, INC.   Year: 1981 Empl: 65   Code–1°:3 2°:VBPJ
   LIPOSOMES FOR RX, DRUG DELIVERY,
   VETERINARY, AGRICULTURE

MOLECULAR GENETICS, INC.   Year: 1979 Empl: 63   Code–1°:A 2°:BKV
   DX; AG DX; BGH; VET DX; SCOURS VX

MONOCLONAL ANTIBODIES, INC. Year: 1979 Empl: 110   Code–1°:K 2°:V
   DX TEST KITS-PREGNANCY, REPRODUCTION,
   INFERTILITY; VET DX

NEOGEN CORP.   Year: 1981 Empl: 40   Code–1°:B 2°:AVFY
   FOOD, PLANT HEALTH, ANIMAL HEALTH, DX

NORDEN LABS   Year: 1983 Empl: 145   Code–1°:Q 2°:V
   VX; HEP B VX; VET VX & RX

PHARMACIA LKB BIOTECH.   Year: 1960   Code–1°:E 2°:GVAJ
   CHROMATOGRAPHY SEPARATION CHEMICALS
   & INSTRUMENTS; ENZYMES; REAGENTS; BIOCHEMICALS

PROBIOLOGICS INTERNATIONAL Year: 1987 Empl: 5   Code–1°:P 2°:AVW
   RX, PROBIOTICS, ANTIBIOTICS FOR
   HUMAN & ANIMAL USE

Listing 5-21 Veterinary Areas

Secondary Focus (Cont.)

PROTATEK INTERNATIONAL   Year: 1984 Empl: 37   Code–1°:K 2°:YJPV
   VIRUS, DX TEST KITS; RX; SPECIALIZED
   MICROBES; VET VX, RX, DX

SYNERGEN, INC.   Year: 1981 Empl: 101   Code–1°:P 2°:VQW
   RX; RESEARCH LUNG DISEASE; FGF; VET VX

SYNGENE PRODUCTS   Year: 1980 Empl: 20   Code–1°:P 2°:VA
   PRODUCTS FOR ANIMALS AND HUMAN HEALTH

SYNTRO CORP.   Year: 1981 Empl: 84   Code–1°:A 2°:V
   ANIMAL VX & DX; ANIMAL HEALTH PRODUCTS

UNIVERSITY GENETICS CO.   Year: 1981 Empl: 59   Code–1°:B 2°:AV
   CLONED OR CULTURED PLANTS;
   MONOCLONAL CELLS & EMBRYOS

LISTING 5-22   COMPANIES IN RESEARCH (W)

Primary Focus

ABC RESEARCH CORP.   Year: 1967 Empl: 82   Code–1°:W 2°:JBHR
   CONTRACT RESEARCH; CUSTOM STARTER
   CULTURES; BACTERIAL PROTEIN; ANTIOXIDANTS

ELCATECH, INC.   Year: 1984 Empl: 4   Code–1°:W 2°:PQXG
   RESEARCH; COAGULATION ASSAY TECHNOLOGY

IMMUNOGEN, INC.   Year: 1981 Empl: 11   Code–1°:W 2°:P
   CONJUGATED ABS FOR TUMORS & DX

LIFE SCIENCES   Year: 1962 Empl: 20   Code–1°:W 2°:XP
   IF; RESEARCH ON VIRUSES, ANIMAL STRAINS

MELOY LABORATORIES, INC.   Year: 1970 Empl: 246   Code–1°:W 2°:EP
   SERVICES; REAGENTS; RESEARCH & DEVELOPMENT; IF

SIBIA   Year: 1981 Empl: 65   Code–1°:W
   RESEARCH IN ANIMAL & HUMAN HEALTH,
   PLANT/AG SCIENCE, BT PROCESS DEVELOPMENT

Secondary Focus

ADVANCED BIOTECHNOLOGIES   Year: 1982 Empl: 17   Code–1°:G 2°:EHW
   MEDIA, SERA, GROWTH FACTORS,
   DNA, RESEARCH SERVICES

AQUASYNERGY, LTD.   Year: 1986 Empl: 5   Code–1°:T 2°:SWF
   CO2 GENERATOR FOR AQUACULTURE PLANT
   PRODUCTION; AQUATIC PLANT/ANIMAL CROPS

BIOSTAR MEDICAL PRODUCTS   Year: 1983 Empl: 30   Code–1°:K 2°:WF
   DX & TEST KITS FOR AUTOIMMUNE
   DISEASES, INSTRUMENTATION

BIOTHERM, INC.   Year: 1985 Empl: 10   Code–1°:K 2°:FDW
   DX, CARDIOVASCULAR DX

CENTRAL BIOLOGICS, INC.   Year: 1982 Empl: 7   Code–1°:Q 2°:AHXW
   VET VX; CUSTOM SERA; CONTRACT RESEARCH

COLLABORATIVE RESEARCH, INC. Year: 1961 Empl: 150   Code–1°:G 2°:WK5
   DX SERVICES (DNA PROBE); IF; RENIN;
   UROKINASE; IL-2; ENZYMES; TPA; REAGENTS

Listing 5-22 Research

Secondary Focus (Cont.)

CREATIVE BIOMOLECULES   Year: 1981 Empl: 70   Code–1°:J 2°:PGW
    PEPTIDES, CONTRACT R&D, TPA, EGF

DNAX RESEARCH INSTITUTE   Year: 1980 Empl: 105   Code–1°:P 2°:XWA
    RESEARCH; MABS; RX

EMTECH RESEARCH   Year: 1983 Empl: 30   Code–1°:B 2°:W
    BIOTECH R&D FOR COSMETICS
    AND AGRICULTURE

IGEN   Year: 1982 Empl: 60   Code–1°:X 2°:KW
    MABS FOR HUMAN PATHOGENS;
    CONTRACT RESEARCH

INCON CORP.   Year: 1983 Empl: 6   Code–1°:S 2°:WT
    R&D ON NEW PLANT/ANIMAL AQUACULTURE CROPS

INTER-AMERICAN RES. ASSOC.   Year: 1979 Empl: 10   Code–1°:G 2°:XWK
    PURIFIED ANTIGENS, ANTISERA, BLOOD PRODS

INT'L BIOTECHNOLOGIES   Year: 1982 Empl: 85   Code–1°:G 2°:FKW
    NUCLEIC ACIDS, CLONING, SEQUENCING,
    GENOME ANALYSES & ELECTROPHORESIS SYSTEMS

MAIZE GENETIC RESOURCES, INC. Year: 1984 Empl: 2   Code–1°:B 2°:W
    PROPRIETARY TECHNOL. FOR CORN
    BREEDING STRESS-RESISTANT VARIETIES

ORGANON TEKNIKA CORP.   Year: 1978 Empl: 270   Code–1°:K 2°:DFWX
    HEALTH CARE/MEDICAL DX

PROBIOLOGICS INTERNATIONAL   Year: 1987 Empl: 5   Code–1°:P 2°:AVW
    RX, PROBIOTICS, ANTIBIOTICS
    FOR HUMAN & ANIMAL USE

SPHINX BIOTECHNOLOGIES   Year: 1987 Empl: 3   Code–1°:P 2°:W
    CELL REGULATION THERAPEUTICS

STRATAGENE, INC.   Year: 1984 Empl: 77   Code–1°:G 2°:W
    BT REAGENTS; MOLECULAR BIOLOGICALS;
    GENE DX; CUSTOM RESEARCH

SYNERGEN, INC.   Year: 1981 Empl: 101   Code–1°:P 2°:VQW
    RX; RESEARCH LUNG DISEASE; FGF; VET VX

TRITON BIOSCIENCES, INC.   Year: 1983 Empl: 180   Code–1°:P 2°:KGW
    RX, B-IF, DX RESEARCH REAGENTS, MAB, RESEARCH

Secondary Focus (Cont.)

WASHINGTON BIOLAB   Year: 1986 Empl: 10   Code–1°:J 2°:GW
    AIDS RX; ENZYMES, RAW MATERIALS FOR BIOTECH.

XENOGEN   Year: 1981 Empl: Code–1°:K 2°:JW
    RDNA RESEARCH FOR DX AND CHEMICALS;
    DX FOR BREAST CANCER, GONORRHEA

ZYMOGENETICS, INC.   Year: 1981 Empl: 95   Code–1°:P 2°:W
    RX, WOUND HEALING, COAGULATION,
    PROTEINS, ANTI-INFECTIVES

Listing 5-23 Immunological Products

LISTING 5-23   COMPANIES IN IMMUNOLOGICAL PRODUCTS (X)

Primary Focus

AMERICAN BIOTECHNOLOGY CO.   Year: 1984 Empl: 5   Code–1°:X 2°:PGHJ
   GROWTH FACTORS, IMMUNOMODULATORS

ANTIBODIES, INC.   Year: 1961 Empl: 35   Code–1°:X 2°:VK
   ANTISERA, MABS, VET DX, HUMAN DX

BERKELEY ANTIBODY CO., INC.   Year: 1983 Empl: 27   Code–1°:X 2°:GHP
   CUSTOM ANTIBODIES, MAB & POLYCLONALS

BETHYL LABS, INC.   Year: 1977 Empl: 8   Code–1°:X 2°:KA
   ANTISERA & CUSTOM ANTISERA SERVICE

BIOPRODUCTS FOR SCIENCE, INC.   Year: 1985 Empl: 19   Code–1°:X 2°:G
   ANTIBODIES; ANTIGEN MARKERS; REAGENTS

CHARLES RIVER BIOTECH.   Year: 1983 Empl: 55   Code–1°:X 2°:HF
   MABS, CELL CULTURE SYSTEMS

DAKO CORP.   Year: 1979 Empl: 50   Code–1°:X 2°:K
   POLYCLONAL AB'S & MABS, CLINICAL
   TEST KITS, PROTEINS

E-Y LABORATORIES, INC.   Year: 1978 Empl: 20   Code–1°:X 2°:KPJ
   ANTISERA; LECTINS; REAGENT; MABS; DX

EARL-CLAY LABORATORIES, INC.   Year: 1984 Empl: 7   Code–1°:X 2°:H
   MABS; ULTRACLONE GROWTH CHAMBERS
   IN TISSUE CULTURE; LYMPHOCYTE STIMULATING ASSAY

IGEN   Year: 1982 Empl: 60   Code–1°:X 2°:KW
   MABS FOR HUMAN PATHOGENS; CONTRACT RESEARCH

IMMUNETECH PHARMACEUT.   Year: 1981 Empl: 45   Code–1°:X 2°:P
   IMMUNE ACTIVE PEPTIDES AS RX

INTEK DIAGNOSTICS, INC.   Year: 1983 Empl: 35   Code–1°:X
   T CELL CLONES, ANTIGENS, FLOW CYTOMETRY

INTER-CELL TECHNOLOGIES, INC.   Year: 1982 Empl: 9   Code–1°:X 2°:GKP
   BIOLOGICALS; IMMUNOLOGICAL REAGENTS; LYMPHOKINES

MONOCLONAL PRODUCTION INT'L Year: 1983 Empl: 4   Code–1°:X
   MABS FORMED IN SHEEP/COW UTERUS

Primary Focus (Cont.)

SERAGEN, INC.   Year: 1979 Empl: 250   Code–1°:X 2°:PG
   CANCER RX; MABS; REAGENTS; ANTISERA

TRANSFORMATION RESEARCH   Year: 1981 Empl: 4   Code–1°:X
   ANTIBODIES, GROWTH FACTORS

Secondary Focus

ALFACELL CORP.   Year: 1981 Empl: 22   Code–1°:G 2°:PX
   TUMOR TOXINS, REAGENTS, SERA

AMTRON   Year: 1981 Empl: 16   Code–1°:V 2°:X
   ANIMAL DISEASE PREVENTATIVES,
   IMMUNOLOGICAL PRODUCTS

BINAX, INC.   Year: 1986 Empl: 17   Code–1°:K 2°:VGX
   MABS, DX TEST KITS, VET DX

BIOPURE   Year: 1984 Empl: 31   Code–1°:O 2°:JPX
   CUSTOM FERMENTATION OF RX,
   MAB, CHEMICALS

BIOTEST DIAGNOSTICS CORP.   Year: 1946 Empl: 10   Code–1°:G 2°:X
   REAGENTS AND MAB MARKERS

CEL-SCI CORP.   Year: 1983 Empl: 2   Code–1°:P 2°:KQX
   RX; LYMPHOKINES; AIDS VX; IL-1, IL-2

CENTRAL BIOLOGICS, INC.   Year: 1982 Empl: 7   Code–1°:Q 2°:AHXW
   VET VX; CUSTOM SERA; CONTRACT RESEARCH

CISTRON BIOTECHNOLOGY, INC.   Year: 1982 Empl: 30   Code–1°:P 2°:XK
   IL-1, IL-2, DX

CYTOGEN CORP.   Year: 1980 Empl: 140   Code–1°:K 2°:PX
   RX; MABS; DX; CANCER RX

DAMON BIOTECH, INC.   Year: 1978 Empl: 101   Code–1°:P 2°:X3
   MABS, DRUG DELIV., CELL ENCAPSULATION
   SYSTEM; AIDS DX; TPA

DNAX RESEARCH INSTITUTE   Year: 1980 Empl: 105   Code–1°:P 2°:XWA
   RESEARCH; MABS; RX

ELCATECH   Year: 1984 Empl: 4   Code–1°:W 2°:PQXG
   RESEARCH; COAGULATION ASSAY TECHNOLOGY

Listing 5-23 Immunological Products

Secondary Focus (Cont.)

ENDOGEN, INC.   Year: 1985 Empl: 7   Code–1°:K 2°:X
   DX ASSAYS (QUANTITATIVE); HUMAN MABS

HAZLETON BIOTECHNOLOGIES   Year: 1983 Empl: 500   Code–1°:G 2°:FX
   LAB REAGENTS AND EQUIPMENT, MEDIA, MABS

IMMUNOSYSTEMS, INC.   Year: 1981 Empl: 4   Code–1°:G 2°:X
   ENZYME IMMUNOASSAYS

IMREG, INC.   Year: 1981 Empl: 50   Code–1°:P 2°:KX
   IMMUNOREGULATORS; CANCER DX;
   AIDS, ARTHRITIS RX

INTER-AMERICAN RES. ASSOC.   Year: 1979 Empl: 10   Code–1°:G 2°:XWK
   PURIFIED ANTIGENS, ANTISERA,
   BLOOD PRODUCTS

INTERNATIONAL ENZYMES, INC. Year: 1983 Empl: 4   Code–1°:G 2°:JX
   ENZYMES; ANTISERA; SERUM-FREE MEDIA SUPPLIES

KIRKEGAARD & PERRY LABS   Year: 1979 Empl: 40   Code–1°:V 2°:AGX
   POULTRY DX; VET DX; IMMUNOLOGICAL REAGENTS

LIFE SCIENCES   Year: 1962 Empl: 20   Code–1°:W 2°:XP
   IF; RESEARCH ON VIRUSES, ANIMAL STRAINS

MICROBIOLOGICAL ASSOC.   Year: 1949 Empl: 180   Code–1°:5 2°:X
   TESTING SERVICE FOR COMPLIANCE
   WITH REGULATORY AGENCIES (FDA/EPA)

ONCOGENE SCIENCE, INC.   Year: 1983 Empl: 65   Code–1°:K 2°:PX
   ONCOGENE PROBES FOR DX AND RX, MABS

ORGANON TEKNIKA CORP.   Year: 1978 Empl: 270   Code–1°:K 2°:DFWX
   HEALTH CARE/MEDICAL DX

PEL-FREEZ BIOLOGICALS, INC.   Year: 1911 Empl: 175   Code–1°:G 2°:JX
   VARIOUS SPECIALTY REAGENTS, MAB SERA

PRAXIS BIOLOGICS   Year: 1983 Empl: 180   Code–1°:Q 2°:XPK
   VX; TEST KITS; RX; DX

PROTEINS INTERNATIONAL, INC. Year: 1983 Empl: 6   Code–1°:K 2°:X
   IMMUNOLOGICAL DX; CUSTOM MABS

RESEARCH & DIAGN. ANTIBODIES Year: 1985   Code–1°:K 2°:XG
   DX KITS, MAB REAGENTS

Secondary Focus (Cont.)

SERONO LABS   Year: 1971 Empl: 198   Code–1°:P 2°:KX
    RX AND DX; FERTILITY RX, GROWTH RX,
    HGH, IMMUNOLOGICAL RX, VIROLOGICAL RX

SOUTHERN BIOTECH. ASSOCIATES Year: 1982 Empl: 12   Code–1°:J 2°:X
    COLLAGEN; COLLAGEN AB'S; CUSTOM
    CONJUGATION AB DEVELOPMENT

SUMMA MEDICAL CORP.   Year: 1978 Empl: 33   Code–1°:G 2°:KX
    HISTOCHEMICAL MARKERS AND DX KITS, CANCER DX

UNITED BIOMEDICAL, INC.   Year: 1983 Empl: 10   Code–1°:K 2°:JXQ
    DX, MABS, SPECIALTY PROTEINS, AIDS VX

XOMA CORP.   Year: 1981 Empl: 149   Code–1°:P 2°:X
    MAB BASED ANTICANCER RX

Listing 5-24 Toxicology

LISTING 5-24 COMPANIES IN TOXICOLOGY (Y)

Primary Focus

ANGENICS   Year: 1980 Empl: 50   Code–1°:Y 2°:K
   SCREENING TESTS FOR MILK AND DRUG ABUSE

APPLIED GENETICS LABS, INC.   Year: 1984 Empl: 11   Code–1°:Y 2°:K
   GENETIC TOX TEST, DNA PROBES

ENVIRONMENTAL DIAGNOSTICS Year: 1983 Empl: 55   Code–1°:Y 2°:KF
   ENVIRONMENTAL AND TOXIN TESTING
   IN FOODS, ANIMALS & HUMANS

IDETEK, INC.   Year: 1983 Empl: 14   Code–1°:Y 2°:MAB
   DX & TOX TEST KITS FOR FOOD & AGRICULTURE

RICERCA, INC.   Year: 1986 Empl: 160   Code–1°:Y 2°:BAO
   FERMENTATION AND RESEARCH

TOXICON   Year: 1978 Empl: 35   Code–1°:Y 2°:5
   TOXICOLOGY SERVICE

Secondary Focus

AGDIA, INC.   Year: 1981 Empl: 8   Code–1°:B 2°:KY
   PLANT AND FOOD TOX DX FOR VIRUSES

AGRI-DIAGNOSTICS ASSOCIATES Year: 1983 Empl: 22   Code–1°:B 2°:Y
   TURF & CROP DISEASE DX KITS,
   FOOD PATHOGEN DX

BIO-RESPONSE, INC.   Year: 1972 Empl: 45   Code–1°:P 2°:HYK
   NON-RDNA TPA; MASS CELL
   CULTURE; DX KITS; MABS

BIOCONTROL SYSTEMS   Year: 1985 Empl: 29   Code–1°:R 2°:YD
   TESTS FOR FOOD, WATER; BIOSENSORS

BIODESIGN, INC.   Year: 1987 Empl: 3   Code–1°:G 2°:PY
   IMMUNOLOGICAL REAGENTS; BIODESIGN
   ANTISERA, PURIFIED BIOLOGICAL COMPOUNDS;
   HAZARDOUS CHEMICALS

BIOSPHERICS, INC.   Year: 1967 Empl: 250   Code–1°:R 2°:YIDF
   WASTEWATER PROCESS & MONITORS & PILOT PLANT

Secondary Focus (Cont.)

BIOTHERAPEUTICS, INC.   Year: 1984 Empl: 120   Code–1°:P 2°:KY
   CANCER RX & DX; BIOLOGICAL
   RESPONSE MODIFIERS

COVALENT TECHNOLOGY CORP.   Year: 1981 Empl: 16   Code–1°:K 2°:Y
   DX FOR HORMONE, DISEASE;
   BIOWAR DX; TROPICAL DISEASE DX

DIAGNON CORP.   Year: 1981 Empl: 70   Code–1°:K 2°:Y
   DX ASSAYS FOR FERRITIN, B12/FOLATE,
   HERPES, VIRAL TREATMENT, TOXIN ASSAYS

INTEGRATED GENETICS, INC.   Year: 1981 Empl: 200   Code–1°:P 2°:KAY
   DX; RX; AIDS DX; FOOD TOXIN DX;
   CANCER DX; TPA; HCG; EPO; HEP-B VX

NEOGEN CORP.   Year: 1981 Empl: 40   Code–1°:B 2°:AVFY
   FOOD, PLANT HEALTH, ANIMAL HEALTH, DX

PROTATEK INTERNATIONAL   Year: 1984 Empl: 37   Code–1°:K 2°:YJPV
   VIRUS, DX TEST KITS; RX; SPECIALIZED
   MICROBES; VET VX, RX, DX

Listing 5-25 Biomaterials

LISTING 5-25 COMPANIES IN BIOMATERIALS (1)

Primary Focus
BIOMATRIX, INC.   Year: 1981 Empl: 50   Code–1°:1
   MATERIAL FOR ARTIFICIAL ORGANS;
   COSMETICS; OPHTHALMICS

BIOPOLYMERS, INC.   Year: 1984 Empl: 30   Code–1°:1 2°:HJ
   CELL & TISSUE, OPHTHALMIC;
   MOLLUSK ADHESIVES

Secondary Focus
COLLAGEN CORP.   Year: 1975 Empl: 225   Code–1°:4 2°:HP1
   MEDICAL DEVICES; COLLAGEN
   IMPLANTS; TISSUE CULTURES

MYCOGEN CORP.   Year: 1982 Empl: 50   Code–1°:B 2°:A1
   PLANT HERBICIDES/PESTICIDES
   BASED ON NATURAL PATHOGENS

LISTING 5-26 COMPANIES IN FUNGI (2)

Primary Focus

MYCOSEARCH, INC.   Year: 1979 Empl: 6   Code–1°:2 2°:OCHR
   CULTURES OF RARE & NEW FUNGI;
   LIVE CULTURES; SPECIALTY CHEMICALS

Secondary Focus

NO COMPANIES LISTED

Listing 5-27 Drug Delivery

LISTING 5-27 COMPANIES IN DRUG DELIVERY (3)

Primary Focus

CYGNUS RESEARCH CORP.   Year: 1985 Empl: 18 Code–1°:3 2°:DKP
 DRUG DELIVERY SYSTEM;
 CONTRACTS TO RX COMPANIES

LIPOSOME COMPANY, INC.   Year: 1981 Empl: 67 Code–1°:3 2°:PQJ
 LIPOSOMES; DRUG DELIVERY; CANCER RX; VX

LIPOSOME TECHNOLOGY, INC. Year: 1981 Empl: 65 Code–1°:3 2°:VBPJ
 LIPOSOMES FOR RX, DRUG DELIVERY,
 VETERINARY, AGRICULTURE

RHOMED, INC.   Year: 1986 Empl: 8 Code–1°:3 2°:PK
 DX, AB DELIVERY SYSTEM FOR
 CANCER DX & RX; IMMUNOASSAYS

Secondary Focus

BIOTECHNOLOGY DEVELOPMENT Year: 1982 Empl: 35 Code–1°:F 2°:PJQ3
 AIDS VX; EQUIPMENT; REAGENTS;
 DRUG DELIVERY SYSTEMS

CALIFORNIA BIOTECHNOLOGY Year: 1981 Empl: 155 Code–1°:P 2°:AKV3
 RX; ANF, EPO, SURFACTANT, HGH, RENIN,
 PTH; VET RX, NASAL DRUG DELIVERY

DAMON BIOTECH, INC.   Year: 1978 Empl: 101 Code–1°:P 2°:X3
 MABS, DRUG DELIVERY, CELL
 ENCAPSULATION SYSTEMS; AIDS DX; TPA

GENETIC THERAPY, INC.   Year: 1986 Empl: 4 Code–1°:P 2°:V3
 ANIMAL/HUMAN GENE THERAPY; RX; RETROVIRAL
 VECTORS; TISSUE-TARGETED DELIVERY SYSTEMS

LISTING 5-28 COMPANIES IN MEDICAL DEVICES (4)

Primary Focus

COLLAGEN CORP.   Year: 1975 Empl: 225   Code–1°:4 2°:HP1
    MEDICAL DEVICES; COLLAGEN
    IMPLANTS; TISSUE CULTURES

Secondary Focus

LIFECORE BIOMEDICAL, INC.   Year: 1975 Empl: 41   Code–1°:J 2°:4
    HYALURONIC ACID; MEDICAL DEVICES

Listing 5-29 Testing/Analytical Services

LISTING 5-29 COMPANIES IN TESTING/ANALYTICAL SERVICES (5)

Primary Focus

MICROBIOLOGICAL ASSOCIATES   Year: 1949 Empl: 180   Code–1°:5 2°:X
  TESTING SERVICE FOR COMPLIANCE
  WITH REGULATORY AGENCIES (FDA/EPA)

TEKTAGEN   Year: 1987 Empl: 3   Code–1°:5
  SERVICE LAB FOR FDA TESTING

VIVIGEN, INC.   Year: 1981 Empl: 70   Code–1°:5 2°:K
  GENETIC TESTING LAB, AMNIOCENTESIS;
  CHROMOSOME ANALYSIS

Secondary Focus

COLLABORATIVE RESEARCH   Year: 1961 Empl: 150   Code–1°:G 2°:WK5
  DX SERVICES (DNA PROBE); IF; RENNIN;
  UROKINASE; IL-2; ENZYMES; TPA; REAGENTS

INTELLIGENETICS, INC.   Year: 1980 Empl: 52   Code–1°:F 2°:5
  GENETIC ENGINEERING SOFTWARE;
  DNA & PROTEIN ANALYSIS

TOXICON   Year: 1978 Empl: 35   Code–1°:Y 2°:5
  TOXICOLOGY SERVICE

TABLE 5-2 COMPANY FOCUS

Code	Classification	Number Primary	Number Secondary	Total Companies
A	Agriculture, Animal	8	28	36
B	Agriculture, Plant	30	18	48
C	Biomass Conversion	1	4	5
D	Biosensors/Bioelectronics	4	9	13
E	Bioseparations	9	14	23
F	Biotechnology Equipment	10	34	44
G	Biotechnology Reagents	38	58	96
H	Cell Culture, General	5	33	38
I	Chemicals, Commodity	3	5	8
J	Chemicals, Specialty	30	51	81
K	Diagnostics, Clinical Human	75	82	157
L	Energy	2	4	6
M	Food Production/Processing	5	20	25
N	Mining	1	1	2
O	Production/Fermentation	4	12	16
P	Therapeutics	62	67	129
Q	Vaccines	5	25	30
R	Waste Disposal/Treatment	9	7	16
S	Aquaculture	2	2	4
T	Marine Natural Products	5	2	7
U	Consulting	0	0	0
V	Veterinary	15	28	43
W	Research	6	23	29
X	Immunological Products	16	33	49
Y	Toxicology	6	12	18
Z	Venture Capital/Financing	0	0	0
1	Biomaterials	2	2	4
2	Fungi	1	0	1
3	Drug Delivery Systems	4	4	8
4	Medical Devices	1	1	2
5	Testing/Analytical Services	3	3	6

Note: The number of companies listing each classification as a primary or a secondary focus is given. The total number of companies represented in this table is 360.

Table 5-3 Ranking by Primary Focus

TABLE 5-3 RANKING BY PRIMARY FOCUS

Code	Classification (1)	Number Primary	Percent of Total (2)
K	Diagnostics, Clinical Human	75	20.8
P	Therapeutics	62	17.2
G	Biotechnology Reagents	38	10.5
B	Agriculture, Plant	30	8.3
J	Chemicals, Specialty	30	8.3
X	Immunological Products	16	4.4
V	Veterinary	15	4.2
F	Biotechnology Equipment	10	2.8
E	Bioseparations	9	2.5
R	Waste Disposal/Treatment	9	2.5
A	Agriculture, Animal	8	2.2
W	Research	6	1.7
Y	Toxicology	6	1.7
H	Cell Culture, General	5	1.4
M	Food Production/Processing	5	1.4
Q	Vaccines	5	1.4
T	Marine Natural Products	5	1.4
D	Biosensors/Bioelectronics	4	1.1
O	Production/Fermentation	4	1.1
3	Drug Delivery Systems	4	1.1
I	Chemicals, Commodity	3	0.8
5	Testing/Analytical Services	3	0.8
L	Energy	2	0.6
S	Aquaculture	2	0.6
1	Biomaterials	2	0.6
C	Biomass Conversion	1	0.3
2	Fungi	1	0.3
4	Medical Devices	1	0.3
N	Mining	1	0.3
U	Consulting	0	
Z	Venture Capital/Financing	0	

Notes:
1. The number of companies listing each classification as a primary focus is given. The total number of companies represented is 360.
2. The percent of all companies with a primary focus in each classification.

TABLE 5-4 RANKING BY COMBINED PRIMARY AND SECONDARY FOCI

Code	Classification (1)	Number Primary	Number Secondary	Number Combined	Percent Total (2)
K	Diagnostics, Clinical Human	75	82	157	43.5
P	Therapeutics	62	67	129	35.7
G	Biotechnology Reagents	38	58	96	26.6
J	Chemicals, Specialty	30	51	81	22.4
X	Immunological Products	16	33	49	13.6
B	Agriculture, Plant	30	18	48	13.3
F	Biotechnology Equipment	10	34	44	12.2
V	Veterinary	15	28	43	11.9
H	Cell Culture, General	5	33	38	10.5
A	Agriculture, Animal	8	28	36	10.0
Q	Vaccines	5	25	30	8.3
W	Research	6	23	29	8.0
M	Food Production/Processing	5	20	25	6.9
E	Bioseparations	9	14	23	6.4
Y	Toxicology	6	12	18	5.0
O	Production/Fermentation	4	12	16	4.4
R	Waste Disposal/Treatment	9	7	16	4.4
D	Biosensors/Bioelectronics	4	9	13	3.6
3	Drug Delivery Systems	4	4	8	2.2
I	Chemicals, Commodity	3	5	8	2.2
T	Marine Natural Products	5	2	7	1.9
5	Testing/Analytical Services	3	3	6	1.7
L	Energy	2	4	6	1.7
C	Biomass Conversion	1	4	5	1.4
1	Biomaterials	2	2	4	1.1
S	Aquaculture	2	2	4	1.1
4	Medical Devices	1	1	2	0.6
N	Mining	1	1	2	0.6
2	Fungi	1	0	1	0.3
U	Consulting	0	0	0	
Z	Venture Capital/Financing	0	0	0	

Notes:

1. The number of companies listing each classification as primary focus, secondary focus, or both combined is given. The total number of companies is 360.

2. The percent of all companies with either a primary or a secondary focus in each classification is shown.

SECTION 6   BIOTECHNOLOGY COMPANY PERSONNEL

INTRODUCTION

The explosive growth in the U.S. biotechnology industry can be best seen in the growth in personnel. This section focuses on industry personnel and the composition of employees in the U.S. biotechnology firms. Listing 6-1 gives the number of employees in the 360 biotechnology companies listed in Section 2. In addition, for 125 of the companies, the company employees engaged in science and technology and those engaged in management or administration are given. Listing 6-2 divides the companies by their primary industry classification, according to our code system, and shows total personnel numbers for each firm. The largest U.S. biotechnology companies, those with at least 100 employees, are shown in Listing 6-3, along with the industry classification and type (i.e., private, public or subsidiary) of each company.

Table 6-1 is an analysis of the average personnel size for companies in each industry classification as well as the total number of employees in the U.S. biotechnology industry for each primary focus classification. Almost 10,000 employees are represented by each of the two most popular industry classifications, therapeutics and diagnostics. In all, about 31,000 employees make up the small firm component of the U.S. biotechnology industry. Table 6-2 breaks down this number by firm type: private, public or subsidiary. Of the three types of firms, subsidiary firms have the largest number of employees, an average of 184, with public and private firms having averages of 150 and 35, respectively.

Figure 6-1 demonstrates the breakdown of types of employees in companies. As would be expected, the largest number of employees is engaged in the science and technical end of the business, an average of 55 percent of the staff, with 40 percent of these holding Ph.D. degrees. The rest of the staff is engaged in management, administration, marketing and production.

Figure 6-2 demonstrates the average and total employee numbers for each of seven key industry classifications. As would be expected, the therapeutics and diagnostics companies have the most employees per company on average as well as per classification.

One key issue is personnel growth. Figure 6-3 demonstrates the explosive growth in personnel over the biotechnology companies' first years. In addition, a significant predicted growth is demonstrated.

Listing 6-1 Company Personnel

LISTING 6-1 COMPANY PERSONNEL

Company	Employees		
	Total	Sci/Tech	Mgt/Admin
A.M. BIOTECHNIQUES, INC.	5	3	2
A/G TECHNOLOGY CORP.	8	3	2
ABC RESEARCH CORP.	82	66	16
ABN	45	43	42
ADVANCED BIOTECHNOLOGIES	17	12	5
ADVANCED GENETIC SCIENCES	130		
ADVANCED MAGNETICS, INC.	40		
ADVANCED MINERAL TECHNOLOGIES	17		
AGDIA, INC.	8	8	2
AGRACETUS	65	40	5
AGRI-DIAGNOSTICS ASSOCIATES	22	19	3
AGRIGENETICS CORP.	750		
AGRITECH SYSTEMS, INC.	90	40	
ALFACELL CORP.	22		
ALLELIC BIOSYSTEMS	7		
ALPHA I BIOMEDICALS, INC.	11	6	5
AMBICO, INC.	25		
AMERICAN BIOCLINICAL	24		
AMERICAN BIOGENETICS CORP.	10	7	3
AMERICAN BIOTECHNOLOGY CO.	5	3	
AMERICAN DIAGNOSTICA, INC.	9		3
AMERICAN LABORATORIES, INC.	26		
AMERICAN QUALEX INTERNATIONAL	10		
AMGEN	340		
AMTRON	16		
AN-CON GENETICS	3		
ANGENICS	50	27	5
ANTIBODIES, INC.	35		
ANTIVIRALS, INC.	5	4	1
APPLIED BIOSYSTEMS, INC.	350		
APPLIED BIOTECHNOLOGY, INC.	45		
APPLIED DNA SYSTEMS, INC.	7	5	2
APPLIED GENETICS LABS, INC.	11		
APPLIED MICROBIOLOGY, INC.	20		
APPLIED PROTEIN TECHNOLOGIES, INC.	5	2	3
AQUASYNERGY, LTD.	5		
ASTRE CORPORATE GROUP	148	144	4
ATLANTIC ANTIBODIES	60		
AUTOMEDIX SCIENCES, INC.	7	2	4
BACHEM BIOSCIENCE, INC.	15	7	7
BACHEM, INC.	45	30	4

Company	Employees		
	Total	Sci/Tech	Mgt/Admin
BEHRING DIAGNOSTICS	100		
BEND RESEARCH, INC.	60		
BENTECH LABORATORIES	7		
BERKELEY ANTIBODY CO., INC.	27		
BETHYL LABS, INC.	8		
BINAX, INC.	17	7	10
BIO HUMA NETICS	15		
BIO TECHNIQUES LABS, INC.	25	17	
BIO-RECOVERY, INC.	49		
BIO-RESPONSE, INC.	45		
BIO-TECHNICAL RESOURCES, INC.	25		
BIOASSAY SYSTEMS CORP.	65		
BIOCHEM TECHNOLOGY, INC.	12	7	3
BIOCONSEP, INC.	6		
BIOCONTROL SYSTEMS	29		
BIODESIGN, INC.	3	2	1
BIOGEN	250		
BIOGENEX LABORATORIES	42	20	20
BIOGROWTH, INC.	16	14	2
BIOKYOWA, INC.	70		
BIOMATERIALS INTERNATIONAL, INC.	30		
BIOMATRIX, INC.	50		
BIOMED RESEARCH LABS, INC.	18		
BIOMEDICAL TECHNOLOGIES	10	6	4
BIOMERICA, INC.	35		
BIONETICS RESEARCH	200	150	50
BIONIQUE LABS, INC.	11	6	
BIOPOLYMERS, INC.	30		
BIOPROBE INTERNATIONAL, INC.	15	7	3
BIOPRODUCTS FOR SCIENCE, INC.	19		
BIOPURE	31		
BIOSCIENCE MANAGEMENT, INC.	8		
BIOSEARCH, INC.	100		
BIOSPHERICS, INC.	250		
BIOSTAR MEDICAL PRODUCTS, INC.	30		10
BIOSYSTEMS, INC.	15		
BIOTECH RESEARCH LABS, INC.	167		
BIOTECHNICA AGRICULTURE	50		
BIOTECHNICA INTERNATIONAL	145	95	45
BIOTECHNOLOGY DEVELOPMENT CORP.	35		
BIOTECHNOLOGY GENERAL CORP.	130		
BIOTEST DIAGNOSTICS CORP.	10		
BIOTHERAPEUTICS, INC.	120	100	20
BIOTHERAPY SYSTEMS, INC.	8		
BIOTHERM	10	5	2

Listing 6-1 Company Personnel

Company	Employees		
	Total	Sci/Tech	Mgt/Admin
BIOTICS RESEARCH CORP.	22		
BIOTROL, INC.	30	16	14
BIOTX (BIOSCIENCES CORP. OF TEXAS)	20	16	4
BMI, INC.	10	5	5
BOEHRINGER MANNHEIM DIAGN.	1,000		
BRAIN RESEARCH, INC.	NA		
CALGENE, INC.	165	102	63
CALIFORNIA BIOTECHNOLOGY, INC.	155	96	59
CALIFORNIA INTEGRATED DIAGNOSTICS	9		6
CALZYME LABORATORIES, INC.	10		
CAMBRIDGE BIOSCIENCE CORP.	97		
CAMBRIDGE RESEARCH BIOCHEMICALS	75	40	35
CARBOHYDRATES INTERNATIONAL	25	20	5
CEL-SCI CORP.	2		
CELLMARK DIAGNOSTICS	30	24	6
CELLULAR PRODUCTS	101		
CENTOCOR	246	150	100
CENTRAL BIOLOGICS, INC.	7	5	2
CETUS CORP.	711		
CHARLES RIVER BIOTECH. SERVICES	55		
CHEMGENES	8		
CHEMICAL DYNAMICS CORP.	45		
CHEMICON INTERNATIONAL, INC.	8		
CHIRON CORP.	300		
CIBA-CORNING DIAGNOSTIC CORP.	750		
CISTRON BIOTECHNOLOGY, INC.	30		
CLINETICS CORP.	10	4	6
CLINICAL SCIENCES, INC.	50		
CLONTECH LABORATORIES, INC.	30		
COAL BIOTECH CORP.	10		
CODON CORP.	120		
COLLABORATIVE RESEARCH, INC.	150	70	
COLLAGEN CORP.	225		30
CONSOLIDATED BIOTECHNOLOGY, INC.	15		
COOPER DEVELOPMENT CO.	900		
COORS BIOTECH PRODUCTS CO.	185		
COVALENT TECHNOLOGY CORP.	16	14	2
CREATIVE BIOMOLECULES	70		
CROP GENETICS INTERNATIONAL CORP.	75	68	7
CYANOTECH CORP.	25		
CYGNUS RESEARCH CORP.	18	12	2
CYTOGEN CORP.	140		
CYTOTECH, INC.	43		
DAKO CORP.	50		
DAMON BIOTECH, INC.	101	51	

Company	Employees		
	Total	Sci/Tech	Mgt/Admin
DEKALB-PFIZER GENETICS	235		
DELTOWN CHEMURGIC CORP.	90		
DIAGNON CORP.	70		
DIAGNOSTIC PRODUCTS CORP.	270		
DIAGNOSTIC TECHNOLOGY, INC.	60		
DIAMEDIX, INC.	35		
DIGENE DIAGNOSTICS, INC.	20	17	
DNA PLANT TECHNOLOGY CORP.	80	70	
DNAX RESEARCH INSTITUTE	105	80	25
E-Y LABORATORIES, INC.	20		7
EARL-CLAY LABORATORIES, INC.	7	3	4
ECOGEN, INC.	75	60	15
ELANEX PHARMACEUTICALS, INC.	10	10	
ELCATECH	4	3	
ELECTRO-NUCLEONICS, INC.	720		
EMBREX, INC.	20	11	9
EMTECH RESEARCH	30		
ENDOGEN, INC.	7	3	4
ENDOTRONICS, INC.	45		
ENVIRONMENTAL DIAGNOSTICS, INC.	55		
ENZO-BIOCHEM, INC.	110		
ENZON	22		
ENZON, INC.	31		
ENZYME BIO-SYSTEMS LTD.	101		
ENZYME CENTER INC.	30		
ENZYME TECHNOLOGY CORP.	26	22	5
EPITOPE, INC.	60	39	21
ESCAGENETICS CORP.	50	31	19
EXOVIR, INC.	5		
FERMENTA ANIMAL HEALTH	359		
FLOW LABORATORIES, INC.	200		
FORGENE, INC.	4	2	2
GAMETRICS, LTD.	20		
GAMMA BIOLOGICALS, INC.	166		
GEN-PROBE, INC.	127	64	63
GENE-TRAK SYSTEMS	196	137	59
GENELABS, INC.	90	72	18
GENENCOR, INC.	130	90	40
GENENTECH	1,500		
GENESIS LABS, INC.	29		
GENETIC DIAGNOSTICS CORP.	17		
GENETIC ENGINEERING, INC.	23		
GENETIC SYSTEMS CORP.	400		
GENETIC THERAPY, INC.	4	3	1
GENETICS INSTITUTE, INC.	302	191	111

Listing 6-1 Company Personnel

Company	Employees Total	Sci/Tech	Mgt/Admin
GENEX, INC.	80		
GENTRONIX LABORATORIES, INC.	6		
GENZYME CORP.	185		
GRANADA GENETICS CORP.	3		
HANA BIOLOGICS, INC.	130	80	50
HAWAII BIOTECHNOLOGY GROUP, INC.	10		
HAZLETON BIOTECHNOLOGIES CO.	500		
HOUSTON BIOTECHNOLOGY, INC.	30		
HUNTER BIOSCIENCES, INC.	8		
HYBRITECH, INC.	708		
HYCLONE LABORATORIES, INC.	107		
HYGEIA SCIENCES	70		
IBF BIOTECHNICS, INC.	5	2	3
ICL SCIENTIFIC, INC.	70		
IDEC PHARMACEUTICAL	45		
IDETEK, INC.	14		
IGEN	60		
IGENE BIOTECHNOLOGY, INC.	30	25	5
IMCERA BIOPRODUCTS, INC.	5	2	3
IMCLONE SYSTEMS, INC.	50	40	10
IMMUCELL CORP.	17	13	1
IMMUNETECH PHARMACEUTICALS	45		
IMMUNEX CORP.	153	124	36
IMMUNEX, INC.	5	3	2
IMMUNO CONCEPTS, INC.	20		
IMMUNO MODULATORS LABS, INC.	25	10	15
IMMUNOGEN, INC.	11		
IMMUNOMED CORP.	17	5	12
IMMUNOMEDICS, INC.	55		
IMMUNOSYSTEMS, INC.	4	2	2
IMMUNOTECH CORP.	15	5	10
IMMUNOVISION, INC.	6	4	2
IMRE CORP.	25		
IMREG, INC.	50		
INCELL CORP.	23		
INCON CORP.	6		
INFERGENE CO.	25	21	
INGENE, INC.	65	55	6
INTEGRATED CHEMICAL SENSORS	8	8	4
INTEGRATED GENETICS, INC.	200		
INTEK DIAGNOSTICS, INC.	35		
INTELLIGENETICS, INC.	52		
INTER-AMERICAN RESEARCH ASSOC.	10		
INTER-CELL TECHNOLOGIES, INC.	9		
INTERFERON SCIENCES, INC.	70	58	

Company	Employees		
	Total	Sci/Tech	Mgt/Admin
INTERNATIONAL BIOTECHNOLOGIES	85		
INTERNATIONAL ENZYMES, INC.	4		
INTERNATIONAL PLANT RESEARCH INST.	50	46	4
INVITRON CORP.	165		
JACKSON IMMUNORESEARCH LABS, INC.	16		
KALLESTAD DIAGNOSTICS	330	58	272
KARYON TECHNOLOGY, LTD.	25		
KIRIN-AMGEN	5		
KIRKEGAARD & PERRY LABS, INC.	40	15	25
LABSYSTEMS, INC.	10	4	6
LEE BIOMOLECULAR RESEARCH LABS	16		
LIFE SCIENCES	20		
LIFE TECHNOLOGIES, INC.	1,300		
LIFECODES CORP.	53		
LIFECORE BIOMEDICAL, INC.	41		
LIPOSOME COMPANY, INC.	67		
LIPOSOME TECHNOLOGY, INC.	65	45	20
LUCKY BIOTECH CORP.	4	2	
LYPHOMED, INC.	1,200		
MAIZE GENETIC RESOURCES, INC.	2	2	
MARCOR DEVELOPMENT CO.	6		
MARICULTURA, INC.	8	5	3
MARINE BIOLOGICALS, INC.	2	2	1
MARTEK	16	14	2
MAST IMMUNOSYSTEMS, INC.	45	23	22
MEDAREX, INC.	3	2	1
MELOY LABORATORIES, INC.	246	177	58
MICROBE MASTERS	37	20	17
MICROBIO RESOURCES	30		
MICROBIOLOGICAL ASSOCIATES, INC.	180		
MICROGENESYS, INC.	32	22	10
MICROGENICS	70		
MOLECULAR BIOLOGY RESOURCES, INC.	10	6	4
MOLECULAR BIOSYSTEMS, INC.	55		
MOLECULAR DIAGNOSTICS, INC.	30	26	
MOLECULAR GENETIC RESOURCES	5		
MOLECULAR GENETICS, INC.	80	75	5
MONOCLONAL ANTIBODIES, INC.	110		
MONOCLONAL PRODUCTION INT'L	4		
MONOCLONETICS INTERNATIONAL, INC.	5		
MULTIPLE PEPTIDE SYSTEMS, INC.	6		1
MUREX CORP.	180	142	38
MYCOGEN CORP.	50		
MYCOSEARCH, INC.	6	6	1
NATIONAL GENO SCIENCES	4		

Listing 6-1 Company Personnel

Company	Employees		
	Total	Sci/Tech	Mgt/Admin
NATIVE PLANTS, INC.	625	125	500
NEOGEN CORP.	40		
NEORX CORP.	194		
NEUREX CORP.	22	16	6
NEUSHUL MARICULTURE	6		
NITRAGIN CO.	80		
NORDEN LABS	145	125	20
NORTH COAST BIOTECHNOLOGY, INC.	2	1	1
NYGENE CORP.	18		
O.C.S. LABORATORIES, INC.	12		
OCEAN GENETICS	35		
OMNI BIOCHEM, INC.	5	3	2
ONCOGENE SCIENCE, INC.	65		
ONCOR, INC.	30	20	
ORGANON TEKNIKA CORP.	270		
PAMBEC LABORATORIES	6	4	2
PEL-FREEZ BIOLOGICALS, INC.	175		
PENINSULA LABORATORIES, INC.	75		
PETROFERM, INC.	68		
PETROGEN, INC.	6		
PHARMACIA LKB BIOTECHNOLOGY, INC.	N A		
PHARMAGENE	3	2	1
PHYTOGEN	20	18	2
PLANT GENETICS, INC.	63	29	34
POLYCELL, INC.	6	3	3
PRAXIS BIOLOGICS	180		
PROBIOLOGICS INTERNATIONAL, INC.	5		1
PROGENX	20	17	3
PROMEGA CORP.	85		
PROTATEK INTERNATIONAL, INC.	37		
PROTEINS INTERNATIONAL INC.	6	4	2
PROVESTA CORP.	50		
QUEST BIOTECHNOLOGY, INC.	9	6	3
QUEUE SYSTEMS, INC.	85		
QUIDEL	325		
R & A PLANT/SOIL, INC.	6	3	3
RECOMTEX CORP.	12		
REPAP TECHNOLOGIES	16		
REPLIGEN CORP.	95		
RESEARCH & DIAGNOSTIC ANTIBODIES	N A		
RHOMED, INC.	8		
RIBI IMMUNOCHEM RESEARCH, INC.	35		
RICERCA, INC.	160		
SCOTT LABORATORIES	375		
SCRIPPS LABORATORIES	60		

Company	Employees		
	Total	Sci/Tech	Mgt/Admin
SEAPHARM, INC.	50		
SENETEK, PLC	14		
SEPRACOR, INC.	40		
SERAGEN, INC.	250		
SERONO LABS	198		
SIBIA	65		
SOUTHERN BIOTECHNOLOGY ASSOC.	12		
SPHINX BIOTECHNOLOGIES	3	3	1
STRATAGENE, INC.	77		
SUMMA MEDICAL CORP.	33		
SUNGENE TECHNOLOGIES CORP.	28		
SYNBIOTICS CORP.	89	62	9
SYNERGEN, INC.	101	75	
SYNGENE PRODUCTS	20		
SYNTHETIC GENETICS, INC.	6	4	2
SYNTRO CORP.	84	60	13
SYVA CO.	1,000		
T CELL SCIENCES, INC.	65		
TECHNICLONE INTERNATIONAL, INC.	15		
TECHNOGENETICS, INC.	30		
TEKTAGEN	3	2	1
THREE-M (3M) DIAGNOSTIC SYSTEMS	130		
TOXICON	35		
TRANSFORMATION RESEARCH, INC.	4		
TRANSGENIC SCIENCES, INC.	3	1	2
TRITON BIOSCIENCES, INC.	180		
UNIGENE LABORATORIES, INC.	30	22	2
UNITED AGRISEEDS, INC.	192		
UNITED BIOMEDICAL, INC.	10		
UNITED STATES BIOCHEMICAL CORP.	NA		
UNIVERSITY GENETICS CO.	59		
UNIVERSITY MICRO REFERENCE LAB	6	3	3
UPSTATE BIOTECHNOLOGY, INC.	4	3	1
VECTOR LABORATORIES, INC.	50		
VEGA BIOTECHNOLOGIES, INC.	60		
VERAX CORP.	80	40	
VIAGENE, INC.	9	7	2
VIRAGEN	15		
VIROSTAT, INC.	1	1	
VIVIGEN, INC.	70	55	15
WASHINGTON BIOLAB	10	8	2
WELGEN MANUFACTURING, INC.	200	120	80
WESTBRIDGE RESEARCH GROUP	15		
XENOGEN	NA		
XOMA CORP.	149		

Listing 6-1 Company Personnel

Company	Employees		
	Total	Sci/Tech	Mgt/Admin
ZOECON CORP.	130	100	10
ZYMED LABORATORIES, INC.	20	12	
ZYMOGENETICS, INC.	95	69	10

Notes:

Total employee data were available from 355 companies. (NA=not available) Sci/Tech represents the number of science- and technology-related employees in the companies, and these figures were available from 142 of the companies. Mgt/Admin represents the number of management- and administration-related employees, and these figures were available from 129 of the companies. Some companies had additional personnel that did not fit into these categories, such as in marketing and production. Data were collected by questionnaire (126 companies responded; mid-1987) and by telephone (approximately 220 companies; mid 1987 to February 1988).

LISTING 6-2 COMPANY PERSONNEL BY PRIMARY CLASSIFICATION

Company Name	Total Employees
Animal Agriculture (A)	
BIO TECHNIQUES LABS, INC.	25
BIOKYOWA, INC.	70
EMBREX, INC.	20
GENETIC ENGINEERING, INC.	23
GRANADA GENETICS CORP.	3
IMMUCELL CORP.	17
MOLECULAR GENETICS, INC.	80
SYNTRO CORP.	84
Plant Agriculture (B)	
ADVANCED GENETIC SCIENCES	130
AGDIA, INC.	8
AGRACETUS	65
AGRI-DIAGNOSTICS ASSOCIATES	22
AGRIGENETICS CORP.	750
BENTECH LABORATORIES	7
BIO HUMA NETICS	15
BIOTECHNICA AGRICULTURE	50
CALGENE, INC.	165
CROP GENETICS INTERNATIONAL CORP.	75
DEKALB-PFIZER GENETICS	235
DNA PLANT TECHNOLOGY CORP.	80
ECOGEN, INC.	75
EMTECH RESEARCH	30
ESCAGENETICS CORP.	50
FORGENE, INC.	4
HAWAII BIOTECHNOLOGY GROUP, INC.	10
INTERNATIONAL PLANT RESEARCH INST.	50
MAIZE GENETIC RESOURCES, INC.	2
MYCOGEN CORP.	50
NATIVE PLANTS, INC.	625
NEOGEN CORP.	40
NITRAGIN CO.	80
PHYTOGEN	20
PLANT GENETICS, INC.	63
R & A PLANT/SOIL, INC.	6
SUNGENE TECHNOLOGIES CORP.	28
UNITED AGRISEEDS, INC.	192
UNIVERSITY GENETICS CO.	59
ZOECON CORP.	130

Listing 6-2 Company Personnel by Primary Classification

Company Name	Total Employees
Biomass Conversion (C)	
REPAP TECHNOLOGIES	16
Biosensors/Bioelectronics (D)	
BIOCHEM TECHNOLOGY, INC.	12
BIOMATERIALS INTERNATIONAL, INC.	30
GENTRONIX LABORATORIES, INC.	6
INTEGRATED CHEMICAL SENSORS	8
Bioseparations (E)	
A/G TECHNOLOGY CORP.	8
BEND RESEARCH, INC.	60
BIOCONSEP, INC.	6
BIOPROBE INTERNATIONAL, INC.	15
COAL BIOTECH CORP.	10
IMRE CORP.	25
PHARMACIA LKB BIOTECHNOLOGY, INC.	N A
SEPRACOR, INC.	40
VIAGENE, INC.	9
Biotechnology Equipment (F)	
AN-CON GENETICS	3
APPLIED BIOSYSTEMS	350
APPLIED PROTEIN TECHNOLOGIES, INC.	5
BIO-RECOVERY, INC.	49
BIOTECHNOLOGY DEVELOPMENT CORP.	35
ENDOTRONICS, INC.	45
INTELLIGENETICS, INC.	52
MULTIPLE PEPTIDE SYSTEMS, INC.	6
NYGENE CORP.	18
VEGA BIOTECHNOLOGIES, INC.	60
Biotechnology Reagents (G)	
ADVANCED BIOTECHNOLOGIES	17
ALFACELL CORP.	22
AMERICAN BIOCLINICAL	24
AMERICAN QUALEX INTERNATIONAL, INC.	10
BIODESIGN, INC.	3
BIOSEARCH, INC.	100
BIOTEST DIAGNOSTICS CORP.	10
CHEMGENES	8
CHEMICAL DYNAMICS CORP.	45
CHEMICON INTERNATIONAL, INC.	8
CLINICAL SCIENCES, INC.	50
COLLABORATIVE RESEARCH, INC.	150
CONSOLIDATED BIOTECHNOLOGY, INC.	15

Company Name	Total Employees
Biotechnology Reagents (Cont.)	
DIAGNOSTIC PRODUCTS CORP.	270
EPITOPE, INC.	60
FLOW LABORATORIES, INC.	200
HAZLETON BIOTECHNOLOGIES CO.	500
IBF BIOTECHNICS, INC.	5
IMMUNOSYSTEMS, INC.	4
INTER-AMERICAN RESEARCH ASSOC.	10
INTERNATIONAL BIOTECHNOLOGIES, INC.	85
INTERNATIONAL ENZYMES, INC.	4
JACKSON IMMUNORESEARCH LABS, INC.	16
MARCOR DEVELOPMENT CO.	6
MOLECULAR BIOLOGY RESOURCES, INC.	10
MOLECULAR GENETIC RESOURCES	5
MONOCLONETICS INTERNATIONAL, INC.	5
PEL-FREEZ BIOLOGICALS, INC.	175
PROMEGA CORP.	85
SCOTT LABORATORIES	375
STRATAGENE, INC.	77
SUMMA MEDICAL CORP.	33
UNITED STATES BIOCHEMICAL CORP.	N A
UNIVERSITY MICRO REFERENCE LAB	6
UPSTATE BIOTECHNOLOGY, INC.	4
VECTOR LABORATORIES, INC.	50
ZYMED LABORATORIES, INC.	20
Cell Culture (H)	
BIONIQUE LABS, INC.	11
HYCLONE LABORATORIES, INC.	107
IMCERA BIOPRODUCTS, INC.	5
KARYON TECHNOLOGY, LTD.	25
QUEUE SYSTEMS, INC.	85
Commodity Chemicals (I)	
AMERICAN BIOGENETICS CORP.	10
BIOASSAY SYSTEMS CORP.	65
GENEX, INC.	80
Specialty Chemicals (J)	
BACHEM BIOSCIENCE, INC.	15
BACHEM, INC.	45
BIOMEDICAL TECHNOLOGIES	7
BIOSCIENCE MANAGEMENT, INC.	8
BIOTECHNICA INTERNATIONAL	145
CALZYME LABORATORIES, INC.	10
CAMBRIDGE RESEARCH BIOCHEMICALS	75

Listing 6-2 Company Personnel by Primary Classification

Company Name	Total Employees
Specialty Chemicals (Cont.)	
CARBOHYDRATES INTERNATIONAL, INC.	25
CREATIVE BIOMOLECULES	70
DELTOWN CHEMURGIC CORP.	90
ENZYME BIO-SYSTEMS, LTD.	101
GENE-TRAK SYSTEMS	196
GENENCOR, INC.	130
INCELL CORP.	23
INFERGENE CO.	25
LIFECORE BIOMEDICAL, INC.	20
MARTEK	16
MICROBIO RESOURCES	30
NORTH COAST BIOTECHNOLOGY, INC.	2
O.C.S. LABORATORIES, INC.	12
OMNI BIOCHEM, INC.	5
PAMBEC LABORATORIES	6
PENINSULA LABORATORIES, INC.	75
PROVESTA CORP.	50
REPLIGEN CORP.	95
SOUTHERN BIOTECHNOLOGY ASSOC.	12
SYNTHETIC GENETICS, INC.	6
VIROSTAT, INC.	1
WASHINGTON BIOLAB	10
WESTBRIDGE RESEARCH GROUP	15
Clinical Diagnostics (K)	
ABN	45
ADVANCED MAGNETICS, INC.	40
AMERICAN DIAGNOSTICA, INC.	9
ANTIVIRALS, INC.	5
ATLANTIC ANTIBODIES	60
BEHRING DIAGNOSTICS	100
BINAX, INC.	17
BIOGENEX LABORATORIES	42
BIOMERICA, INC.	35
BIONETICS RESEARCH	200
BIOSTAR MEDICAL PRODUCTS, INC.	30
BIOTECH RESEARCH LABS, INC.	167
BIOTHERM	10
BIOTX (BIOSCIENCES CORP. OF TEXAS)	20
BMI, INC.	10
BOEHRINGER MANNHEIM DIAGNOSTICS	1,000
BRAIN RESEARCH, INC.	N A
CALIFORNIA INTEGRATED DIAGNOSTICS	9
CAMBRIDGE BIOSCIENCE CORP.	97
CELLMARK DIAGNOSTICS	30

Company Name	Total Employees
Clinical Diagnostics (Cont.)	
CELLULAR PRODUCTS, INC.	101
CENTOCOR	246
CIBA-CORNING DIAGNOSTIC CORP.	750
CLINETICS CORP.	10
CLONTECH LABORATORIES, INC.	30
COVALENT TECHNOLOGY CORP.	16
CYTOGEN CORP.	140
CYTOTECH, INC.	43
DIAGNON CORP.	70
DIAGNOSTIC TECHNOLOGY, INC.	60
DIAMEDIX, INC.	35
DIGENE DIAGNOSTICS, INC.	20
ELECTRO-NUCLEONICS, INC.	720
ENDOGEN, INC.	7
ENZO-BIOCHEM, INC.	110
GAMMA BIOLOGICALS, INC.	166
GEN-PROBE, INC.	127
GENELABS, INC.	90
GENESIS LABS, INC.	29
GENETIC DIAGNOSTICS CORP.	17
GENETIC SYSTEMS CORP.	400
HYGEIA SCIENCES	70
ICL SCIENTIFIC, INC.	70
IMCLONE SYSTEMS, INC.	50
IMMUNEX, INC.	5
IMMUNO CONCEPTS, INC.	20
IMMUNOTECH CORP.	15
KALLESTAD DIAGNOSTICS	330
LABSYSTEMS, INC.	10
LIFE TECHNOLOGIES, INC.	1,300
LIFECODES CORP.	53
MAST IMMUNOSYSTEMS, INC.	45
MICROGENICS	70
MOLECULAR BIOSYSTEMS, INC.	55
MOLECULAR DIAGNOSTICS, INC.	30
MONOCLONAL ANTIBODIES, INC.	110
MUREX CORP.	180
ONCOGENE SCIENCE, INC.	65
ONCOR, INC.	30
ORGANON TEKNIKA CORP.	270
POLYCELL, INC.	6
PROGENX	20
PROTATEK INTERNATIONAL, INC.	37
PROTEINS INTERNATIONAL, INC.	6
QUEST BIOTECHNOLOGY, INC.	9

Listing 6-2 Company Personnel by Primary Classification

Company Name	Total Employees
Clinical Diagnostics (Cont.)	
QUIDEL	325
RECOMTEX CORP.	12
RESEARCH & DIAGNOSTIC ANTIBODIES	N A
SENETEK PLC	14
SYVA CO.	1,000
TECHNICLONE INTERNATIONAL, INC.	15
TECHNOGENETICS, INC.	30
THREE-M (3M) DIAGNOSTIC SYSTEMS	130
UNITED BIOMEDICAL, INC.	10
XENOGEN	N A
Energy (L)	
PETROFERM, INC.	68
PETROGEN, INC.	6
Food Production/Processing (M)	
BIO-TECHNICAL RESOURCES, INC.	25
BIOTICS RESEARCH CORP.	22
COORS BIOTECH PRODUCTS CO.	185
CYANOTECH CORP.	25
IGENE BIOTECHNOLOGY, INC.	30
Production/Fermentation (O)	
BIOPURE	31
ENZYME CENTER INC.	30
ENZYME TECHNOLOGY CORP.	26
VERAX CORP.	80
Therapeutics (P)	
ALPHA I BIOMEDICALS, INC.	11
AMERICAN LABORATORIES, INC.	26
AMGEN	340
APPLIED DNA SYSTEMS, INC.	7
AUTOMEDIX SCIENCES, INC.	7
BIO-RESPONSE, INC.	45
BIOGEN	250
BIOGROWTH, INC.	16
BIOTECHNOLOGY GENERAL CORP.	130
BIOTHERAPEUTICS, INC.	120
BIOTHERAPY SYSTEMS, INC.	8
CALIFORNIA BIOTECHNOLOGY, INC.	155
CEL-SCI CORP.	2
CETUS CORP.	711
CHIRON CORP.	300
CISTRON BIOTECHNOLOGY, INC.	30

Company Name	Total Employees
Therapeutics (Cont.)	
COOPER DEVELOPMENT CO.	900
DAMON BIOTECH, INC.	101
DNAX RESEARCH INSTITUTE	105
ELANEX PHARMACEUTICALS, INC.	10
ENZON	22
ENZON, INC.	31
EXOVIR, INC.	5
GENENTECH	1,500
GENETIC THERAPY, INC.	4
GENETICS INSTITUTE, INC.	302
GENZYME CORP.	185
HANA BIOLOGICS, INC.	130
HOUSTON BIOTECHNOLOGY, INC.	30
HYBRITECH, INC.	708
IDEC PHARMACEUTICAL	45
IMMUNEX CORP.	153
IMMUNO MODULATORS LABS, INC.	25
IMMUNOMEDICS, INC.	55
IMMUNOVISION, INC.	6
IMREG, INC.	50
INGENE, INC.	65
INTEGRATED GENETICS, INC.	200
INTERFERON SCIENCES, INC.	70
INVITRON CORP.	165
KIRIN-AMGEN	5
LEE BIOMOLECULAR RESEARCH LABS	16
LUCKY BIOTECH CORP.	4
LYPHOMED, INC.	1,200
MEDAREX, INC.	3
NATIONAL GENO SCIENCES	4
NEORX CORP.	194
NEUREX, CORP.	22
PROBIOLOGICS INTERNATIONAL, INC.	5
SCRIPPS LABORATORIES	60
SERONO LABS	198
SPHINX BIOTECHNOLOGIES	3
SYNERGEN, INC.	101
SYNGENE PRODUCTS	20
T CELL SCIENCES, INC.	65
TRITON BIOSCIENCES, INC.	180
UNIGENE LABORATORIES, INC.	30
VIRAGEN	15
WELGEN MANUFACTURING, INC.	200
XOMA CORP.	149
ZYMOGENETICS, INC.	95

Listing 6-2 Company Personnel by Primary Classification

Company Name	Total Employees
Vaccines (Q)	
CENTRAL BIOLOGICS, INC.	7
IMMUNOMED CORP.	17
NORDEN LABS	145
PRAXIS BIOLOGICS	180
Waste Disposal/Treatment (R)	
ADVANCED MINERAL TECHNOLOGIES	17
ASTRE CORPORATE GROUP	148
BIOCONTROL SYSTEMS	29
BIOSPHERICS, INC.	250
BIOSYSTEMS, INC.	15
BIOTROL, INC.	30
HUNTER BIOSCIENCES	8
MICROBE MASTERS	37
Aquaculture (S)	
INCON CORP.	6
NEUSHUL MARICULTURE	6
Marine Natural Products (T)	
AQUASYNERGY, LTD.	5
MARICULTURA, INC.	8
MARINE BIOLOGICALS, INC.	2
OCEAN GENETICS	35
SEAPHARM, INC.	50
Veterinary Areas (V)	
A.M. BIOTECHNIQUES, INC.	5
AGRITECH SYSTEMS, INC.	90
ALLELIC BIOSYSTEMS	7
AMBICO, INC.	25
AMTRON	16
APPLIED BIOTECHNOLOGY	45
APPLIED MICROBIOLOGY, INC.	20
BIOMED RESEARCH LABS, INC.	18
CODON CORP.	120
FERMENTA ANIMAL HEALTH	359
GAMETRICS, LTD.	20
KIRKEGAARD & PERRY LABS, INC.	40
RIBI IMMUNOCHEM RESEARCH, INC.	35
SYNBIOTICS CORP.	89
TRANSGENIC SCIENCES, INC.	3

Company Name	Total Employees
Research (W)	
ABC RESEARCH CORP.	82
ELCATECH	4
IMMUNOGEN, INC.	11
LIFE SCIENCES	20
MELOY LABORATORIES, INC.	246
SIBIA	65
Immunological Products (X)	
AMERICAN BIOTECHNOLOGY CO.	5
ANTIBODIES, INC.	35
BERKELEY ANTIBODY CO., INC.	27
BETHYL LABS, INC.	8
BIOPRODUCTS FOR SCIENCE, INC.	19
CHARLES RIVER BIOTECH. SERVICES	55
DAKO CORP.	50
E-Y LABORATORIES, INC.	20
EARL-CLAY LABORATORIES, INC.	7
IGEN	60
IMMUNETECH PHARMACEUTICALS	45
INTEK DIAGNOSTICS, INC.	35
INTER-CELL TECHNOLOGIES, INC.	9
MONOCLONAL PRODUCTION INT'L	4
SERAGEN, INC.	250
TRANSFORMATION RESEARCH, INC.	4
Toxicology (Y)	
ANGENICS	50
APPLIED GENETICS LABS, INC.	11
ENVIRONMENTAL DIAGNOSTICS, INC.	55
IDETEK, INC.	14
RICERCA, INC.	160
TOXICON	35
Biomaterials (1)	
BIOMATRIX, INC.	50
BIOPOLYMERS, INC.	30
Fungi (2)	
MYCOSEARCH, INC.	6
Drug Delivery (3)	
CYGNUS RESEARCH CORP.	18
LIPOSOME COMPANY, INC.	67
LIPOSOME TECHNOLOGY, INC.	65
RHOMED, INC.	8

Listing 6-2 Company Personnel by Primary Classification

Company Name	Total Employees
Medical Devices (4)	
COLLAGEN CORP.	225
Testing/Analytical Services (5)	
MICROBIOLOGICAL ASSOCIATES, INC.	180
TEKTAGEN	3
VIVIGEN, INC.	70

Notes:

Companies are listed by their primary classification. Total company personnel is given, where available. (NA=not available)

LISTING 6-3 THE LARGEST BIOTECHNOLOGY COMPANIES (1)

Rank	Company	Employees	Code (2)	Financing (3)
1	GENENTECH	1,500	P	PUB
2	LIFE TECHNOLOGIES, INC.	1,300	K	PUB
3	LYPHOMED, INC.	1,200	P	PUB
4	BOEHRINGER MANNHEIM DIAGN.	1,000	K	SUB
5	SYVA CO.	1,000	K	SUB
6	COOPER DEVELOPMENT CO.	900	P	PUB
7	AGRIGENETICS CORP.	750	B	SUB
8	CIBA-CORNING DIAGNOSTIC CORP.	750	K	SUB
9	ELECTRO-NUCLEONICS, INC.	720	K	PUB
10	CETUS CORP.	711	P	PUB
11	HYBRITECH, INC.	708	P	SUB
12	NATIVE PLANTS, INC.	625	B	PUB
13	HAZLETON BIOTECHNOLOGIES CO.	500	G	SUB
14	GENETIC SYSTEMS CORP.	400	K	SUB
15	SCOTT LABORATORIES	375	G	SUB
16	FERMENTA ANIMAL HEALTH	359	V	SUB
17	APPLIED BIOSYSTEMS, INC.	350	F	PUB
18	AMGEN	340	P	PUB
19	KALLESTAD DIAGNOSTICS	330	K	PRI
20	QUIDEL	325	K	PRI
21	GENETICS INSTITUTE, INC.	302	P	PUB
22	CHIRON CORP.	300	P	PUB
23	DIAGNOSTIC PRODUCTS CORP.	270	G	PUB
24	ORGANON TEKNIKA CORP.	270	K	SUB
25	BIOGEN	250	P	PUB
26	BIOSPHERICS, INC.	250	R	PUB
27	SERAGEN, INC.	250	X	PRI
28	CENTOCOR	246	K	PUB
29	MELOY LABORATORIES, INC.	246	W	SUB
30	DEKALB-PFIZER GENETICS	235	B	PRI
31	COLLAGEN CORP.	225	4	PUB
32	BIONETICS RESEARCH	200	K	SUB
33	FLOW LABORATORIES, INC.	200	G	SUB
34	INTEGRATED GENETICS, INC.	200	P	PUB
35	WELGEN MANUFACTURING, INC.	200	P	PRI
36	SERONO LABS	198	P	PRI
37	GENE-TRAK SYSTEMS	196	J	PUB
38	NEORX CORP.	194	P	PRI
39	UNITED AGRISEEDS, INC.	192	B	SUB
40	COORS BIOTECH PRODUCTS CO.	185	M	SUB
41	GENZYME CORP.	185	P	PUB
42	MICROBIOLOGICAL ASSOCIATES, INC.	180	5	SUB
43	MUREX CORP.	180	K	PRI

Listing 6-3 The Largest Biotechnology Companies

Rank	Company	Employees	Code (2)	Financing (3)
44	PRAXIS BIOLOGICS, INC.	180	Q	PUB
45	TRITON BIOSCIENCES, INC.	180	P	SUB
46	PEL-FREEZ BIOLOGICALS, INC.	175	G	PRI
47	BIOTECH RESEARCH LABS, INC.	167	K	PUB
48	GAMMA BIOLOGICALS, INC.	166	K	PUB
49	CALGENE, INC.	165	B	PUB
50	INVITRON CORP.	165	P	PRI
51	RICERCA, INC.	160	Y	SUB
52	CALIFORNIA BIOTECHNOLOGY, INC.	155	P	PUB
53	IMMUNEX CORP.	153	P	PUB
54	COLLABORATIVE RESEARCH, INC.	150	G	PUB
55	XOMA CORP.	149	P	PUB
56	ASTRE CORPORATE GROUP	148	R	PRI
57	BIOTECHNICA INTERNATIONAL	145	J	PUB
58	NORDEN LABS	145	Q	SUB
59	CYTOGEN CORP.	140	K	PUB
60	ADVANCED GENETIC SCIENCES	130	B	PUB
61	BIOTECHNOLOGY GENERAL CORP.	130	P	PUB
62	GENENCOR, INC.	130	J	PRI
63	HANA BIOLOGICS, INC.	130	P	PUB
64	THREE-M (3M) DIAGNOSTIC SYSTEMS	130	K	SUB
65	ZOECON CORP.	130	B	SUB
66	GEN-PROBE, INC.	127	K	PUB
67	BIOTHERAPEUTICS, INC.	120	P	PUB
68	CODON CORP.	120	V	PRI
69	ENZO-BIOCHEM, INC.	110	K	PUB
70	MONOCLONAL ANTIBODIES, INC.	110	K	PUB
71	HYCLONE LABORATORIES, INC.	107	H	PRI
72	DNAX RESEARCH INSTITUTE	105	P	SUB
73	CELLULAR PRODUCTS	101	K	PUB
74	DAMON BIOTECH, INC.	101	P	SUB
75	ENZYME BIO-SYSTEMS, LTD.	101	J	SUB
76	SYNERGEN, INC.	101	P	PUB
77	BEHRING DIAGNOSTICS	100	K	SUB
78	BIOSEARCH, INC.	100	G	SUB

Notes:
1. These are the largest companies, in personnel number, of the 360 biotechnology companies in Listing 2-2, and data were collected by questionnaire or telephone between mid-1987 and February 1988. Companies on this list have 100 or more employees.
2. Primary industry code. See industry classification codes in Tables 1-4 or 5-1.
3. PRI=Privately held, PUB=Public corporation, SUB=Subsidiary.

TABLE 6-1 PERSONNEL BY PRIMARY CLASSIFICATION

Code	Classification	Number Primary (1)	Average Employees (2)	Total Employees (3)
A	Agriculture, Animal	8	40	322
B	Agriculture, Plant	30	116	3,465
C	Biomass Conversion	1	16	16
D	Biosensors/Bioelectronics	4	14	56
E	Bioseparations	9	22	194
F	Biotechnology Equipment	10	63	626
G	Biotechnology Reagents	38	70	2,626
H	Cell Culture, General	5	47	233
I	Chemicals, Commodity	2	38	75
J	Chemicals, Specialty	30	44	1,320
K	Diagnostics, Clinical Human	75	130	9,728
L	Energy	2	37	74
M	Food Production/Processing	5	57	287
N	Mining	1	10	10
O	Production/Fermentation	4	42	167
P	Therapeutics	65	150	9,731
Q	Vaccines	5	76	381
R	Waste Disposal/Treatment	9	68	608
S	Aquaculture	2	6	12
T	Marine Natural Products	5	14	70
U	Consulting**	0		
V	Veterinary	15	56	836
W	Research	6	71	428
X	Immunological Products	16	40	634
Y	Toxicology	6	54	325
Z	Venture Capital/Financing**	0		
1	Biomaterials	2	40	80
2	Fungi	1	6	6
3	Drug Delivery Systems	4	40	158
4	Medical Devices	1	23	23
5	Testing/Analytical Services	3	84	253

Notes:

**Only companies using the new biotechnologies were sampled for this book. Consulting and venture capital firms working with biotechnology were not included.

1. The number of companies with a primary focus in the indicated category are given.

2. Average number of employees for companies with indicated primary focus.

3. Total number of employees in all companies with indicated primary focus.

Table 6-2 Personnel by Company Type

TABLE 6-2 PERSONNEL BY COMPANY TYPE

| Type (1) | n | Employees (2) | | |
		Mean	Median	Sum
All Companies	352	90.3	30	31,811
Private Companies	197	34.5	17	6,794
Public Companies	101	149.5	65	15,099
Subsidiary Companies	54	183.7	85	9,918

Notes:
1. Of the 360 companies we have included in our database as U.S. biotechnology firms (see Listing 2-1), 354 have provided data on their current personnel size. Of these, more than half are privately held, and the rest have public stock traded or are subsidiaries of other companies.
2. The mean and median values are given for each type category. The sum represents the total number of personnel in the companies in each category. Although the private firms represent more than 55 percent of the number of companies, they only represent about 20 percent of the personnel.

FIGURE 6-1 TYPES OF EMPLOYEES IN THE COMPANIES

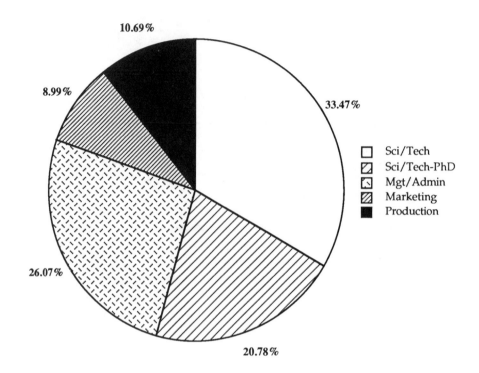

These data are the average employee composition of 133 biotechnology companies from questionnaire data taken between February 1987 and February 1988. The categories are science/technology employees without Ph.D.(Sci/Tech), science/technology employees with Ph.D. (Sci/Tech-PhD), management and administrative staff (Mgt/Admin), marketing, and production. The total science/technology employees represent 55 percent and include the Sci/Tech section and S/T with Ph.D. section. Percentages of employees in each category are shown. The average company in this sample was founded in 1981 and has 86 employees.

Figure 6-2 Personnel Numbers by Industry Classification

FIGURE 6-2 PERSONNEL NUMBERS BY INDUSTRY CLASSIFICATION

A. Average employee numbers in biotechnology companies with the indicated primary industry classification (1).

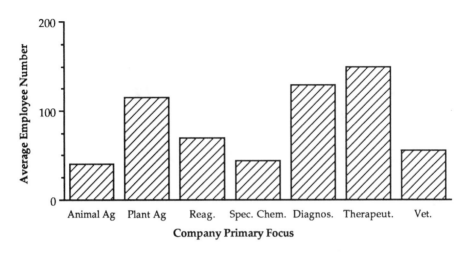

B. Total personnel for the biotechnology companies in each classification (2).

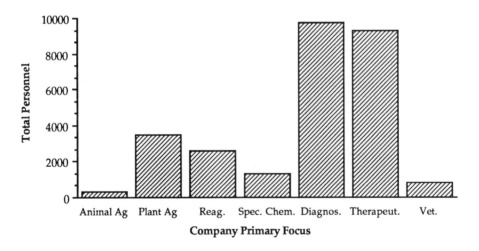

Notes:
1. The indicated classifications are: animal agriculture (Code A, 8 companies), plant agriculture (B, 30), biotechnology reagents (G, 38), specialty chemicals (J, 30), clinical diagnostics (K, 75), therapeutics (P, 62), and veterinary areas (V, 15).
2. Total personnel are the average employee number multiplied by the number of companies with a primary focus in the indicated industry class.

FIGURE 6-3 PERSONNEL GROWTH

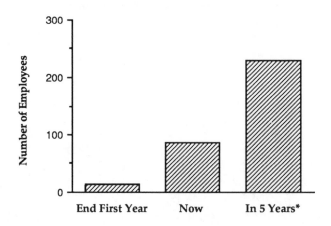

*expected

Managers of 300 U.S. biotechnology companies were asked, in a questionnaire in mid-1987, how many employees their company had at the end of its first year, at present, and how many employees they expected to have at the end of five years. Questionnaires were returned by 135 of the firms contacted, and 85 percent of the respondents were from company CEO's or presidents. The employee numbers at the end of the first year ranged from 1 to 160 and averaged 13.5. These companies represented 86 employees on average in mid-1987 and ranged from 2 to 1,000. This represented an average of 145 percent yearly growth in the biotechnology industry over the companies' first years (the average firm was six years old). Expected growth, to an average of 230 employees (ranging from 5 to 2,000 employees), is projected to be at an average rate of 27.4 percent per year.

SECTION 7   FINANCING OF BIOTECHNOLOGY COMPANIES

INTRODUCTION

The financing availability and the range of alternatives for U.S. biotechnology companies are believed by many to be primarily responsible for the explosive growth of these firms in the United States, which has not been seen in other countries. To date, most biotechnology companies remain privately held, but about 30 percent have made public stock offerings and another 20 percent are subsidiaries of other companies. Listing 7-1 shows the type of each company as private, public or subsidiary, and gives information on equity held, parent firm, or year of public offering, where available. Listings 7-2 and 7-3 demonstrate equity held in U.S. biotechnology companies by other U.S. companies. Listing 7-4 gives equity held in U.S. biotechnology companies by foreign firms. Listing 7-5 demonstrates the same data sorted by country of the equity purchaser.

In a questionnaire-based study with direct information from almost 140 company executives, we were able to gather financing data from their firms. Although the individual firm financing, R&D budget and expected revenue data are confidential, Figure 7-1 demonstrates the primary funding source (i.e., the source of at least 50 percent of company funding) for the firms.

Figure 7-2 shows the average financing of firms for seven key industry classifications. Average financing ranged from less than $10 million for reagents-oriented companies to $40 million on average for therapeutics-oriented companies. Financing data, extrapolated to the 360 biotechnology firms in this Guide, amounts to more than $7 billion by mid 1987.

Listing 7-1 Financing of Companies

LISTING 7-1 FINANCING OF COMPANIES

Company	Year	Financing (1)
A.M. BIOTECHNIQUES, INC.	1980	PRI
A/G TECHNOLOGY CORP.	1981	PRI
ABC RESEARCH CORP.	1967	PRI
ABN	1981	PUB; FORMERLY AMERICAN BIONETICS
ADVANCED BIOTECHNOLOGIES	1982	PRI
ADVANCED GENETIC SCIENCES	1979	PUB 1983; EQ-ROHM & HAAS, KARLSHAMNS OLIEFABRIKER (7.8%) PLANT GENETIC SYSTEMS (33%)
ADVANCED MAGNETICS, INC.	1981	PUB
ADVANCED MINERAL TECHNOL.	1982	PRI
AGDIA, INC.	1981	PRI
AGRACETUS	1981	PRI; JV OF W.R. GRACE, A.E. STALEY CO., & CETUS
AGRI-DIAGNOSTICS ASSOCIATES	1983	PRI; JV OF KOPPERS (60%), DNAP (40%)
AGRIGENETICS CORP.	1975	SUB OF LUBRIZOL
AGRITECH SYSTEMS, INC.	1984	PRI
ALFACELL CORP.	1981	PUB
ALLELIC BIOSYSTEMS	1984	PRI
ALPHA I BIOMEDICALS, INC.	1982	PUB
AMBICO, INC.	1974	PRI
AMERICAN BIOCLINICAL	1977	PRI
AMERICAN BIOGENETICS CORP.	1984	PRI
AMERICAN BIOTECHNOLOGY CO.	1984	PRI
AMERICAN DIAGNOSTICA, INC.	1983	PRI
AMERICAN LABORATORIES, INC.	1962	PRI
AMERICAN QUALEX INT'L, INC.	1981	PRI
AMGEN	1980	PUB; EQ-FERMENTA (7.95%), BECTON DICKINSON (20%), SMITHKLINE
AMTRON	1981	PRI; FORMERLY CALLED INTRON
AN-CON GENETICS	1982	SUB OF UNIVERSITY GENETICS
ANGENICS	1980	PRI
ANTIBODIES, INC.	1961	PRI
ANTIVIRALS, INC.	1980	PRI; EQ-DUPONT
APPLIED BIOSYSTEMS, INC.	1981	PUB; EQ-BECTON DICKINSON (10%)
APPLIED BIOTECHNOLOGY, INC.	1982	PRI; EQ-DUPONT, PRUTEC
APPLIED DNA SYSTEMS, INC.	1982	PUB 1984
APPLIED GENETICS LABS, INC.	1984	PRI
APPLIED MICROBIOLOGY, INC.	1983	PUB; EQ-AMERICAN CYANAMID
APPLIED PROTEIN TECHNOL.	1984	PRI
AQUASYNERGY, LTD.	1986	PRI
ASTRE CORPORATE GROUP	1967	PRI
ATLANTIC ANTIBODIES	1972	SUB OF INCSTAR
AUTOMEDIX SCIENCES, INC.	1978	PUB 1987
BACHEM BIOSCIENCE, INC.	1987	PRI
BACHEM, INC.	1971	PRI
BEHRING DIAGNOSTICS	1952	SUB OF AMERICAN HOECHST CORP.
BEND RESEARCH, INC.	1975	PRI
BENTECH LABORATORIES	1984	PRI
BERKELEY ANTIBODY CO., INC.	1983	PRI
BETHYL LABS, INC.	1977	PRI
BINAX, INC.	1986	PRI

Company	Year	Financing (1)
BIO HUMA NETICS	1984	PRI; FORMERLY SUNBURST MINING CO.
BIO TECHNIQUES LABS, INC.	1982	PRI
BIO-RECOVERY, INC.	1983	PRI
BIO-RESPONSE, INC.	1972	PUB 1982
BIO-TECHNICAL RESOURCES, INC.	1962	PRI
BIOCHEM TECHNOLOGY, INC.	1977	PRI; EQ-BERWIND CORP.
BIOCONSEP, INC.	1983	PRI
BIOCONTROL SYSTEMS	1985	PRI; EQ-FLOW INDUSTRIES
BIODESIGN, INC.	1987	PRI
BIOGEN	1979	PUB; EQ-MONSANTO, SCHERING-PLOUGH (20%), INCO
BIOGENEX LABORATORIES	1981	PRI
BIOGROWTH, INC.	1985	PRI
BIOKYOWA, INC.	1983	SUB OF KYOWA HAKKO KOGYO-JAPAN
BIOMATERIALS INT'L, INC.	1981	PRI; EQ-HOSPITAL CORP. OF AMERICA
BIOMATRIX, INC.	1981	PRI
BIOMED RESEARCH LABS, INC.	1974	PRI
BIOMEDICAL TECHNOLOGIES	1981	PRI
BIOMERICA, INC.	1971	PUB 1983; FORMERLY NMS PHARMACEUT.
BIONETICS RESEARCH	1985	SUB OF ORGANON TEKNIKA, NETHERLANDS
BIONIQUE LABS, INC.	1983	PRI
BIOPOLYMERS, INC.	1984	PRI
BIOPROBE INTERNATIONAL, INC.	1983	PRI; EQ-TIE CORP, STEINER DIAMOND
BIOPRODUCTS FOR SCIENCE, INC.	1985	PRI; SAME OWNER OWNS UNITED VACCINES
BIOPURE	1984	PRI; FORMERLY TUFTS NEW ENGLAND ENZYME CENTER
BIOSCIENCE MANAGEMENT, INC.	1984	PRI
BIOSEARCH, INC.	1977	SUB OF NEW BRUNSWICK SCI.
BIOSPHERICS, INC.	1967	PUB; EQ-MONTEDISON, FERUZZI GROUP
BIOSTAR MEDICAL PRODUCTS, INC.	1983	PRI; EQ-KEMA NOBEL, SKANDIGEN
BIOSYSTEMS, INC.	1976	PRI
BIOTECH RESEARCH LABS, INC.	1973	PUB 1986; EQ-DUPONT (7%), FUJIZOKI, ETHYL (1.8%)
BIOTECHNICA AGRICULTURE	1987	SUB OF BIOTECHNICA INT'L
BIOTECHNICA INTERNATIONAL	1981	PUB 1983; EQ-SEAGRAMS (31.6%), STATE FARM (26%)
BIOTECHNOLOGY DEVEL. CORP.	1982	PUB
BIOTECHNOLOGY GENERAL CORP.	1980	PUB 1983; EQ-PHARMACIA (4.5%)
BIOTEST DIAGNOSTICS CORP.	1946	SUB OF BIOTEST SERUM INST. (WGER)
BIOTHERAPEUTICS, INC.	1984	PUB
BIOTHERAPY SYSTEMS, INC.	1984	PUB; EQ-DAMON BIOTECH (80%)
BIOTHERM	1985	PRI
BIOTICS RESEARCH CORP.	1975	PRI
BIOTROL, INC.	1985	PRI
BIOTX	1985	PRI
BMI, INC.	1984	PRI
BOEHRINGER MANNHEIM DIAGN.	1975	SUB OF BOEHRINGER MANNHEIM BIOCHEM.
BRAIN RESEARCH, INC.	1968	PRI; EQ-DREXEL BURNHAM LAMBERT (9%)
CALGENE, INC.	1980	PUB; EQ-ALLIED CHEMICAL, MITSUI, RHONE-POULENC, CONTINENTAL GRAIN
CALIFORNIA BIOTECHNOLOGY	1981	PUB; EQ-AMERICAN HOME PRODUCTS (15%)
CALIFORNIA INTEGRATED DIAGN.	1981	SUB OF INFRAGENE; EQ-SYNCOR
CALZYME LABORATORIES, INC.	1983	PRI

Listing 7-1 Financing of Companies

Company	Year	Financing (1)
CAMBRIDGE BIOSCIENCE CORP.	1981	PUB 1983; EQ-JOHNSON & JOHNSON
CAMBRIDGE RESEARCH BIOCHEM.	1980	SUB OF CAMBRIDGE RESEARCH BIOCHEMICALS, LTD.; EQ- MILLIGEN (11%)
CARBOHYDRATES INT'L, INC.	1987	SUB OF MEDICARB, SWEDEN
CEL-SCI CORP.	1983	PUB; FORMERLY CALLED INTERLEUKIN-2
CELLMARK DIAGNOSTICS	1986	SUB OF ICI AMERICAS
CELLULAR PRODUCTS	1982	PUB 1983
CENTOCOR	1979	PUB 1982; EQ-FMC
CENTRAL BIOLOGICS, INC.	1982	PRI
CETUS CORP.	1971	PUB; EQ-AMOCO, CHEVRON, SQUIBB (5%)
CHARLES RIVER BIOTECH. SERVICES	1983	SUB OF BAUSCH & LOMB
CHEMGENES	1981	PRI
CHEMICAL DYNAMICS CORP.	1972	PRI
CHEMICON INTERNATIONAL, INC.	1981	PRI
CHIRON CORP.	1981	PUB 1983; EQ-JOHNSON & JOHNSON
CIBA-CORNING DIAGNOSTIC CORP.	1985	SUB; JV OF CIBA-GEIGY & CORNING GLASS
CISTRON BIOTECHNOLOGY, INC.	1982	PUB; EQ-DUPONT, WARNER LAMBERT
CLINETICS CORP.	1979	PRI
CLINICAL SCIENCES, INC.	1971	PUB
CLONTECH LABORATORIES, INC.	1984	PRI
COAL BIOTECH CORP.	1984	PRI
CODON CORP.	1980	PRI; EQ-CELANESE
COLLABORATIVE RESEARCH, INC.	1961	PUB; EQ-DOW CHEMICAL (5%), GREEN CROSS
COLLAGEN CORP.	1975	PUB; EQ-MONSANTO (23.6%)
CONSOLIDATED BIOTECHNOLOGY	1983	PRI
COOPER DEVELOPMENT CO.	1980	PUB
COORS BIOTECH PRODUCTS CO.	1984	SUB OF ADOLPH COORS CO.
COVALENT TECHNOLOGY CORP.	1981	SUB OF KMS INDUSTRIES
CREATIVE BIOMOLECULES	1981	PRI; EQ-LUBRIZOL
CROP GENETICS INT'L CORP.	1981	PUB 1987
CYANOTECH CORP.	1983	PRI; EQ-DAIKYO OIL, CONTINENTAL CORP., COSMO OIL
CYGNUS RESEARCH CORP.	1985	PRI; EQ-ELF AQUATAINE
CYTOGEN CORP.	1980	PUB 1986; EQ-LEDERLE (15%), KODAK, AMPERSAND
CYTOTECH, INC.	1982	PRI
DAKO CORP.	1979	SUB OF DAKOPATTS A/S (DENMARK)
DAMON BIOTECH, INC.	1978	SUB OF DAMON CORP.; PUBLIC 1983
DEKALB-PFIZER GENETICS	1982	PRI; JV OF PFIZER (30%), DEKALB (70%)
DELTOWN CHEMURGIC CORP.	1968	PRI; SPLIT OFF FROM DEL FOODS
DIAGNON CORP.	1981	PUB
DIAGNOSTIC PRODUCTS CORP.	1972	PUB
DIAGNOSTIC TECHNOLOGY, INC.	1980	PRI
DIAMEDIX, INC.	1986	PRI
DIGENE DIAGNOSTICS, INC.	1984	PRI
DNA PLANT TECHNOLOGY CORP.	1981	PUB; EQ-CAMPBELL SOUP CO.
DNAX RESEARCH INSTITUTE	1980	SUB OF SCHERING PLOUGH
E-Y LABORATORIES, INC.	1978	PRI
EARL-CLAY LABORATORIES, INC.	1984	PUB
ECOGEN, INC.	1983	PUB 1986; EQ-AMERICAN CYANAMID
ELANEX PHARMACEUTICALS, INC.	1984	PRI
ELCATECH	1984	PRI
ELECTRO-NUCLEONICS, INC.	1960	PUB; EQ-PHARMACIA (15.8%)
EMBREX, INC.	1985	PRI; EQ-AM. CYANAMID, PLANT RESOURCES

Company	Year	Financing (1)
EMTECH RESEARCH	1983	PRI; PREVIOUSLY SUB OF ENZYME MEDICAL TECHNOLOGIES
ENDOGEN, INC.	1985	PRI
ENDOTRONICS, INC.	1981	PUB; EQ-CELANESE
ENVIRONMENTAL DIAGN., INC.	1983	PUB
ENZO-BIOCHEM, INC.	1976	PUB; EQ-JOHNSON & JOHNSON (14.6%)
ENZON	1965	PUB; EQ-KODAK; FORMERLY DIAGNOSTIC DATA, INC.
ENZON, INC.	1981	PUB 1984
ENZYME BIO-SYSTEMS, LTD.	1983	SUB OF CPC, INTERNATIONAL
ENZYME CENTER INC.	1978	PRI
ENZYME TECHNOLOGY CORP.	1981	SUB OF GREAT LAKES CHEMICAL CORP.
EPITOPE, INC.	1979	PUB; EQ-SYNTEX
ESCAGENETICS CORP.	1986	PUB
EXOVIR, INC.	1981	PUB
FERMENTA ANIMAL HEALTH	1986	SUB OF FERMENTA AB, SWEDEN
FLOW LABORATORIES, INC.	1961	SUB OF FLOW GENERAL, INC.
FORGENE, INC.	1986	PRI
GAMETRICS, LTD.	1974	PRI
GAMMA BIOLOGICALS, INC.	1969	PUB 1983
GEN-PROBE, INC.	1984	PUB 1987; EQ-HYBRITECH, LILLY
GENE-TRAK SYSTEMS	1981	PUB
GENELABS, INC.	1984	PRI
GENENCOR, INC.	1982	PRI; JV OF CORNING, GENENTECH; EQ-A.E. STALEY (33%), KODAK (16%)
GENENTECH	1976	PUB 1980; EQ-ALFA LAVAL, LUBRIZOL, MONSANTO, BOEHR. INGLEHEIM
GENESIS LABS, INC.	1984	PRI
GENETIC DIAGNOSTICS CORP.	1981	PUB
GENETIC ENGINEERING, INC.	1980	PUB; FOUNDERS OWN 60% EQUITY
GENETIC SYSTEMS CORP.	1980	SUB OF BRISTOL-MYERS
GENETIC THERAPY, INC.	1986	PRI
GENETICS INSTITUTE, INC.	1981	PUB 1986; EQ-BAXTER-TRAVENOL
GENEX, INC.	1977	PUB; EQ-MONSANTO
GENTRONIX LABORATORIES, INC.	1972	PRI; FORMERLY CALLED EMV ASSOC., INC.
GENZYME CORP.	1981	PUB; EQ-ROTHSCHILD FUND, COLE YEAGER & WOOD
GRANADA GENETICS CORP.	1979	SUB OF GRANADA CORP.
HANA BIOLOGICS, INC.	1979	PUB 1986; EQ-FUJIZOKI, RECORDATI
HAWAII BIOTECHNOLOGY GROUP	1982	PRI; EQ-HAWAII SUGARPLANTERS (10%)
HAZLETON BIOTECHNOLOGIES CO.	1983	SUB OF HAZELTON LABORATORIES CORP
HOUSTON BIOTECHNOLOGY, INC.	1984	PRI
HUNTER BIOSCIENCES	1987	PRI
HYBRITECH, INC.	1978	SUB OF ELI LILLY
HYCLONE LABORATORIES, INC.	1975	PRI
HYGEIA SCIENCES	1980	PUB 1986; EQ-DENNISON; FORMERLY BTC DIAGNOSTICS
IBF BIOTECHNICS, INC.	1987	SUB OF IBF, SA; EQ-INSTITUT MERIEUX (40%), RHONE-POULENC (40%), BANEXI (20%)
ICL SCIENTIFIC, INC.	1981	SUB OF HYCOR BIOMED
IDEC PHARMACEUTICAL	1986	PRI
IDETEK, INC.	1983	PRI; EQ-NOVO INDUSTRI (20%)
IGEN	1982	PRI; EQ-GRYPHON VENTURES
IGENE BIOTECHNOLOGY, INC.	1981	PUB; FORMERLY IGI BIOTECHNOLOGY, INC.

Listing 7-1 Financing of Companies

Company	Year	Financing (1)
IMCERA BIOPRODUCTS, INC.	1986	SUB OF INT'L MINERALS & CHEMICALS
IMCLONE SYSTEMS, INC.	1984	PRI
IMMUCELL CORP.	1982	PUB
IMMUNETECH PHARMACEUTICALS	1981	PRI
IMMUNEX CORP.	1981	PUB
IMMUNEX, INC.	1981	PRI
IMMUNO CONCEPTS, INC.	1981	PRI
IMMUNO MODULATORS LABS, INC.	1981	PRI; EQ-VENTREX (25%), GRANADA
IMMUNOGEN, INC.	1981	PRI
IMMUNOMED CORP.	1979	PRI
IMMUNOMEDICS, INC.	1983	PUB; EQ-JOHNSON & JOHNSON (9%)
IMMUNOSYSTEMS, INC.	1981	PRI
IMMUNOTECH CORP.	1980	PRI
IMMUNOVISION, INC.	1985	PRI
IMRE CORP.	1981	PUB; JV (RESEARCH) WITH PHARMACIA
IMREG, INC.	1981	PUB 1981
INCELL CORP.	1982	PRI
INCON CORP.	1983	PRI
INFERGENE CO.	1984	PUB; EQ-MOLSON CO.
INGENE, INC.	1980	PUB; EQ-ATLANTIC RICHFIELD, BEATRICE
INTEGRATED CHEMICAL SENSORS	1984	PRI
INTEGRATED GENETICS, INC.	1981	PUB
INTEK DIAGNOSTICS, INC.	1983	PRI
INTELLIGENETICS, INC.	1980	PRI; JV OF AMOCO (60%), INTELLICORP (40%)
INTER-AMERICAN RES. ASSOC.	1979	PRI
INTER-CELL TECHNOLOGIES, INC.	1982	PRI
INTERFERON SCIENCES, INC.	1981	SUB OF NAT'L PATENTS DEVELOPMENT (79%) EQ-ANHEUSER BUSCH
INT'L BIOTECHNOLOGIES, INC.	1982	SUB OF EASTMAN KODAK
INT'L ENZYMES, INC.	1983	PRI
INT'L PLANT RESEARCH INST.	1978	PRI; EQ-BIO-RAD (70%)
INVITRON CORP.	1984	PRI; EQ-MONSANTO
JACKSON IMMUNORESEARCH LABS	1982	PRI
KALLESTAD DIAGNOSTICS	1967	PRI; DIAGNOSTIC DIV OF ERBAMONT, INC.
KARYON TECHNOLOGY, LTD.	1984	PRI; EQ-TEXTRON
KIRIN-AMGEN	1984	PRI; JV OF KIRIN, AMGEN
KIRKEGAARD & PERRY LABS, INC.	1979	PRI
LABSYSTEMS, INC.	1983	SUB OF LABSYSTEMS OY-FINLAND
LEE BIOMOLEC. RESEARCH LABS	1980	PRI
LIFE SCIENCES	1962	PUB 1968
LIFE TECHNOLOGIES, INC.	1983	PUB; EQ-DEXTER CORP. (62%)
LIFECODES CORP.	1982	SUB OF QUANTUM CORP. 1/88
LIFECORE BIOMEDICAL, INC.	1975	PUB; EQ-COOPER CO. (22%)
LIPOSOME COMPANY, INC.	1981	PUB 1986; EQ-LILLY, MERRILL LYNCH, NIPPON, YAMANOUCHI
LIPOSOME TECHNOLOGY, INC.	1981	PRI
LUCKY BIOTECH CORP.	1984	SUB OF LUCKY LTD., S. KOREA
LYPHOMED, INC.	1981	PUB; EQ-FUJISAWA (22%)
MAIZE GENETIC RESOURCES	1984	PRI
MARCOR DEVELOPMENT CO.	1977	PRI
MARICULTURA, INC.	1984	PRI
MARINE BIOLOGICALS, INC.	1981	PRI
MARTEK	1985	PRI
MAST IMMUNOSYSTEMS, INC.	1979	PRI

Company	Year	Financing (1)
MEDAREX, INC.	1987	PRI; JV OF ESSEX CHEM., DARTMOUTH COLL.
MELOY LABORATORIES, INC.	1970	SUB OF RORER GROUP, INC.
MICROBE MASTERS	1982	PRI
MICROBIO RESOURCES	1981	PRI
MICROBIOLOGICAL ASSOC., INC.	1949	PRI
MICROGENESYS, INC.	1983	PRI
MICROGENICS	1981	PRI; EQ-YAMANOUCHI
MOLECULAR BIOLOGY RESOURCES	1986	PRI
MOLECULAR BIOSYSTEMS, INC.	1981	PUB; EQ-DUPONT (8%)
MOLECULAR DIAGNOSTICS, INC.	1981	SUB OF BAYER AG (60%)
MOLECULAR GENETIC RESOURCES	1983	PRI
MOLECULAR GENETICS, INC.	1979	PUB 1982; EQ-AMERICAN CYANAMID (20%), MARTIN MARIETTA (21%)
MONOCLONAL ANTIBODIES, INC.	1979	PUB; EQ-DIAGNOSTIC PRODUCTS (6.6%)
MONOCLONAL PRODUCTION INT'L	1983	PRI
MONOCLONETICS INT'L, INC.	1984	PRI
MULTIPLE PEPTIDE SYSTEMS	1986	PRI
MUREX CORP.	1984	PRI; EQ-PILOT LABS, SYNTEX
MYCOGEN CORP.	1982	PUB-1987
MYCOSEARCH, INC.	1979	PRI
NATIONAL GENO SCIENCES	1980	PRI
NATIVE PLANTS, INC.	1973	PUB; EQ-ELF AQUATAINE, KYOWA HAKKO, SANDOZ, SUMITOMO, MARTIN MARIETTA, MCCORMICK (2.5%)
NEOGEN CORP.	1981	PRI
NEORX CORP.	1984	PRI; EQ-KODAK (40%), GENETICS INSTITUTE, MALINKRODT
NEUREX CORP.	1986	PRI; VENTURE CAPITAL
NEUSHUL MARICULTURE	1978	PRI; EQ-M&S NEUSHUL
NITRAGIN COMPANY	1898	SUB OF ALLIED CORP
NORDEN LABS	1983	SUB OF SMITHKLINE BECKMAN
NORTH COAST BIOTECHNOLOGY	1986	PRI
NYGENE CORP.	1985	PRI
O.C.S. LABORATORIES, INC.	1983	PRI
OCEAN GENETICS	1981	PRI
OMNI BIOCHEM, INC.	1986	PRI
ONCOGENE SCIENCE, INC.	1983	PUB 1986; EQ-BECTON-DICKINSON (20%), PFIZER
ONCOR, INC.	1983	PUB
ORGANON TEKNIKA CORP.	1978	SUB OF ORGANON TEKNIKA-NETHERLANDS
PAMBEC LABORATORIES	1987	PRI
PEL-FREEZ BIOLOGICALS, INC.	1911	PRI
PENINSULA LABORATORIES, INC.	1971	PRI
PETROFERM, INC.	1977	SUB OF PETROLEUM FERM. NV
PETROGEN, INC.	1980	PRI
PHARMACIA LKB BIOTECHNOLOGY	1960	SUB OF PHARMACIA AB, UPPSALA, SWEDEN
PHARMAGENE	1986	PRI
PHYTOGEN	1980	PRI
PLANT GENETICS, INC.	1981	PUB 1987; EQ-KIRIN BREWERY, MCCORMICK, TWYFORD
POLYCELL, INC.	1983	SUB OF QUEST BIOTECHNOLOGY
PRAXIS BIOLOGICS	1983	PUB 1986; EQ-BRISTOL-MYERS (10.6%)
PROBIOLOGICS INT'L, INC.	1987	PRI
PROGENX	1987	PRI

Listing 7-1 Financing of Companies

Company	Year	Financing (1)
PROMEGA CORP.	1978	PRI; FORMERLY CALLED BIOTEC, INC.
PROTATEK INTERNATIONAL, INC.	1984	PRI
PROTEINS INTERNATIONAL, INC.	1983	PRI
PROVESTA CORP.	1975	SUB OF PHILLIPS PETROLEUM
QUEST BIOTECHNOLOGY, INC.	1986	PUB
QUEUE SYSTEMS, INC.	1980	PRI
QUIDEL	1981	PRI; EQ-BECTON DICKINSON, DIATEK, MAATSCHAPPIJ
R & A PLANT/SOIL, INC.	1978	PRI
RECOMTEX CORP.	1983	PRI
REPAP TECHNOLOGIES	1981	SUB OF REPAP ENTERPRISES, CANADA
REPLIGEN CORP.	1981	PUB 1986; EQ-C. ITOH
RESEARCH & DIAGN. ANTIBODIES	1985	PRI
RHOMED, INC.	1986	PRI
RIBI IMMUNOCHEM RESEARCH	1981	PUB; EQ-EXECUTIVE LIFE INS. CO.
RICERCA, INC.	1986	SUB OF FERMENTA; FORMERLY SDS BIOTECH
SCOTT LABORATORIES	1981	SUB OF MICROBIOLOGICAL SCIENCES, INC.
SCRIPPS LABORATORIES	1980	PRI; NO CONNECTION WITH SCRIPPS INST.
SEAPHARM, INC.	1983	PRI
SENETEK PLC	1983	PUB 1983
SEPRACOR, INC.	1984	PRI; EQ-ALCOA, AMERICAN CYANAMID
SERAGEN, INC.	1979	PRI; EQ-BOSTON UNIV.
SERONO LABS	1971	PRI; U.S. AFFILIATE OF ARES APPLIED RESEARCH, SWITZ
SIBIA	1981	PRI; EQ-PHILLIPS PETROLEUM (37%), SALK INSTITUTE (63%)
SOUTHERN BIOTECHNOL. ASSOC.	1982	PRI
SPHINX BIOTECHNOLOGIES	1987	PRI
STRATAGENE, INC.	1984	PRI 82%; VEN CAP 18%
SUMMA MEDICAL CORP.	1978	PUB
SUNGENE TECHNOLOGIES CORP.	1981	PRI; MANY INVESTORS
SYNBIOTICS CORP.	1982	PUB; EQ-INT'L MIN. & CHEM. (6.42%), SMITHKLINE (6.12%)
SYNERGEN, INC.	1981	PUB 1987; EQ-ELI LILLY
SYNGENE PRODUCTS	1980	SUB OF TECHAMERICA GROUP
SYNTHETIC GENETICS, INC.	1985	PRI
SYNTRO CORP.	1981	PUB; EQ-CPC INT'L, LUBRIZOL, NIPPON OIL & FATS
SYVA CO.	1966	SUB OF SYNTEX
T CELL SCIENCES, INC.	1984	PUB; EQ-AETNA, SYNTEX
TECHNICLONE INT'L, INC.	1981	PUB
TECHNOGENETICS, INC.	1983	PUB 1984
TEKTAGEN	1987	PRI
THREE-M (3M) DIAGNOS. SYSTEMS	1981	SUB OF 3M; FORMER NAME WAS AXONICS
TOXICON	1978	PRI
TRANSFORMATION RESEARCH, INC.	1981	PRI
TRANSGENIC SCIENCES, INC.	1988	PRI
TRITON BIOSCIENCES, INC.	1983	SUB OF SHELL OIL CO.
UNIGENE LABORATORIES, INC.	1980	PUB 1987; EQ-ERBAMONT, SIGMA TAU
UNITED AGRISEEDS, INC.	1981	SUB OF DOW CHEMICAL; EQ-GENETICS INSTITUTE (29%)
UNITED BIOMEDICAL, INC.	1983	PRI
UNITED STATES BIOCHEMICAL	1973	PRI
UNIVERSITY GENETICS CO.	1981	PUB

Company	Year	Financing (1)
UNIVERSITY MICRO REFERENCE LAB	1980	PRI
UPSTATE BIOTECHNOLOGY, INC.	1986	PRI
VECTOR LABORATORIES, INC.	1976	PRI
VEGA BIOTECHNOLOGIES, INC.	1979	PUB 1983; EQ-DUPONT
VERAX CORP.	1978	PRI; EQ-GENENTECH, PHARMACIA, SANDOZ, ELI LILLY
VIAGENE, INC.	1987	PRI; VENTURE CAPITAL
VIRAGEN	1980	PUB; FORMED BY MEDICORE
VIROSTAT, INC.	1985	PRI
VIVIGEN, INC.	1981	PUB 1983
WASHINGTON BIOLAB	1986	PRI
WELGEN MANUFACTURING, INC.	1988	PRI; JV OF GENETICS INSTITUTE, WELLCOME BIOTECHNOLOGY
WESTBRIDGE RESEARCH GROUP	1982	PRI
XENOGEN	1981	PRI; EQ-JOHNSON & JOHNSON
XOMA CORP.	1981	PUB 1986
ZOECON CORP.	1968	SUB OF SANDOZ; PREVIOUSLY OWNED BY OCCIDENTAL PETROLEUM
ZYMED LABORATORIES, INC.	1980	PRI
ZYMOGENETICS, INC.	1981	PRI; EQ-NOVO INDUSTRI, EISAI CO., CABLE & HOWSE

Note:

1.   All companies were classified PRI=Private, PUB=Public or SUB=Subsidiary as main financing type. EQ indicates equity holders.   JV indicates a new joint venture company formed. Where available, other information, such as year that company went public, former company name, or previous ownership is shown. Data were collected from company contacts and literature between February 1987 and March 1988 and were generated from the **Companies** database.

Listing 7-2 Companies with Equity Held

LISTING 7-2 COMPANIES WITH EQUITY HELD

Biotechnology Firm (1)	Equity Purchaser (U.S.)	Year (2)	Details (2)
ADVANCED GENETIC SCIENCES	ROHM & HAAS	79	$12,000,000
AGRIGENETICS	KELLOGG CO.	82	$10,000,000
AMERICAN BIONUCLEAR	SYNTEX	84	17%
AMGEN	BECTON DICKINSON		20%
AMGEN	SMITHKLINE	86	$5,000,000
ANTIVIRALS, INC.	DUPONT		
APPLIED BIOSYSTEMS, INC.	BECTON DICKINSON	84	10%
APPLIED BIOTECHNOLOGY, INC.	DUPONT		
AUTOMATED DIAGNOSTICS	UNIVERSITY GENETICS	87	
BIOCHEM TECHNOLOGY, INC.	BERWIND CORP.		
BIOCONTROL SYSTEMS	FLOW INDUSTRIES		
BIOGEN	MONSANTO	81	
BIOGEN	SCHERING-PLOUGH	81	20%
BIOMATERIALS INTERNATIONAL	HOSP. CORP. OF AMERICA		
BIOTECH RESEARCH LABS	DUPONT	84	7%
BIOTECH RESEARCH LABS	ETHYL CORP.		1.80%
BIOTECHNICA INTERNATIONAL	STATE FARM INSURANCE	88	26%
BIOTHERAPY SYSTEMS	DAMON BIOTECH	84	80%
BRAIN RESEARCH, INC.	DREXEL BURNHAM LAMBERT		9%
CALGENE	ALLIED CHEMICAL		
CALGENE	CONTINENTAL GRAIN		
CALIFORNIA BIOTECHNOLOGY	AMERICAN HOME PRODUCTS	85	15%
CALIFORNIA INTEGRATED DIAGN.	SYNCOR		
CAMBRIDGE BIOSCIENCE CORP.	JOHNSON & JOHNSON		
CAMBRIDGE RESEARCH BIOCHEM.	MILLIGEN (MILLIPORE)	86	11%
CENTOCOR	FMC		
CETUS	AMOCO	82	
CETUS	SQUIBB	87	5%
CETUS	CHEVRON		
CHIRON	JOHNSON & JOHNSON	85	$8,000,000
CODON CORP.	CELANESE	87	$12,000,000
COLLABORATIVE RESEARCH	DOW CHEMICAL	81	5%
COLLAGEN	MONSANTO	81	23.60%
CREATIVE BIOMOLECULES	LUBRIZOL		
CYANOTECH CORP.	THE CONTINENTAL CORP.		
CYTOGEN	AMPERSAND		
CYTOGEN	EASTMAN KODAK	86	$15,000,000
CYTOGEN	LEDERLE	83	15%
DIAGNOSTIC, INC.	RORER	84	10%
DIAGNOSTIC REAGENT TECHNOL.	VENTREX	83	25%
DNA PLANT TECHNOLOGY	CAMPBELL SOUP CO.	81	$10,000,000
ECOGEN	AMERICAN CYANAMID	86	
ENDOTRONICS	CELANESE	86	$15,000,000
ENDOTRONICS	VENTURE FUNDING, LTD	87	
ENZO	JOHNSON & JOHNSON	82	14.60%
ENZON	EASTMAN KODAK	87	$15,000,000
EPITOPE	SYNTEX		
GEN-PROBE	HYBRITECH	84	$2,000,000
GENENCOR	A.E. STALEY MFG. CO.	84	33%
GENENCOR	EASTMAN KODAK	87	16%

Biotechnology Firm (1)	Equity Purchaser (U.S.)	Year (2)	Details (2)
GENENTECH	LUBRIZOL		
GENENTECH	MONSANTO	81	
GENETIC SYSTEMS CORP.	SYNTEX	85	18%
GENETICS INSTITUTE	BAXTER TRAVENOL	82	$4,900,000
GENEX	MONSANTO	81	
HAWAII BIOTECHNOLOGIES	HAWAII SUGARPLANTERS		10%
HEM RESEARCH	DUPONT	87	
HYGEIA SCIENCES	DENNISON MFG. CO.		
IGEN	GRYPHON VENTURES	87	$2,000,000
IMCERA BIOPRODUCTS	INT'L. MINERALS & CHEMICALS		
IMMUNO MODULATORS LABS	GRANADA CORP.		
IMMUNO MODULATORS LABS	VENTREX	83	25%
IMMUNOMEDICS, INC.	JOHNSON & JOHNSON		
INGENE	ATLANTIC RICHFIELD		
INGENE	BEATRICE CO.		
INT'L PLANT RESEARCH INST.	BIO-RAD LABORATORIES	83	70%
INTERFERON SCIENCES	ANHEUSER-BUSCH		
INVITRON	MONSANTO		
KARYON TECHNOLOGY	TEXTRON		
LIPOSOME COMPANY	ELI LILLY		
LIPOSOME COMPANY	MERRILL LYNCH		
MEDI-CONTROL CORP	SEARLE	86	10%
MOLECULAR BIOSYSTEMS	DUPONT	85	8%
MOLECULAR GENETICS	AMERICAN CYANAMID	81	20%
MOLECULAR GENETICS	MARTIN MARIETTA	82	21%
MONOCLONAL ANTIBODIES	DIAGNOSTIC PRODUCTS	87	6.60%
MUREX	SYNTEX		
NEORX	EASTMAN KODAK	86	40%
NEORX	GENETICS INSTITUTE	87	$7,500,000
NEUROGENETIC	BIOTECHNOLOGY GENERAL	87	$2,000,000
NATIVE PLANTS, INC.	MARTIN MARIETTA	82	
NATIVE PLANTS, INC.	MC CORMICK	85	2.50%
ONCOGENE SCIENCE	BECTON DICKINSON		20%
ONCOGENE SCIENCE	PFIZER	86	
PLANT GENETICS	MC CORMICK		
PRAXIS BIOLOGICS	BRISTOL-MYERS	87	10.60%
PROTEIN DATABASES	MILLIPORE	87	
QUIDEL	BECTON DICKINSON	85	
RIBI IMMUNOCHEM	EXECUTIVE LIFE INS. CO.		
SEPRACOR	ALCOA		
SEPRACOR	AMERICAN CYANAMID		
SERAGEN	BOSTON UNIVERSITY	87	
SIBIA	PHILLIPS PETROLEUM		
SYNBIOTICS	INT'L MINERALS & CHEMICAL	87	6.42%
SYNBIOTICS	SMITHKLINE	87	6.12%
SYNERGEN	ELI LILLY	84	$7,000,000
SYNTRO CORP.	LUBRIZOL		
SYVA	SYNTEX		
T CELL SCIENCES	AETNA LIFE AND CASUALTY		
T CELL SCIENCES	SYNTEX	85	$2,000,000
UNIGENE LABORATORIES	ERBAMONT		
UNITED AGRISEEDS	GENETICS INSTITUTE	85	29%
VEGA BIOTECHNOLOGIES	DUPONT	86	

Listing 7-2 Companies with Equity Held

Biotechnology Firm (1)	Equity Purchaser (U.S.)	Year (2)	Details (2)
VERAX CORP.	ELI LILLY		
VIRATEK	EASTMAN KODAK	84	$8,400,000
XENOGEN	JOHNSON & JOHNSON		

Notes:

1. Most biotechnology firms listed appear in the directory section (Listing 2-2). Data are from **Actions** database and questionnaires returned by 136 companies, and represent equity purchase actions and not necessarily equity currently held. These data may therefore vary from those in Listing 7-1, generated from the **Companies** database.

2. Details for year, amount, and percent of purchase are given, where available.

LISTING 7-3 U.S. PURCHASERS OF EQUITY

Equity Purchaser (U.S.)	Biotechnology Firm (1)	Year (2)	Details (2)
A.E. STALEY MFG. CO.	GENENCOR	84	33%
AETNA LIFE AND CASUALTY	T CELL SCIENCES		
ALCOA	SEPRACOR		
ALLIED CHEMICAL	CALGENE		
AMERICAN CYANAMID	ECOGEN	86	
AMERICAN CYANAMID	MOLECULAR GENETICS	81	20%
AMERICAN CYANAMID	SEPRACOR		
AMERICAN HOME PRODUCTS	CALIFORNIA BIOTECHNOLOGY	85	15%
AMOCO	CETUS	82	
AMPERSAND	CYTOGEN		
ANHEUSER-BUSCH	INTERFERON SCIENCES		
ATLANTIC RICHFIELD	INGENE		
BAXTER TRAVENOL	GENETICS INSTITUTE	82	$4,900,000
BEATRICE CO.	INGENE		
BECTON DICKINSON	AMGEN		20%
BECTON DICKINSON	APPLIED BIOSYSTEMS	84	10%
BECTON DICKINSON	ONCOGENE SCIENCE		20%
BECTON DICKINSON	QUIDEL	85	
BERWIND CORP.	BIOCHEM TECHNOLOGY, INC.		
BIO-RAD LABORATORIES	INT'L PLANT RESEARCH INST.	83	70%
BIOTECHNOLOGY GENERAL	NEUROGENETIC	87	$2,000,000
BOSTON UNIVERSITY	SERAGEN	87	
BRISTOL-MYERS	PRAXIS BIOLOGICS, INC.	87	10.6%
CAMPBELL SOUP CO.	DNA PLANT TECHNOLOGY	81	$10,000,000
CELANESE	CODON	87	$12,000,000
CELANESE	ENDOTRONICS	86	$15,000,000
CHEVRON	CETUS		
CONTINENTAL GRAIN	CALGENE		
DAMON BIOTECH	BIOTHERAPY SYSTEMS	84	80%
DENNISON MFG. CO.	HYGEIA SCIENCES		
DIAGNOSTIC PRODUCTS	MONOCLONAL ANTIBODIES	87	6.6%
DOW CHEMICAL	COLLABORATIVE RESEARCH	81	5%
DREXEL BURNHAM LAMBERT	BRAIN RESEARCH, INC.		9%
DUPONT	ANTIVIRALS, INC.		
DUPONT	APPLIED BIOTECHNOLOGY		
DUPONT	BIOTECH RESEARCH LABS	84	7%
DUPONT	HEM RESEARCH	87	
DUPONT	MOLECULAR BIOSYSTEMS	85	8%
DUPONT	VEGA BIOTECHNOLOGIES	86	
EASTMAN KODAK	CYTOGEN	86	$15,000,000
EASTMAN KODAK	ENZON	87	$15,000,000
EASTMAN KODAK	GENENCOR	87	16%
EASTMAN KODAK	NEORX	86	40%
EASTMAN KODAK	VIRATEK	84	$8,400,000
ELI LILLY	LIPOSOME COMPANY		
ELI LILLY	SYNERGEN	84	$7,000,000
ELI LILLY	VERAX CORP		
ERBAMONT	UNIGENE LABORATORIES		
ETHYL CORP.	BIOTECH RESEARCH LABS		1.8%
EXECUTIVE LIFE INS. CO.	RIBI IMMUNOCHEM		

Listing 7-3 U.S. Purchasers of Equity

Equity Purchaser (U.S.)	Biotechnology Firm (1)	Year (2)	Details (2)
FLOW INDUSTRIES	BIOCONTROL SYSTEMS		
FMC	CENTOCOR		
GENETICS INSTITUTE	NEORX	87	$7,500,000
GENETICS INSTITUTE	UNITED AGRISEEDS	85	29%
GRANADA CORP	IMMUNO MODULATORS LABS		
GRYPHON VENTURES	IGEN	87	$2,000,000
HAWAII SUGARPLANTERS	HAWAII BIOTECHNOLOGIES		10%
HOSP. CORP. OF AMERICA	BIOMATERIALS INT'L		
HYBRITECH	GEN-PROBE	84	$2,000,000
INT'L MINERALS & CHEMICALS	SYNBIOTICS	87	6.4%
INT'L MINERALS & CHEMICALS	IMCERA BIOPRODUCTS		
JOHNSON & JOHNSON	CAMBRIDGE BIOSCIENCE CORP.		
JOHNSON & JOHNSON	CHIRON	85	$8,000,000
JOHNSON & JOHNSON	ENZO	82	14.60%
JOHNSON & JOHNSON	IMMUNOMEDICS, INC.		
JOHNSON & JOHNSON	XENOGEN		
KELLOGG COMPANY	AGRIGENETICS	82	$10,000,000
LEDERLE	CYTOGEN	83	15%
LUBRIZOL	CREATIVE BIOMOLECULES		
LUBRIZOL	GENENTECH		
LUBRIZOL	SYNTRO CORP.		
MARTIN MARIETTA	MOLECULAR GENETICS	82	21%
MARTIN MARIETTA	NATIVE PLANTS, INC.	82	
MC CORMICK	NATIVE PLANTS, INC.	85	2.5%
MC CORMICK	PLANT GENETICS		
MERRILL LYNCH	LIPOSOME COMPANY		
MILLIGEN (MILLIPORE)	CAMBRIDGE RES. BIOCHEM.	86	11%
MILLIPORE	PROTEIN DATABASES	87	
MONSANTO	BIOGEN	81	
MONSANTO	COLLAGEN	81	23.60%
MONSANTO	GENENTECH	81	
MONSANTO	GENEX	81	
MONSANTO	INVITRON		
MONTEDISON	SRI INTERNATIONAL	87	$25,000,000
PFIZER	ONCOGENE SCIENCE	86	
PHILLIPS PETROLEUM	SIBIA		
ROHM & HAAS	ADVANCED GENETIC SCIENCES	79	$12,000,000
RORER	DIAGNOSTIC, INC.	84	10%
SCHERING-PLOUGH	BIOGEN	81	20%
SEARLE	MEDI-CONTROL CORP.	86	10%
SMITHKLINE	AMGEN	86	$5,000,000
SMITHKLINE	SYNBIOTICS	87	6.1%
SQUIBB	CETUS	87	5%
STATE FARM INSURANCE	BIOTECHNICA INTERNATIONAL	88	26%
SYNCOR	CALIFORNIA INTEGRATED DIAG.		
SYNTEX	AMERICAN BIONUCLEAR	84	17%
SYNTEX	EPITOPE		
SYNTEX	MUREX		
SYNTEX	SYVA		
SYNTEX	T CELL SCIENCES	85	$2,000,000
TEXTRON	KARYON TECHNOLOGY		
THE CONTINENTAL CORP.	CYANOTECH CORP.		
UNIVERSITY GENETICS	AUTOMATED DIAGNOSTICS	87	

Equity Purchaser (U.S.)	Biotechnology Firm (1)	Year (2)	Details (2)
VENTREX	DIAGNOSTIC REAGENT TECHNOL.	83	25%
VENTREX	IMMUNO MODULATORS LABS	83	25%
VENTURE FUNDING, LTD	ENDOTRONICS	87	

Notes:

1.   Most biotechnology firms listed appear in the directory section (Listing 2-2). Data are from **Actions** database and questionnaires returned by 136 companies, and represent equity purchase actions and not necessarily equity currently held. These data may therefore vary from those in Listing 7-1, generated from the **Companies** database.

2.   Details for year, amount, and percent of purchase are given, where available.

Listing 7-4 Foreign Equity

LISTING 7-4 FOREIGN EQUITY IN U.S. BIOTECHNOLOGY COMPANIES

Biotechnology Firm (1)	Equity Purchaser (Foreign)	Country	Yr.	Details (2)
ADVANCED GENETIC SCIENCES	KARLSHAMNS OLJEFABRIKER	SWE	86	7.80%
ADVANCED GENETIC SCIENCES	PLANT GENETIC SYSTEMS	BELG	83	33%
AGRIGENETICS	HOFFMANN-LA ROCHE	SWI	82	
AMGEN	AB FERMENTA	SWE	86	7.95%
APPLIED BIOTECHNOLOGY	PRUTEC, LTD.	UK	83	
BIOKYOWA, INC.	KYOWA HAKKO KOGYO	JAPAN		
BIOSPHERICS	FERUZZI GROUP	ITALY		
BIOSPHERICS	MONTEDISON	ITALY		
BIOSTAR MEDICAL PRODUCTS	KEMA NOBEL AB	SWE		
BIOSTAR MEDICAL PRODUCTS	SKANDIGEN AB	SWE	86	$2,300,000
BIOTECH RESEARCH LABS	FUJIZOKI	JAPAN	82	$500,000
BIOTECHNICA INTERNATIONAL	SEAGRAMS	CAN	84	11.60%
BIOTECHNICA INTERNATIONAL	SEAGRAMS	CAN	86	20%
BIOTECHNOLOGY GENERAL	PHARMACIA	SWE	85	4.50%
CALGENE	MITSUI	JAPAN		
CALGENE	RHONE-POULENC	FRANCE		
COLLABORATIVE RESEARCH	GREEN CROSS	JAPAN	81	14,000 Sh.
CYANOTECH	DAIKYO OIL CO.	JAPAN	86	$500,000
CYGNUS	ELF AQUATAINE	FRANCE		
ELECTRO-NUCLEONICS	PHARMACIA	SWE	86	15.80%
GENENTECH	ALFA LAVAL	SWE		
GENENTECH	BOEHRINGER INGELHEIM	WGER	85	$40,000,000
GENENTECH	JAPANESE INVESTORS GRP.	JAPAN	81	145,000 Sh.
GENZYME	ROTHSCHILD FUND	UK		
HANA BIOLOGICS	FUJIZOKI	JAPAN	82	
HANA BIOLOGICS	RECORDATI	ITALY	82	
IBF BIOTECHNICS	BANEXI	FRANCE		20%
IBF BIOTECHNICS	INST. MERIEUX	FRANCE		40%
IBF BIOTECHNICS	RHONE-POULENC	FRANCE		40%
IDETEK	NOVO INDUSTRI	DEN.	87	20%
INFERGENE	MOLSON	CAN		
INTERNATIONAL GENETIC SCI.	BIO-LOGICALS	CAN	85	$2,000,000
JOHNSON & JOHNSON	CELLTECH	UK	87	
LIPOSOME COMPANY	NIPPON LIFE	JAPAN		
LIPOSOME COMPANY	YAMANOUCHI	JAPAN		
LYPHOMED	FUJISAWA	JAPAN	84	22%
MICROGENICS	YAMANOUCHI	JAPAN		
MOLECULAR DIAGNOSTICS	BAYER	WGER	82	60%
MUREX	PILOT LABORATORIES	CAN	86	$5,000,000
NATIVE PLANTS, INC.	ELF AQUATAINE	FRANCE		
NATIVE PLANTS, INC.	KYOWA HAKKO KOGYO	JAPAN		
NATIVE PLANTS, INC.	SANDOZ	SWI		
NATIVE PLANTS, INC.	SUMITOMO	JAPAN		
PLANT CELL RESEARCH INST.	MONTEDISON	ITALY	87	$6,000,000
PLANT GENETICS	KIRIN BREWERY	JAPAN	84	$1,000,000
PLANT GENETICS	TWYFORD SEED CO.	UK		
QUIDEL	MAATSCHAPPIJ VOOR IND.	NETH	88	$4,000,000
REPLIGEN	C. ITOH	JAPAN	83	30,000 Sh.
SYNTRO	NIPPON OIL AND FATS	JAPAN		

Biotechnology Firm (1)	Equity Purchaser (Foreign)	Country	Yr.	Details (2)
UNIGENE LABORATORIES	SIGMA TAU	ITALY		
VERAX CORP.	SANDOZ	SWI		
ZYMOGENETICS	EISAI	JAPAN		
ZYMOGENETICS	NOVO INDUSTRI	DEN		
ZYMOS	NOVO INDUSTRI	DEN	82	

Notes:
1. Most biotechnology firms listed appear in the directory section (Listing 2-2). Data are from **Actions** database and questionnaires returned by 136 companies, and represent equity purchase actions and not necessarily equity currently held. These data may therefore vary from those in Listing 7-1, generated from the **Companies** database.
2. Details for year, amount, shares purchased, and percent of purchase are given, where available.

Listing 7-5 Foreign Equity, by Country

LISTING 7-5 FOREIGN EQUITY, BY COUNTRY (1)

Country	Equity Purchaser	Biotechnology Firm (1)	Year	Details (2)
BELG	PLANT GENETIC SYSTEMS	ADVANCED GENETIC SCIENCES	83	33%
CAN	BIO-LOGICALS	INTERNATIONAL GENETIC SCI.	85	$2,000,000
CAN	MOLSON	INFERGENE		
CAN	PILOT LABORATORIES	MUREX	86	$5,000,000
CAN	SEAGRAMS	BIOTECHNICA INT'L	84	11.60%
CAN	SEAGRAMS	BIOTECHNICA INT'L	86	20%
DEN	NOVO INDUSTRI	IDETEK	87	20%
DEN	NOVO INDUSTRI	ZYMOGENETICS		
DEN	NOVO INDUSTRI	ZYMOS	82	
FRANCE	BANEXI	IBF BIOTECHNICS		20%
FRANCE	ELF AQUATAINE	CYGNUS		
FRANCE	ELF AQUATAINE	NATIVE PLANTS, INC.		
FRANCE	INST. MERIEUX	IBF BIOTECHNICS		40%
FRANCE	RHONE-POULENC	CALGENE		
FRANCE	RHONE-POULENC	IBF BIOTECHNICS		40%
ITALY	FERUZZI GROUP	BIOSPHERICS		
ITALY	MONTEDISON	BIOSPHERICS		
ITALY	MONTEDISON	PLANT CELL RESEARCH INST.	87	$6,000,000
ITALY	RECORDATI	HANA BIOLOGICS	82	
ITALY	SIGMA TAU	UNIGENE LABORATORIES		
JAPAN	C. ITOH	REPLIGEN	83	30,000 Sh.
JAPAN	DAIKYO OIL CO.	CYANOTECH	86	$500,000
JAPAN	EISAI	ZYMOGENETICS		
JAPAN	FUJISAWA	LYPHOMED	84	22%
JAPAN	FUJIZOKI	BIOTECH RESEARCH LABS	82	$500,000
JAPAN	FUJIZOKI	HANA BIOLOGICS	82	
JAPAN	GREEN CROSS	COLLABORATIVE RESEARCH	81	14,000 Sh.
JAPAN	JAPANESE INVESTORS GROUP	GENENTECH	81	145,000 Sh.
JAPAN	KIRIN BREWERY	PLANT GENETICS	84	$1,000,000
JAPAN	KYOWA HAKKO KOGYO	BIOKYOWA, INC.		
JAPAN	KYOWA HAKKO KOGYO	NATIVE PLANTS, INC.		
JAPAN	MITSUI	CALGENE		
JAPAN	NIPPON LIFE	LIPOSOME COMPANY		
JAPAN	NIPPON OIL AND FATS	SYNTRO		
JAPAN	SUMITOMO	NATIVE PLANTS, INC.		
JAPAN	YAMANOUCHI	LIPOSOME COMPANY		
JAPAN	YAMANOUCHI	MICROGENICS		
NETH	MAATSCHAPPIJ VOOR IND.	QUIDEL	88	$4,000,000
SWE	ALFA LAVAL	GENENTECH		
SWE	KEMA NOBEL AB	BIOSTAR MEDICAL PRODUCTS		
SWE	AB FERMENTA	AMGEN	86	7.95%
SWE	KARLSHAMNS OLJEFABRIKER	ADVANCED GENETIC SCIENCES	86	7.80%
SWE	PHARMACIA	BIOTECHNOLOGY GENERAL	85	4.50%
SWE	PHARMACIA	ELECTRO-NUCLEONICS	86	15.80%
SWE	SKANDIGEN AB	BIOSTAR MEDICAL PRODUCTS	86	$2,300,000
SWI	HOFFMANN-LA ROCHE	AGRIGENETICS	82	
SWI	SANDOZ	NATIVE PLANTS, INC.		
SWI	SANDOZ	VERAX CORP.		
UK	CELLTECH	JOHNSON & JOHNSON	87	

Country	Equity Purchaser	Biotechnology Firm (1)	Year	Details (2)
UK	PRUTEC, LTD.	APPLIED BIOTECHNOLOGY	83	
UK	ROTHSCHILD FUND	GENZYME		
UK	TWYFORD SEED CO.	PLANT GENETICS		
WGER	BAYER	MOLECULAR DIAGNOSTICS	82	60%
WGER	BOEHRINGER INGELHEIM	GENENTECH	85	$40,000,000

Notes:

1. Most biotechnology firms listed appear in the directory section (Listing 2-2). Data are from **Actions** database and questionnaires returned by 136 companies, and represent equity purchase actions and not necessarily equity currently held. These data may therefore vary from those in Listing 7-1, generated from the **Companies** database.

2. Details for year, amount, shares purchased, and percent of purchase are given, where available.

Figure 7-1 Primary Financing Sources Over Time

FIGURE 7-1 PRIMARY FINANCING SOURCES OVER TIME

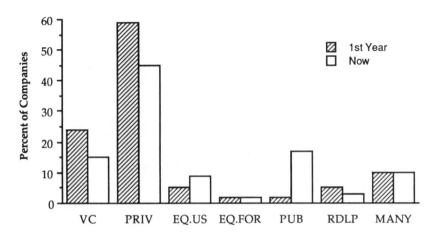

A total of 138 U.S. biotechnology companies were examined for the primary source of their financing at the end of their first year and in mid-1987 (Now), when the average company was 6 years old. Primary source is defined for this study as the source of more than 50 percent of total company financing. Financing sources shown above are venture capital (VC), private funds (PRIV), equity from U.S. companies (EQ.US), equity from foreign companies (EQ.FOR), public stock offerings (PUB), research and development limited partnerships (RDLP) or multiple sources with no one source accounting for at least 50 percent of the total (MANY).

FIGURE 7-2 FINANCING OF COMPANIES BY INDUSTRY CLASSIFICATION

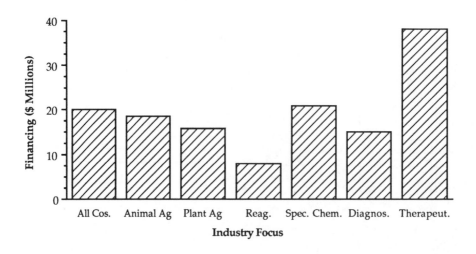

Average financing of 121 U.S. biotechnology companies and average financing of these companies by their primary focus is shown. Values include all financing from all sources from their start to the present (mid 1987 for most companies in this study). Data came from all 121 companies, 6 animal agriculture companies, 8 plant agriculture companies, 3 reagents companies, 3 specialty chemicals companies, 19 diagnostics companies and 28 therapeutics companies.

SECTION 8 COMPANY REVENUES

INTRODUCTION

We were able to ascertain biotechnology company revenues for the current (1987-1988) fiscal year from 139 companies. Although individual company revenue data are kept confidential for most firms, we have analyzed these data by industry groupings. Table 8-1 shows that the average firm has expected revenues of $11 million, and expected revenue data from private, public and subsidiary firms are also shown. Extrapolated to the entire set of 360 U.S. biotechnology firms, total revenues are expected to be about $4 billion for the current fiscal year.

Figure 8-1 shows average revenues for seven key industry classifications. They range from about $5 million for veterinary or reagents companies to about $20 million for diagnostics or therapeutics companies. Figure 8-2 is a graphical representation of the data in Table 8-1, both mean and median data for all companies together as well as breakdowns for private, public and subsidiary companies.

Table 8-1 Revenues of Companies by Type

TABLE 8-1 REVENUES OF COMPANIES BY TYPE

	n	Expected Revenues	Range Low	Range High	Median
All Companies	139	$11,000,000	$75,000	$230,500,000	$2,700,000
Private Companies	73	$2,620,000	$75,000	$15,000,000	$1,400,000
Public Companies	50	$18,000,000	$500,000	$230,500,000	$6,000,000
Subsidiary Companies	16	$28,200,000	$350,000	$100,000,000	$8,500,000

Note:
The mean expected revenues for the current fiscal year (1987-1988) are shown for 139 U.S. biotechnology companies and separately for these that are public, private or subsidiaries. In addition, the range and median values are provided.

FIGURE 8-1 REVENUES OF COMPANIES BY INDUSTRY CLASSIFICATION

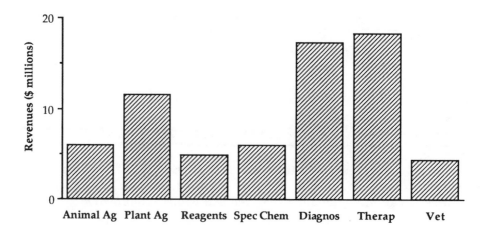

Expected annual revenues for 1987-1988 for U.S. biotechnology companies with the indicated industry focus are shown. Data are the mean values for 4 animal agriculture companies, 17 plant agriculture companies, 13 reagents companies, 14 specialty chemicals/enzymes companies, 30 diagnostics companies, 33 pharmaceutical companies, and 6 veterinary-related companies. Other industry classifications are not shown as there were not enough data to be meaningful. Data are from questionnaire and telephone contacts with company executives. Expected revenues for all 139 companies in the sample averaged $11,000,000.

Figure 8-2 Revenues of Companies

FIGURE 8-2 REVENUES OF COMPANIES

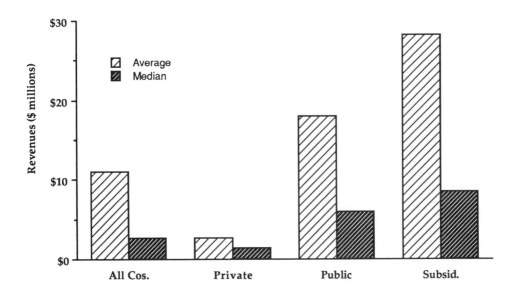

The average and median expected revenues for the current fiscal year (1987-1988) are shown, as given by managers of 139 U.S. biotechnology companies. In addition, data are given for 73 private companies, 50 public companies and 16 subsidiary companies.

SECTION 9   RESEARCH AND DEVELOPMENT

INTRODUCTION

Industry Changes

As can be seen in Figure 6-3 and its accompanying text, biotechnology firms have been undergoing tremendous growth. However, most of the products that will emanate from the advances in the new technology are still in the development phase and can be considered potential products only. As firms grow in size and begin to have a product line, it is likely that their initial focus on research and development must broaden to downstream activities. It is therefore likely that the managers of biotechnology firms realize the need for a shift in emphasis from science to business. We tested this hypothesis in a questionnaire-based study of U.S. biotechnology companies (1).

Questionnaires were sent in mid-1987 to managers of 300 of the U.S. biotechnology companies listed in this Guide. The response rate was 45 percent, with 85 percent of the questionnaires answered by the companies' top management. Respondents were questioned about the relative proportion of science to business in their total operations and the importance of research, development, production and marketing. Ratings were given for the end of the first year of the company's operation, at the present time, and for what was expected in five years. We were also able to look at these ratings by company size and company industry focus.

Proportion of Science to Business

Seen in Figure 9-1, at the end of their first year the companies weighted the relative proportion of science/technology to business in their total operations as almost three to one. Currently, the proportion is almost equal, with a 57 to 43 emphasis on science over business. The view toward the future furthers this trend toward business activities with a 55 to 45 percent emphasis in favor of business.

Importance of the Firms' Activities

In the companies' first year of operation, research was rated as being higher in importance than product development, manufacturing, or marketing. (See Figure 9-2.) The ratings have changed over time and are expected to change further. At present, product development is rated more important than the other three functions, with research significantly reduced in importance. Manufacturing and marketing are now elevated in importance to about the same level as research. Marketing is expected to be of highest importance to the company in five years, followed by product development and manufacturing. Research, most important to the firms early in their histories, is projected to be the least important activity in five years.

Further Analysis

The relative proportion of science to business appeared independent of company size over the three time points. The breakdown of firms by industry area reveals some differences. Companies working with therapeutics had the highest relative proportion of science to business in their first year, at present or projected for five years from now (at 81 percent, 65 percent and 48 percent, respectively). In contrast, agricultural biotechnology companies (plant and animal combined) had the lowest relative proportion of science to business (at 72 percent, 51 percent and 40 percent). Further, therapeutics companies rated research more important than did agricultural or diagnostics companies at all three time points.

The importance of manufacturing increased over time in the therapeutics and agricultural companies, as it did in all companies combined. In contrast, diagnostics companies reported the largest increase in the importance of manufacturing between their first year and the present, with no significant increase expected for the next five years. Marketing importance increased significantly for *all* firms over time. It is clearly rated the most important company function expected in five years from now.

Internal and External Activities

The biotechnology firm managers were also asked what percentage of research, development, marketing and manufacturing they expected to conduct externally to their company, as opposed to in-house. Large companies look to the outside for marketing help now and expect to do so in the future. In contrast, small companies have about twice as much external research activities than do the large firms at all three time points.

Therapeutics companies currently look to the outside for marketing activities twice as much as agricultural or diagnostic firms at present and expect this trend to continue over the next five years. Although all firms have about the same external manufacturing activities at present, the therapeutics firms expect to do one-third of their manufacturing externally, or about 50 percent more than other firms.

Conclusion

One final observation is that these data reflect the maturing of a very young "industry." The predicted future emphasis, from mostly science to more business than science, is more typical of a more developed industry. As more products of biotechnology reach the market, the shift in emphasis from R&D to business is likely to continue as predicted.

Note:

1. For a more detailed description of the shifting emphasis from science to business in the U.S. biotechnology industry, see: M.D. Dibner, W.F. Hamilton and J. Vila. The Maturing of Biotech Companies: Shifting Emphasis from Science to Business. Bio/Technology 6: 276-279, March 1988.

FIGURE 9-1 SHIFTING EMPHASIS FROM R&D TO BUSINESS

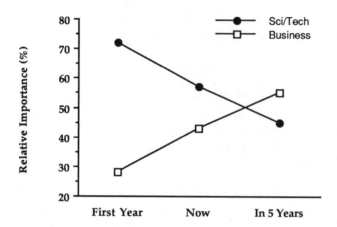

The importance of science to business is shown over time. Responders rated the relative proportion of science/technology to business in their total operation in their first year, now, and expected in five years. Ratings were as percent of total; science plus business equals 100 percent at each time point. There is a significant increase in proportion of business activities over the three time points (and a concomitant decrease in proportion of science activities) at p<0.001.

Figure 9-2 Relative Importance of Company Activities

FIGURE 9-2 RELATIVE IMPORTANCE OF COMPANY ACTIVITIES

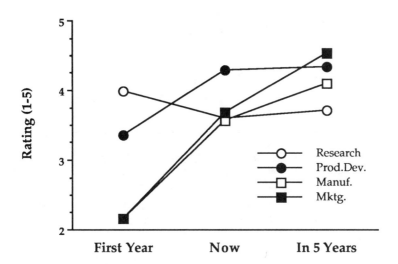

Biotechnology company executives rated the importance of research, product development, manufacturing, and marketing activities from 1 (lowest importance) to 5 (highest importance) at the three time points. The importance of research significantly decreased in importance (p<0.01), whereas the other three activities increased in importance (p<0.001) between the companies' first year and the present. Marketing and manufacturing are predicted to significantly increase in importance over the next five years (p<0.001).

TABLE 9-1 R&D BUDGET OF BIOTECHNOLOGY COMPANIES

	R&D Budget ($ millions) (1)	R&D Budget per Employee (2)	R&D Budget as % Revenues (3)
Average	4.64	$46,431	42%
Low	0.01	$370	1.33%
High	120.00	$266,687	3,333%
Median	1.70	$32,600	63%

Notes:
1.   Data represent the average, median and range of data of R&D budgets of 140 U.S. biotechnology companies for the period of mid-1987 to mid-1988.
2.   The R&D budget per company employee is given for 130 companies.
3.   The R&D budget as a percent of expected revenues is shown with data from 111 companies. Not all data were available from all companies.

Figure 9-3 R&D Budget by Company Type

FIGURE 9-3 R&D BUDGET BY COMPANY TYPE

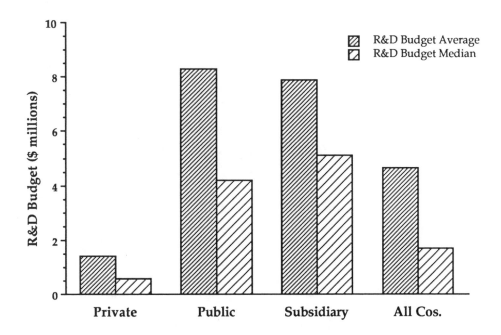

The average and median R&D budget is given for U.S. biotechnology companies in millions of dollars. Data include 68 privately held companies, 47 public companies and 15 subsidiary companies. R&D budgets ranged from $10,000 to $1.5 million for the private companies, $100,000 to $120 million for the public companies, and $750,000 to $30 million for the subsidiary companies.

FIGURE 9-4 R&D BUDGET PER EMPLOYEE

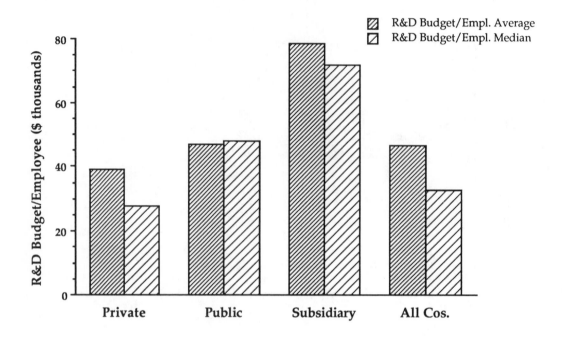

The R&D budget per employee in thousands of dollars is shown for 68 private, 47 public and 15 subsidiary U.S. biotechnology companies. Average and median figures are shown for the data taken between mid 1987 and February 1988. Data range from $370 per employee to $187,500 per employee.

Figure 9-5 R&D Budget by Industry Classification

FIGURE 9-5 R&D BUDGET BY INDUSTRY CLASSIFICATION

A.

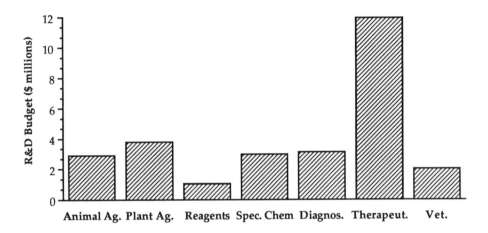

B.

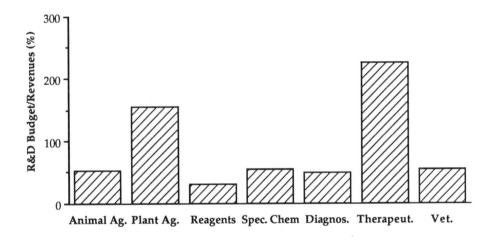

A. The average R&D budget in millions of dollars is shown for U.S. biotechnology companies specializing in animal agriculture (4 companies), plant agriculture (17), reagents (13), specialty chemicals (14), diagnostics (30), therapeutics (33) or veterinary areas (6). Data are from mid-1987 to February 1988.

B. The average R&D budget, as a percent of revenues, is shown for the same groups of companies.

SECTION 10 PATENTS AND PRODUCTS

INTRODUCTION

The backlog of biotechnology-related patents at the U.S. Patent Office has been a prominent issue. This section demonstrates that patent applications pending from the U.S. biotechnology firms alone may have reached 10,000 in number by the end of 1987. This figure does not include patent applications submitted by foreign firms or by the large U.S. corporations listed in Section 11. Our figures were obtained from top executives of 111 U.S. biotechnology companies in mid 1987. Figure 10-1 shows average data on patent applications for all firms, the private, public and subsidiary firm types, and for five key industry classifications. Individual company patent data is confidential, but these analyses show the average numbers by category.

Figure 10-2 shows the average number of biotechnology-related patents granted each year from the responses of the 111 companies. This figure demonstrates that there is a clear increase in patents each year.

Figure 10-3 shows the number of biotechnology-related products reaching the market each year on average taken from data provided by 100 U.S. biotechnology firms in mid 1987. These data are taken for all company classes and average figures may be greater than expected due to the large number of products of companies providing biotechnology reagents, specialty chemicals and monoclonal antibodies. For example, to date there are only six therapeutic products on the market that are made by new biotechnological methods: human insulin, alpha interferon, OKT3 antibody, human growth hormone, hepatitis B vaccine, and tissue plasminogen activator.

Figure 10-1 Patent Applications by Company Type

FIGURE 10-1 PATENT APPLICATIONS BY COMPANY TYPE

A. The mean number of patent applications on file as of mid-1987 for 111 U.S. biotechnology companies is shown. Data are from 56 private, 40 public and 15 subsidiary companies. The average company had 27.3 biotechnology-related patent applications pending in the U.S. Patent Office. Extrapolated for the whole industry, this number represents almost 10,000 patent applications for the small U.S. biotechnology companies. The patent applications from the large or foreign corporations would add an additional number.

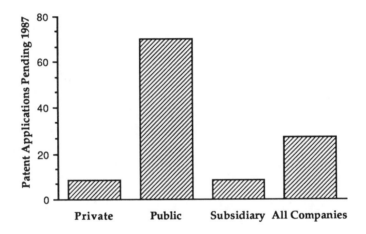

B. The figure below shows the average patent applications for these companies with focus in animal agriculture, plant agriculture, specialty chemicals, diagnostics or therapeutics.

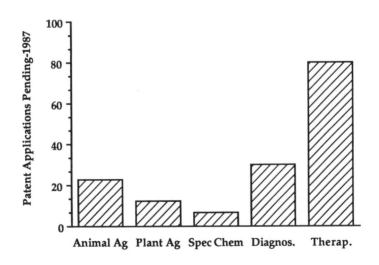

FIGURE 10-2 PATENTS PER YEAR

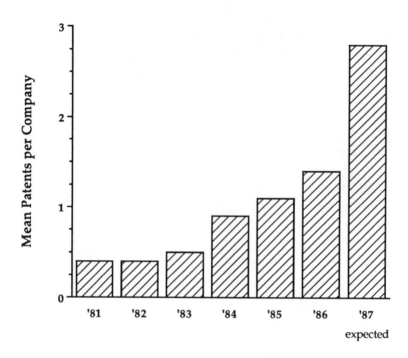

The mean number of U.S. patents granted per year is shown for 111 U.S. biotechnology companies in mid-1987. Note the significant growth in patents in recent years to six times the rate in 1982. The average company in this study was founded in 1981.

Figure 10-3 Products per Year

FIGURE 10-3 PRODUCTS PER YEAR

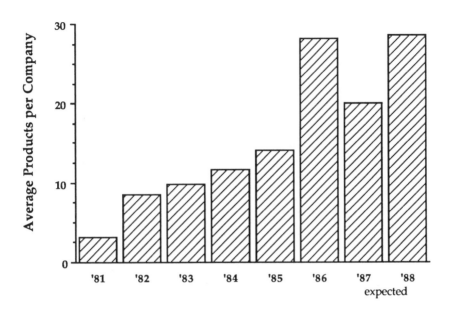

The number of products per company, on average, is given for 100 companies from 1981 to 1988. These are all products, and do not just represent a major product, such as a novel therapeutic, reaching the market. The data were collected by question-naire in conjunction with the Office of Technology Assessment (see Appendix E). The total number of products between 1981 and 1988 (expected) for the average com-pany is 87.6 products. This number varies widely by industry focus: plant agricul-ture companies had 4.9 products on average during that period, therapeutic compa-nies had 18.6, diagnostic companies had 75.2 and reagent companies had 492.6 prod-ucts. Interestingly, the small companies, with less than 20 employees, had 154.3 products cumulatively between 1981 and 1988 whereas the medium companies (21 to 100 employees) had only 30.2 products and the large companies (100-plus employ-ees) had 99.7 products.

SECTION 11 LARGE U.S. CORPORATIONS IN BIOTECHNOLOGY

Section 11   Large U.S. Corporations

INTRODUCTION

The U.S. biotechnology industry is composed of the new biotechnology firms, detailed in the preceding sections of this Guide, and major corporations with biotechnology programs, detailed in this section. The literature has indicated more than 100 corporations that have entered the biotechnology arena, and of these we have been able to identify about 50 large U.S. corporations, operating at about 60 sites, that have significant programs in biotechnology. We contacted these corporations and were able to obtain the data provided in Listings 11-1 and 11-2. For at least half of these corporations, their biotechnology efforts were scattered in many groups and departments. Most could or would not provide information on the number of biotechnology-related employees in their company, and we often had to turn to a number of sources and contacts to put together an overview of corporate biotechnology. Just as our contacts had trouble providing biotechnology-related personnel data, only a few of the corporations could provide biotechnology-related revenue and R&D data. One company, Johnson and Johnson, has its biotechnology efforts spread over five sites. In addition, as shown in Listing 11-3, quite a few U.S. corporations have biotechnology-related subsidiaries.

Although the information is diffuse, many of these corporations have significant biotechnology efforts, with more resources devoted to biotechnology than the average biotechnology firm. Table 11-1 analyzes the average large biotechnology firm and the average large pharmaceutical corporation. The differences in size and available resources are noteworthy.

Listing 11-1 Directory of Large Corporations

LISTING 11-1 DIRECTORY OF LARGE CORPORATIONS

ABBOTT LABORATORIES
ROUTES IL 43 AND IL 137
ABBOTT PARK, IL 60064
Telephone: 312-937-0201
Products: HEP B & AIDS DX; ENZYMES, UROKINASE;
INSECTICIDES; LAB PRODS; DX EQUIP.
Total Corporation Employees: 37,000

CEO: ROBERT SCHOELLMAN
President: JACK SCHULER
R&D Dir: DWAYNE BURNHAM
Started: 1914   Class Code: P

AMERICAN CYANAMID
AGRICULTURAL DIVISION
P.O. BOX 400
PRINCETON, NJ 08540
Telephone: 609-799-0400
Products: ANIMAL GROWTH FACTORS; BOVINE
SOMATOTROPIN; BOVINE MASTITIS RX;
HERBICIDES; PESTICIDES
Site Employees: 630
Total Corporation Employees: 34,000

CEO: GEORGE J. SELLA, JR.
President: GEORGE J. SELLA, JR.
Started: 1907   Class Code: B

AMERICAN CYANAMID
LEDERLE LABORATORIES
MIDDLETOWN ROAD
PEARL RIVER, NY 10965
Telephone: 914-732-5000
Products: RX; ANTICANCER RX; CARDIOVASCULAR RX;
ANTIBIOTICS; VX; VITAMINS; ANTI-FUNGALS
Site Employees: 6,000
Total Corporation Employees: 34,000

CEO: GEORGE SELLA
President: GEORGE SELLA
R&D Dir: ARNOLD R. ARONSKY
Started: 1907   Class Code: P

AMERICAN HOME PRODUCTS
WYETH-AYERST LABORATORIES
P.O. BOX 8299
PHILADELPHIA, PA 19101
Telephone: 215-341-2003
Products: ETHICAL RX; PENICILLIN; VX; INFANT
FORMULAS FROM WHEY; BICILLIN
Site Employees: 1,760
Total Corporation Employees: 23,000

CEO: DR. BERNARD CANAVAN
R&D Dir: DR. H.P. AGERSBORG
Started: 1931   Class Code: P

AMOCO
RESEARCH CENTER
PO BOX 400, MF F-2
NAPERVILLE, IL 60566
Telephone: 312-420-5111
Products: SPECIALTY CHEMICALS; ENERGY/BIOMASS;
DATABANKS/PROGRAMS; FERMENTATION
Site Employees: 1,800
Total Corporation Employees: 47,000

CEO: RICHARD M. MORROW
President: H. LAURENCE FULLER
R&D Dir: DR. GAR ROYER
Started: 1977   Class Code: L

ANHEUSER-BUSCH, INC.
ONE BUSCH PLACE
ST. LOUIS, MO 63188
Telephone: 314-577-2000
Products: BEER; YEAST & FERMENTATION TECHNOL.;
             YEAST TO PRODUCE RDNA IF
Total Corporation Employees: 41,805

CEO: AUGUST BUSCH III
President: DENNIS LONG
Class Code: M

BATELLE MEMORIAL INSTITUTE
505 KING AVE.
COLUMBUS, OH 43201
Telephone: 614-424-6424
Products: DX; VET VX; SEPARATION TECHNIQUES;
             CONVERSION TECHNIQUES
Site Employees: 3,000
Total Corporation Employees: 7,200

CEO: DR. DOUGLAS OLESON
President: DR. DOUGLAS OLESON
R&D Dir: DR. ANNA D. BARKER
Started: 1929   Class Code: K

BAXTER HEALTHCARE CORP.
ONE BAXTER PARKWAY
DEERFIELD, IL 60015
Telephone: 312-948-2000
Products: FACTOR VIII; AIDS DX; MABS; TPA; IF-G; IL-2;
             ALPHA-FETOPROTEIN RIA KIT
Total Corporation Employees: 60,000

CEO: VERNON R. LOUCKS, JR.
President: WILBUR H. GANTZ
Started: 1931   Class Code: P

BAYER USA
MILES LABS BT PRODUCT DIVISION
1127 MYRTLE ST.
ELKHART, IN 46514
Telephone: 219-264-8111
Products: INDUSTRIAL ENZYMES; CITRIC ACID;
             INGREDIENTS FOR FERMENTATION
Site Employees: 400
Total Corporation Employees: 173,000

CEO: CONRAD WEISS
President: DR. KLAUS H. RISSE
R&D Dir: DR. RONALD WEISS
Started: 1884   Class Code: J

BECTON DICKINSON & CO.
MEDICAL SECTOR
1 BECTON DR.
FRANKLIN LAKES, NJ 07417
Telephone: 201-848-6800
Products: MEDICAL DEVICES, MAB-BASED REAGENTS
Site Employees: 650
Total Corporation Employees: 19,900

CEO: WESLEY J. HOWE
President: WESLEY J. HOWE
R&D Dir: DR. HERBERT MOROTE
Started: 1898   Class Code: 4

BECTON DICKINSON & CO.
BECTON DICKINSON IMMUNOCYTOMETRY
2375 GARCIA
MOUNTAIN VIEW, CA 94043
Telephone: 415-968-7744
Products: MABS; CELL SORTER REAGENTS
Site Employees: 550
Total Corporation Employees: 19,900

CEO: WESLEY J. HOWE
President: DR. NAGESH MHATRE
R&D Dir: DR. NOEL WARNER
Started: 1979   Class Code: F

Listing 11-1 Directory of Large Corporations

BIO-RAD LABORATORIES, INC.
2200 WRIGHT AVE.
RICHMOND, CA 94804
Telephone: 415-232-7000
Products: REAGENTS/RESEARCH MATERIALS; GENE
             MACHINES; SEPARATION EQUIP/TECHNOL.
Site Employees: 230
Total Corporation Employees: 1,500

CEO: DAVID SCHWARTZ
President: DAVID SCHWARTZ
R&D Dir: RUSSELL FROST
Started: 1957   Class Code: F

BRISTOL-MYERS CO.
(SEE ALSO GENETIC SYSTEMS IN LISTING 2-2)
P.O. BOX 667, THOMPSON RD.
SYRACUSE, NY 13201
Telephone: 315-432-2000
Products: CANCER, VIRAL RX; OTC HEALTH CARE,
             PROSTHETIC DEVICES
Total Corporation Employees: 35,000

CEO: RICHARD L. GELB
R&D Dir: WILLIAM R. MILLER
Biotech Started: 1975   Class Code: P

BURROUGHS WELLCOME CO.
P.O. BOX 1887
GREENVILLE, NC 27835
Telephone: 919-830-6476
Products: RX--ANTIBACTERIALS, ANTIHISTAMINES,
             CARDIOVASCULAR, ANTICANCER, IF
Site Employees: 1,552
Total Corporation Employees: 18,000

CEO: T.E. HAIGLER, JR.
President: T.E. HAIGLER, JR.
R&D Dir: DR. JOHN BETTIS
Started: 1880   Class Code: P

BURROUGHS WELLCOME CO.
3030 CORNWALLIS RD.
RESEARCH TRIANGLE PARK, NC 27709
Telephone: 919-248-3000
Products: RX; ANTI-VIRALS, ANTI-DEPRESSANTS, ANTI-
             CANCER, CARDIOVASCULAR, ANTI-INFECTIVES
Site Employees: 1,500
Total Corporation Employees: 18,000

CEO: T.E. HAIGLER, JR.
President: T.E. HAIGLER, JR.
R&D Dir: DR. HOWARD SCHAEFFER
Started: 1880   Class Code: P

CAMPBELL SOUP CO.
CAMPBELL INSTITUTE OF R & D
CAMPBELL PLACE
CAMDEN, NJ 08103
Telephone: 609-342-4899
Products: RDNA DISEASE-FREE TOMATOES, MUSHROOMS
Total Corporation Employees: 49,226

CEO: R.G. MCGOVERN
President: R.G. MCGOVERN
Started: 1869   Class Code: M

CIBA-GEIGY CORP.
BIOTECHNOLOGY RESEARCH DIVISION
P.O. BOX 12257
RESEARCH TRIANGLE PARK, NC 27709
Telephone: 919-549-8164
Products: IMPROVED CROP PLANTS; PLANT
             TISSUE CULTURE
Site Employees: 75   Site Biotech Employees: 75
Total Corporation Employees: 82,000

CEO: ALBERT BODMER
President: ALBERT BODMER
R&D Dir: DR. MARY-DELL CHILTON
Started: 1983   Class Code: B

CIBA-GEIGY CORP.
PHARMACEUTICALS DIVISION
556 MORRIS AVE.
SUMMIT, NJ 07901
Telephone: 201-277-5000
Products: DRUG DELIVERY SYSTEMS; VX
Site Employees: 2,200
Total Corporation Employees: 82,000

CEO: ALBERT BODMER
President: ALBERT BODMER
Started: 1938   Class Code: P

DOW CHEMICAL CO.
MERRELL DOW RESEARCH INSTITUTE
2110 E. GALBRAITH RD.
CINCINNATI, OH 45215
Telephone: 513-948-9111
Products: MAB PRODUCTION; REGULATION OF GENE
             EXPRESSION; RDNA PROTEIN FOR
             STRUCTURE ANALYSIS
Total Corporation Employees: 50,000

CEO: FRANK POPOFF
President: FRANK POPOFF
R&D Dir: RICHARD JACKSON
Started: 1985   Class Code: P

DOW CHEMICAL CO.
BIOPRODUCTS LAB
1701 BLDG.
MIDLAND, MI 48674
Telephone: 517-636-1066
Products: CANCER IMMUNOTHERAPY,
             BIOLOGICAL PEST CONTROL
Site Employees: 35   Site Biotech Employees: 32
Total Corporation Employees: 50,000

CEO: FRANK POPOFF
President: FRANK POPOFF
R&D Dir: DR. CLIFF THOMPSON
Started: 1979   Class Code: P

E I DU PONT DE NEMOURS & CO
MEDICAL PRODUCTS DEPT.
1007 MARKET ST.
WILMINGTON, DE 19898
Telephone: 302-744-1000
Products: DX; EQUIPMENT; DNA PROBES; CELL
             CULTURE PRODUCTS; AIDS DX; IF; MABS
Total Corporation Employees: 140,000

CEO: RICHARD E. HECKERT
President: E.S. WOOLARD, JR.
R&D Dir: DR. DAVID JACKSON
Started: 1804   Class Code: F

EASTMAN KODAK
343 STATE STREET
ROCHESTER, NY 14650
Telephone: 716-458-1000
Products: ANIMAL NUTRITION; AGRICULTURAL
             TECHNOLOGY; PHARMACEUTICALS
Total Corporation Employees: 124,350

CEO: COLBY H. CHANDLER
President: KAY R. WHITMORE
R&D Dir: EUGENE H. CORDES,VP
Started: 1984   Class Code: A

ELI LILLY & CO.
(SEE ALSO HYBRITECH, INC. IN LISTING 2-2)
LILLY CORPORATE CENTER
INDIANAPOLIS, IN 46285
Telephone: 317-261-2000
Products: RX; HGH; PROINSULIN; ANTIBIOTICS; DX;
             AGRICHEMICALS
Total Corporation Employees: 29,000

CEO: RICHARD D. WOOD
President: RICHARD D. WOOD
R&D Dir: D.W. GRIMM
Started: 1975   Class Code: P

Listing 11-1 Directory of Large Corporations

GLAXO, INC.
FIVE MOORE DR.
RESEARCH TRIANGLE PARK, NC 27709
Telephone: 919-248-2100
Products: RX
Site Employees: 1,135
Total Corporation Employees: 33,000

CEO: JOSEPH J. RUVANE, JR.
President: DR. ERNEST MARIO
R&D Dir: DR. PEDRO CUATRECASAS
Started: 1983   Class Code: P

W.R. GRACE & CO.
BIOTECH RESEARCH DEPT.
7379 ROUTE 32
COLUMBIA, MD 21004
Telephone: 301-531-4000
Products: SPECIALTY CHEMICALS; MABS FOR
             RESEARCH; GENETIC ENGINEERING
Site Employees: 75
Total Corporation Employees: 40,000

CEO: J. PETER GRACE
President: J. PETER GRACE
R&D Dir: DR. F. VAN REMOORTERE
Started: 1977   Class Code: A

HOECHST CELANESE CORP.
RT. 202-206 N.
SOMERVILLE, NJ 08876
Telephone: 201-231-2000
Products: ENZYMES, RESEARCH ON MICRO-
             ORGANISMS AS CATALYSTS
Total Corporation Employees: 23,000

CEO: DR. ERNEST H. DREW
President: DR. ERNEST H. DREW
R&D Dir: DR. LEON STARR
Started: 1987   Class Code: J

HOFFMANN-LA ROCHE, INC.
THE ROCHE INSTITUTE OF MOLECULAR BIOLOGY
340 KINGSLAND AVE.
NUTLEY, NJ 07110
Telephone: 201-235-5000
Products: IF-2A; IL-2; DX
Site Employees: 6,000
Total Corporation Employees: 13,000

CEO: IRWIN LERNER
President: IRWIN LERNER
R&D Dir: DR. RONALD KUNTZMAN
Started: 1969   Class Code: P

ICI AMERICAS, INC.
P.O. BOX 208
GOLDSBORO, NC 27533
Telephone: 919-731-5201
Products: SEED BREEDING; INSECTICIDES;
             CHEMOTHERAPY
Site Employees: 150   Site Biotech Employees: 20
Total Corporation Employees: 200,000

CEO: DR. D. CUNTHWAITE
President: D.C. WALKER
R&D Dir: DR. A.A. AKHAVEIN
Started: 1926   Class Code: B

INTERNATIONAL MINERALS & CHEMICALS
2315 SAUNDERS RD.
NORTHBROOK, IL 60062
Telephone: 312-564-8600
Products: ANIMAL GROWTH, HEALTH & NUTRITION
Total Corporation Employees: 11,000

CEO: GEORGE D. KENNEDY
President: GEORGE D. KENNEDY
R&D Dir: M.B. INGLE
Started: 1970   Class Code: A

JOHNSON & JOHNSON
501 GEORGE ST.
NEW BRUNSWICK, NJ 08903
Telephone: 201-524-0400
Products: DX; MABS; IMMUNOTECHNOLOGY
Total Corporation Employees: 78,000

CEO: JAMES E. BURKE
President: DAVID R. CLARE
R&D Dir: ROBERT GUSSIN
Started: 1886   Class Code: P

JOHNSON & JOHNSON
ORTHO PHARMACEUTICALS
P.O. BOX 300, U.S. ROUTE 202
RARITAN, NJ 08869
Telephone: 201-218-6000
Products: RX; DX; EPO, IL-2, CANCER MABS; HEP B
             VX; RX DELIVERY SYSTEMS; HOME
             PREGNANCY TESTS
Site Employees: 2,600
Total Corporation Employees: 78,000

CEO: JAMES E. BURKE
President: DAVID R. CLARE
R&D Dir: DENNIS N. LONGSTREET
Started: 1886   Class Code: P

JOHNSON & JOHNSON
ORTHO DIAGNOSTICS SYSTEMS, INC.
U.S. ROUTE 202
RARITAN, NJ 08869
Telephone: 201-218-1300
Products: DX; HERPES 1 & 2, CHLAMYDIA TESTS;
             HERPES DNA PROBE DX, BIOCLONE (TM)
Site Employees: 659
Total Corporation Employees: 78,000

CEO: JAMES E. BURKE
President: DAVID R. CLARE
R&D Dir: DR. JACK GOLDSTEIN
Started: 1886   Class Code: K

JOHNSON & JOHNSON
BIOTECHNOLOGY CENTER
P.O. BOX 8289
LA JOLLA, CA 92038
Telephone: 619-450-2000
Products: DX; SYNTHETIC VX; SOME VET PRODUCTS
Site Employees: 70   Site Biotech Employees: 53
Total Corporation Employees: 78,000

CEO: JAMES E. BURKE
President: DAVID R. CLARE
R&D Dir: DR. NADAV FRIEDMANN
Started: 1886   Class Code: K

JOHNSON & JOHNSON
ETHICON, INC.
ROUTE 22 WEST
SOMERVILLE, NJ 08876
Telephone: 201-218-0707
Products: GROWTH FACTORS
Site Employees: 2,500
Total Corporation Employees: 78,000

CEO: JAMES E. BURKE
President: DAVID R. CLARE
R&D Dir: RICHARD KRONENTHAL
Started: 1886   Class Code: P

LUBRIZOL ENTERPRISES, INC.
29400 LAKELAND BLVD.
WICKLIFFE, OH 44092
Telephone: 216-943-4200
Products: VENTURE DEVELOPMENT CO.;
             RAW MATERIALS
Site Employees: 4,200

CEO: LESTER E. COLEMAN
President: W.G. BARES
R&D Dir: GEORGE R. HILL
Started: 1928   Class Code: B

Listing 11-1 Directory of Large Corporations

MERCK & COMPANY, INC.
P.O. BOX 2000
RAHWAY, NJ 07005
Telephone: 201-574-4000
Products: RX; ANTIBIOTICS; VITAMINS; HEP B VX;
            VET RX; SPECIALTY CHEMICALS; PESTICIDES
Total Corporation Employees: 31,000

CEO: DR. P. ROY VAGELOS
President: DR. P. ROY VAGELOS
R&D Dir: DR. ROBERT GERETY
Started: 1900   Class Code: P

MONSANTO CO.
LIFE SCIENCES RESEARCH CENTER
700 CHESTERFIELD VILLAGE PKWY.
ST. LOUIS, MO 63198
Telephone: 314-694-1000
Products: ASPARTAME; INDUSTRIAL CHEMICALS;
            AG CHEMICALS; RX
Site Employees: 800
Total Corporation Employees: 52,000

CEO: RICHARD J. MAHONEY
President: EARLE HARBISON
R&D Dir: DR. H SCHNEIDERMAN
Started: 1978   Class Code: I

MONSANTO CO.
G.D. SEARLE & CO.
4711 GOLF RD.
SKOKIE, IL 60076
Telephone: 312-982-7000
Products: TPA; ATRIAL PEPTIDES; IF-B; ASPARTAME
Site Employees: 9,500
Total Corporation Employees: 52,000

CEO: RICHARD J. MAHONEY
President: EARLE HARBISON
R&D Dir: DR. ALAN R. TIMMS
Searle Started: 1968   Class Code: P

PFIZER, INC.
235 EAST 42ND ST.
NEW YORK, NY 10017
Telephone: 212-573-2323
Products: SORBOSE; XANTHAN; ANTIBIOTICS;
            TETRACYCLINE
Total Corporation Employees: 40,000

CEO: EDMUND T. PRATT, JR.
President: GERALD LAUBACH
R&D Dir: BARRY M. BLOOM
Started: 1849   Class Code: P

PHILLIPS PETROLEUM
(SEE ALSO PROVESTA CORP. IN LISTING 2-2)
PHILLIPS BLDG.
BARTLESVILLE, OK 74004
Telephone: 918-661-3947
Products: FERMENTATION BASED FLAVORS; DX;
            NUTRITIONAL YEAST; ENZYMES; VET VX
Total Corporation Employees: 20,000

CEO: C.J. SILAS
President: GLENN COX
R&D Dir: DR. EARNEST ZUECH
Started: 1917   Class Code: L

PIONEER HI-BRED INTERNATIONAL, INC.
700 CAPITAL SQUARE, 400 LOCUST
DES MOINES, IA 50309
Telephone: 515-245-3500
Products: HERBICIDE RESISTANT CORN; IMPROVED
            SEED PRODUCTS, CORN, SUNFLOWER
Site Employees: 500   Site Biotech Employees: 135
Total Corporation Employees: 5,000

CEO: THOMAS N. URBAN
President: THOMAS N. URBAN
R&D Dir: DR. DONALD DUVICK
Started: 1977   Class Code: B

PROCTER & GAMBLE
MIAMI VALLEY LABS
P.O. BOX 39175
CINCINNATI, OH 45247
Telephone: 513-562-1100
Products: DETERGENT PRODUCTS WITH ENZYMES
Total Corporation Employees: 10,000

CEO: JOHN G. SMALE
President: JOHN G. SMALE
R&D Dir: CHARLES BROADDUS
Started: 1982   Class Code: J

QUANTUM CHEMICAL CORP.
(SEE ALSO LIFECODES CORP. IN LISTING 2-2)
99 PARK AVE.
NEW YORK, NY 10016
Telephone: 212-949-5000
Products: POLYETHELYNE RESINS; OLEO CHEMICALS;
        PROPANE MARKETING
Total Corporation Employees: 11,000

CEO: JOHN HOYT STOOKEY
R&D Dir: JEROME KREKELER
Started: 1978   Class Code: L

ROHM & HAAS CO.
INDEPENDENCE MALL WEST
PHILADELPHIA, PA 19105
Telephone: 215-592-3000
Products: HYBRID SEED TECH; PLANT SCIENCE
Total Corporation Employees: 12,000

CEO: VINCENT L. GREGORY, JR.
President: JOHN P. MULRONEY
R&D Dir: ROBERT E. NAYLOR
Started: 1909   Class Code: B

RORER GROUP, INC., THE
RORER CENTER RESEARCH (MELOY)
800 BUSINESS CENTER DRIVE
HORSHAM, PA 19044
Telephone: 215-956-5000
Products: RX
Total Corporation Employees: 8,000

CEO: ROBERT CAWTHORN
President: ROBERT CAWTHORN
R&D Dir: DR. ALAIN SCHREIBER
Started: 1910   Class Code: P

SANDOZ PHARMACUETICALS CORP.
59 ROUTE 10
E. HANOVER, NJ 07936
Telephone: 201-503-7500
Products: CANCER RX; INSECTICIDE; CORN; KPA
Total Corporation Employees: 37,000

CEO: DANIEL C. WAGNIERE
President: FRED HASSAN
R&D Dir: DR. DAVID WINTER
Biotech Started: 1984   Class Code: P

SHELL OIL CO.
(SEE ALSO TRITON BIOSCIENCES IN LISTING 2-2)
1 SHELL PLACE
HOUSTON, TX 77001
Telephone: 713-241-6161
Products: PETROLEUM RELATED PRODUCTS
Total Corporation Employees: 32,000

CEO: JOHN F. BOOKOUT
President: JOHN F. BOOKOUT
R&D Dir: DR. JOHN COLE, VP
Started: 1912   Class Code: L

E.R. SQUIBB & SONS, INC.
P.O. BOX 4000
PRINCETON, NJ 08540
Telephone: 609-921-4000
Products: STEROID OXIDATIONS; ANTIBIOTICS;
        INSULIN
Total Corporation Employees: 23,700

CEO: RICHARD M. FURLAND
President: RICHARD M. FURLAND
R&D Dir: RICHARD SYKES
Class Code: P

Listing 11-1 Directory of Large Corporations

SYNTEX CORP.
3401 HILLVIEW AVE.
PALO ALTO, CA 94604
Telephone: 415-855-5050
Products: RX; DX PRODUCTS; DNA RESEARCH;
             NEOPLASM DX; MABS
Site Employees: 3,500
Total Corporation Employees: 9,000

CEO: ALBERT BOWERS
President: PAUL FREIMAN
R&D Dir: DR. JOHN FRIED
Started: 1944   Class Code: K

UPJOHN CO.
7000 PORTAGE RD.
KALAMAZOO, MI 49001
Telephone: 616-323-4000
Products: ANTIBIOTICS; TETRACYCLINE; SEEDS; BGH
Site Employees: 7,500   Site Biotech Employees: 117
Total Corporation Employees: 20,371

CEO: DR. THEODORE COOPER
President: LAWRENCE C. HOFF
R&D Dir: RALPH CHRISTOFFERSON
Started: 1886   Class Code: P

WEYERHAEUSER CO.
255 S. 336TH ST.
TACOMA, WA 98477
Telephone: 206-924-2345
Products: TISSUE CULTURE OF FLOWERS
             AND PLANTS
Site Employees: 2,500
Total Corporation Employees: 43,000

CEO: GEORGE WEYERHAEUSER
President: JACK CREIGHTON
R&D Dir: DR. N.E. JOHNSON
Started: 1900   Class Code: B

Notes:

Corporations with significant programs in biotechnology are listed. Multiple sites are shown, where applicable. R&D Director is shown for site.

LISTING 11-2 CORPORATIONS LISTED BY INDUSTRY CLASSIFICATION

Animal Agriculture (A)

AMERICAN CYANAMID
AGRICULTURAL DIVISION

Products: ANIMAL GROWTH FACTORS; BOVINE SOMATOTROPIN; BOVINE MASTITIS RX; HERBICIDES; PESTICIDES

W.R. GRACE & CO.
RESEARCH DEPT.

Products: SPECIALTY CHEMICALS; MAB FOR BIOTECH RESEARCH, GENETIC ENGINEERING

INTERNATIONAL MINERALS & CHEMICALS

Products: ANIMAL GROWTH, HEALTH & NUTRITION

EASTMAN KODAK

Products: ANIMAL NUTRITION; AGRICULTURAL TECHNOL; PHARMACEUTICALS

ELI LILLY & CO.
(*SEE ALSO HYBRITECH, INC.)

Products: RX; HGH; PROINSULIN; ANTIBIOTICS; DX; AGRICHEMICALS

PFIZER, INC.

Products: SORBOSE; XANTHAN; ANTIBIOTICS; TETRACYCLINE

SYNTEX CORP.

Products: RX; DX PRODUCTS; DNA RESEARCH; NEOPLASMA DX; MABS

UPJOHN CO.

Products: ANTIBIOTICS; TETRACYCLINE; SEEDS; BGH

Plant Agriculture (B)

ABBOTT LABORATORIES

Products: HEP B & AIDS DX; ENZYMES, UROKINASE; INSECTICIDES; LAB PRODS; DX EQUIPMENT

AMERICAN CYANAMID
AGRICULTURAL DIVISION

Products: ANIMAL GROWTH FACTORS; BOVINE SOMATOTROPIN; BOVINE MASTITIS RX; HERBICIDES; PESTICIDES

CAMPBELL SOUP CO.
CAMPBELL INSTITUTE OF R & D

Products: RDNA DISEASE-FREE TOMATOES, MUSHROOMS

CIBA-GEIGY CORP.
BIOTECHNOLOGY RESEARCH DIV.

Products: IMPROVED CROP PLANTS; PLANT TISSUE CULTURE

DOW CHEMICAL CO.
BIOPRODUCTS LAB

Products: CANCER IMMUNOTHERAPY, BIOLOGICAL PEST CONTROL

E.I. DU PONT DE NEMOURS & CO. Products: DX; EQUIPMENT; DNA PROBES; CELL CULTURE PRODUCTS; AGRICHEMICALS; AIDS DX; IF; MABS

ICI AMERICAS, INC.

Products: SEED BREEDING, INSECTICIDES, CHEMOTHERAPY

LUBRIZOL ENTERPRISES, INC. Products: VENTURE DEVELOPMENT CO.; RAW MATERIALS

Listing 11-2 Corporations Listed by Industry Classification

Plant Agriculture (Cont.)

MERCK & COMPANY, INC.
Products: RX; ANTIBIOTICS; VITAMINS; HEP B VX; VET RX, SPECIALTY CHEMICALS, PESTICIDES

MONSANTO CO.
LIFE SCIENCES RESEARCH CTR.
Products: ASPARTAME; INDUSTRIAL CHEMICALS; AG CHEMICALS; RX

PFIZER, INC.
Products: SORBOSE; XANTHAN; ANTIBIOTICS; TETRACYCLINE

PIONEER HI-BRED INT'L, INC.
Products: HERBICIDE RESISTANT CORN, IMPROVED SEED PRODUCTS, CORN, SUNFLOWER

ROHM & HAAS CO.
Products: HYBRID SEED TECH; PLANT SCIENCE RESEARCH

SANDOZ PHARMACUETICALS
Products: CANCER RX; INSECTICIDE; CORN HYBRIDS; KPA

UPJOHN CO.
Products: ANTIBIOTICS; TETRACYCLINE; SEEDS; BGH

WEYERHAEUSER CO.
Products: TISSUE CULTURE OF FLOWERS AND PLANTS

Biomass Conversion (C)

AMOCO
RESEARCH CENTER
Products: SPECIALTY CHEMICALS; ENERGY/BIOMASS; DATABANKS/PROGRAMS; FERMENTATION

Biotechnology Equipment (F)

BECTON DICKINSON & CO.
IMMUNOCYTOMETRY
Products: MABS; CELL SORTER REAGENTS

BIO-RAD LABORATORIES, INC.
Products: REAGENTS/RESEARCH MATERIALS; GENE MACHINES; SEPARATION EQUIP/TECHNOL.

E.I. DU PONT DE NEMOURS & CO.
Products: DX; EQUIPMENT; DNA PROBES; CELL CULTURE PRODUCTS; AGRICHEMICALS; AIDS DX; IF; MABS

Specialty Chemicals (J)

ABBOTT LABORATORIES
Products: HEP B & AIDS DX; ENZYMES, UROKINASE; INSECTICIDES; LAB PRODS; DX EQUIPMENT

AMOCO
RESEARCH CENTER
Products: SPECIALTY CHEMICALS; ENERGY/BIOMASS; DATABANKS/PROGRAMS; FERMENTATION

BAYER USA
MILES LABS
BIOTECHNOLOGY PRODUCTS DIV.
Products: INDUSTRIAL ENZYMES; CITRIC ACID; INGREDIENTS FOR FERMENTATION

HOECHST CELANESE CORP.
Products: ENZYMES, RESEARCH ON MICROORGANISMS AS CATALYSTS

PROCTER & GAMBLE
MIAMI VALLEY LABS
Products: DETERGENT PRODUCTS WITH ENZYMES

Specialty Chemicals (Cont.)

QUANTUM CHEMICAL CORP.
(*SEE ALSO LIFECODES CORP.)

Products: POLYETHELYNE RESINS, OLEO CHEMICALS, PROPANE

ROHM & HAAS CO.

Products: HYBRID SEED TECH; PLANT SCIENCE RESEARCH

Diagnostics (K)

ABBOTT LABORATORIES

Products: HEP B & AIDS DX; ENZYMES, UROKINASE; INSECTICIDES; LAB PRODS; DX EQUIPMENT

ARES-SERONO GROUP
SERONO LABS

Products: INFERTILITY PRODS FROM NATURAL SUBSTANCES; DX, RX

BATELLE MEMORIAL INST.

Products: DX; VET VX; SEPARATION TECHNIQUES; CONVERSION TECHNIQUES

BAXTER HEALTH CARE CORP.

Products: FACTOR VIII; AIDS DX; MABS; TPA; IF-G; IL-2; ALPHA-FETOPROTEIN RIA KIT

HOFFMANN-LA ROCHE, INC.
ROCHE INSTITUTE OF
MOLECULAR BIOLOGY

Products: IF-2A; IL-2; DX

JOHNSON & JOHNSON
BIOTECHNOLOGY CENTER

Products: DX; SYNTHETIC VX; SOME VET PRODUCTS

JOHNSON & JOHNSON
ORTHO DIAGNOSTICS SYSTEMS

Products: DX; HERPES 1 & 2 CHLAMYDIA TESTS, HERPES DNA PROBE DX, BIOCLONE (TM)

JOHNSON & JOHNSON
ORTHO PHARMACEUTICALS

Products: RX; DX; EPO, IL-2, CANCER MABS; HEP B VX; RX DELIVERY SYSTEMS; HOME PREGNANCY TESTS

ELI LILLY & CO.
(*SEE ALSO HYBRITECH, INC.)

Products: RX; HGH; PROINSULIN; ANTIBIOTICS; DX; AGRICHEMICALS

PHILLIPS PETROLEUM
(*SEE ALSO PROVESTA CORP.)

Products: FERMENTATION BASED FLAVORS, NUTRITIONAL YEAST, ENZYMES, VET VX, DX

SHELL OIL CO.
(*SEE ALSO TRITON BIOSCIENCES)

Products: PETROLEUM RELATED PRODUCTS, PETROCHEMICALS & PRODUCTS

SYNTEX CORP.

Products: RX; DX PRODUCTS; DNA RESEARCH; NEOPLASMA DX; MABS

Energy (L)

AMOCO
RESEARCH CENTER

Products: SPECIALTY CHEMICALS; ENERGY/BIOMASS; DATABANKS/PROGRAMS; FERMENTATION

PHILLIPS PETROLEUM
(*SEE ALSO PROVESTA CORP.)

Products: FERMENTATION BASED FLAVORS, NUTRITIONAL YEAST, ENZYMES, VET VX, DX

QUANTUM CHEMICAL CORP.
(*SEE ALSO LIFECODES CORP.)

Products: POLYETHELYNE RESINS, OLEO CHEMICALS, PROPANE

Listing 11-2 Corporations Listed by Industry Classification

Energy (Cont.)

SHELL OIL CO.   Products: PETROLEUM RELATED PRODUCTS,
(*SEE ALSO TRITON BIOSCIENCES)   PETROCHEMICALS & PRODUCTS

Food Production/Processing (M)

ANHEUSER-BUSCH, INC.   Products: BEER; YEAST & FERMENTATION TECHNOLOGY;
YEAST TO PRODUCE RDNA IF

CAMPBELL SOUP CO.   Products: RDNA DISEASE-FREE TOMATOES, MUSHROOMS
CAMPBELL INSTITUTE OF R & D

ICI AMERICAS, INC.   Products: SEED BREEDING, INSECTICIDES,
CHEMOTHERAPY

MONSANTO CO.   Products: ASPARTAME; INDUSTRIAL CHEMICALS; AG
LIFE SCIENCES RESEARCH CTR.   CHEMICALS; RX

PFIZER, INC.   Products: SORBOSE; XANTHAN; ANTIBIOTICS;
TETRACYCLINE

PIONEER HI-BRED INT'L, INC.   Products: HERBICIDE RESISTANT CORN, IMPROVED SEED
PRODUCTS, CORN, SUNFLOWER

Production/Fermentation (O)

AMOCO   Products: SPECIALTY CHEMICALS; ENERGY/BIOMASS;
RESEARCH CENTER   DATABANKS/PROGRAMS; FERMENTATION

ANHEUSER-BUSCH, INC.   Products: BEER; YEAST & FERMENTATION TECHNOLOGY;
YEAST TO PRODUCE RDNA IF

BAYER USA   Products: INDUSTRIAL ENZYMES; CITRIC ACID;
MILES LABS   INGREDIENTS FOR FERMENTATION
BIOTECHNOLOGY PRODUCTS DIV.

E.I. DU PONT DE NEMOURS & CO.   Products: DX; EQUIPMENT; DNA PROBES; CELL CULTURE
PRODUCTS; AGRICHEMICALS; AIDS DX; IF; MABS

PHILLIPS PETROLEUM   Products: FERMENTATION BASED FLAVORS,
(*SEE ALSO PROVESTA CORP.)   NUTRITIONAL YEAST, ENZYMES, VET VX, DX

E.R. SQUIBB & SONS, INC.   Products: STEROID OXIDATIONS; ANTIBIOTICS; INSULIN

Therapeutics (P)

ABBOTT LABORATORIES   Products: HEP B & AIDS DX; ENZYMES, UROKINASE;
INSECTICIDES; LAB PRODS; DX EQUIPMENT

AMERICAN CYANAMID   Products: RX; ANTICANCER RX; CARDIOVASCULAR RX;
LEDERLE LABORATORIES   ANTIBIOTICS; VX; VITAMINS; ANTI-FUNGALS

AMERICAN HOME PRODUCTS   Products: ETHICAL RX; PENICILLIN; VX; INFANT
WYETH-AYERST LABORATORIES   FORMULAS FROM WHEY; BICILLIN

Therapeutics (Cont.)

BAXTER HEALTH CARE CORP.   Products: FACTOR VIII; AIDS DX; MABS; TPA; IF-G; IL-2; ALPHA-FETOPROTEIN RIA KIT

BECTON DICKINSON & CO.   Products: MEDICAL DEVICES, MAB BASED REAGENTS
MEDICAL SECTOR

BRISTOL-MYERS CO.   Products: CANCER, VIRAL RX, OTC HEALTH CARE,
(*SEE ALSO GENETIC SYSTEMS)   PROSTHETIC DEVICES

BURROUGHS WELLCOME CO.   Products: RX--ANTIBACTERIALS, ANTIHISTAMINES, CARDIOVASCULAR, ANTICANCER, IF, ANTI-CANCER, ANTI-VIRALS, ANTI-DEPRESSANTS

CIBA-GEIGY CORP.   Products: DRUG DELIVERY SYSTEMS; VX
PHARMACEUTICALS DIVISION

DOW CHEMICAL CO.   Products: CANCER IMMUNOTHERAPY, BIOLOGICAL PEST
BIOPRODUCTS LAB   CONTROL

DOW CHEMICAL CO.   Products: MAB PRODUCTION; REGULATION OF GENE
MERRELL DOW RESEARCH INST.   EXPRESSION; RDNA PROTEIN FOR STRUCTURE ANALYSIS

E.I. DU PONT DE NEMOURS & CO   Products: DX; EQUIPMENT; DNA PROBES; CELL CULTURE PRODUCTS; AGRICHEMICALS; AIDS DX; IF; MABS

GLAXO, INC.   Products: RX

HOFFMANN-LA ROCHE, INC.   Products: IF-2A; IL-2; DX
ROCHE INSTITUTE OF
MOLECULAR BIOLOGY

ICI AMERICAS, INC.   Products: SEED BREEDING, INSECTICIDES, CHEMOTHERAPY

JOHNSON & JOHNSON   Products: GROWTH FACTORS
ETHICON, INC.

JOHNSON & JOHNSON   Products: RX; DX; EPO, IL-2, CANCER MABS; HEP B VX; RX
ORTHO PHARMACEUTICALS   DELIVERY SYSTEMS; HOME PREGNANCY TESTS

EASTMAN KODAK   Products: ANIMAL NUTRITION; AGRICULTURAL TECHNOL; PHARMACEUTICALS

ELI LILLY & CO.   Products: RX; HGH; PROINSULIN; ANTIBIOTICS; DX;
(*SEE ALSO HYBRITECH, INC.)   AGRICHEMICALS

MERCK & COMPANY, INC.   Products: RX; ANTIBIOTICS; VITAMINS; HEP B VX; VET RX, SPECIALTY CHEMICALS, PESTICIDES

MONSANTO CO.   Products: TPA; ATRIAL PEPTIDES; IF-B; ASPARTAME
G.D. SEARLE & CO.

Listing 11-2 Corporations Listed by Industry Classification

Therapeutics (Cont.)

MONSANTO CO.
LIFE SCIENCES RESEARCH CTR.

Products: ASPARTAME; INDUSTRIAL CHEMICALS; AG CHEMICALS; RX

PFIZER, INC.

Products: SORBOSE; XANTHAN; ANTIBIOTICS; TETRACYCLINE

RORER GROUP, INC.
RORER CENTER RESEARCH (MELOY)

Products: RX

SANDOZ PHARMACUETICALS

Products: CANCER RX; INSECTICIDE; CORN HYBRIDS; KPA

E.R. SQUIBB & SONS, INC.

Products: STEROID OXIDATIONS; ANTIBIOTICS; INSULIN

UPJOHN CO.

Products: ANTIBIOTICS; TETRACYCLINE; SEEDS; BGH

Vaccines (Q)

AMERICAN CYANAMID
LEDERLE LABORATORIES

Products: RX; ANTICANCER RX; CARDIOVASCULAR RX; ANTIBIOTICS; VX; VITAMINS; ANTI-FUNGALS

AMERICAN HOME PRODUCTS
WYETH-AYERST LABORATORIES

Products: ETHICAL RX; PENICILLIN; VX; INFANT FORMULAS FROM WHEY; BICILLIN

CIBA-GEIGY CORP.
PHARMACEUTICALS DIVISION

Products: DRUG DELIVERY SYSTEMS; VX

JOHNSON & JOHNSON
ORTHO PHARMACEUTICALS

Products: RX; DX; EPO, IL-2, CANCER MABS; HEP B VX; RX DELIVERY SYSTEMS; HOME PREGNANCY TESTS

MERCK & COMPANY, INC.

Products: RX; ANTIBIOTICS; VITAMINS; HEP B VX; VET RX, SPECIALTY CHEMICALS, PESTICIDES

Veterinary (V)

AMERICAN CYANAMID
AGRICULTURAL DIVISION

Products: ANIMAL GROWTH FACTORS; BOVINE SOMATOTROPIN; BOVINE MASTITIS RX; HERBICIDES; PESTICIDES

INTERNATIONAL MINERALS & CHEMICALS

Products: ANIMAL GROWTH, HEALTH & NUTRITION

EASTMAN KODAK

Products: ANIMAL NUTRITION; AGRICULTURAL TECHNOLOGY

ELI LILLY & CO.
(*SEE ALSO HYBRITECH, INC.)

Products: RX; HGH; PROINSULIN; ANTIBIOTICS; DX; AGRICHEMICALS; PHARMACEUTICALS

Medical Devices (4)

BECTON DICKINSON & CO.
MEDICAL SECTOR

Products: MEDICAL DEVICES, MAB-BASED REAGENTS

BRISTOL-MYERS CO.
(*SEE ALSO GENETIC SYSTEMS)

Products: CANCER, VIRAL RX, OTC HEALTH CARE, PROSTHETIC DEVICES

Medical Devices (Cont.)

E.I. DU PONT DE NEMOURS & CO Products: DX; EQUIPMENT; DNA PROBES; CELL CULTURE
PRODUCTS; AGRICHEMICALS; AIDS DX; IF; MABS

JOHNSON & JOHNSON   Products: GROWTH FACTORS
ETHICON, INC.

Notes:

*See also the biotechnology company indicated, in Listing 2-2.

Corporations from Listing 11-1 with primary or secondary focus in the indicated areas are presented. Not all industry classes are represented.

Listing 11-3 Corporations with Biotechnology Subsidiaries

LISTING 11-3 CORPORATIONS WITH BIOTECHNOLOGY SUBSIDIARIES

Corporation	Biotechnology Subsidiary
ADOLPH COORS CO.	COORS BIOTECH PRODUCTS CO.
ALLIED CORP.	NITRAGIN CO.
AMERICAN HOECHST CORP.	BEHRING DIAGNOSTICS
BAUSCH & LOMB	CHARLES RIVER BIOTECH. SERVICES
BAYER AG (WGER)	MOLECULAR DIAGNOSTICS, INC.
BIOTECHNICA INT'L	BIOTECHNICA AGRICULTURE
BIOTEST SERUM INST. (WGER)	BIOTEST DIAGNOSTICS CORP.
BOEHRINGER MANNHEIM (WGER)	BOEHRINGER MANNHEIM DIAGN.
BRISTOL-MYERS	GENETIC SYSTEMS CORP.
CAMBRIDGE RES. BIOCHEM. (UK)	CAMBRIDGE RES. BIOCHEMICALS, INC.
CIBA-GEIGY & CORNING GLASS	CIBA-CORNING DIAGNOSTIC CORP.
CPC INTERNATIONAL	ENZYME BIO-SYSTEMS, LTD.
DAKOPATTS A/S (DENMARK)	DAKO CORP.
DAMON CORP.	DAMON BIOTECH, INC.
DARYL LABS	MICROBIOLOGICAL ASSOCIATES, INC.
DOW CHEMICAL	UNITED AGRISEEDS, INC.
EASTMAN KODAK	INT'L BIOTECHNOLOGIES, INC.
ELI LILLY	HYBRITECH, INC.
FERMENTA AB (SWE)	RICERCA, INC.
FERMENTA AB (SWE)	FERMENTA ANIMAL HEALTH
FLOW GENERAL, INC.	FLOW LABORATORIES, INC.
GRANADA CORP.	GRANADA GENETICS CORP.
GREAT LAKES CHEMICAL CORP.	ENZYME TECHNOLOGY CORP.
HAZELTON LABORATORIES	HAZLETON BIOTECHNOLOGIES CO.
HYCOR BIOMED	ICL SCIENTIFIC, INC.
IBF, SA	IBF BIOTECHNICS, INC.
ICI AMERICAS	CELLMARK DIAGNOSTICS
INCSTAR	ATLANTIC ANTIBODIES
INFRAGENE	CALIFORNIA INTEGRATED DIAGN.
INT'L MINERALS & CHEMICALS	IMCERA BIOPRODUCTS, INC.
KMS INDUSTRIES	COVALENT TECHNOLOGY CORP.
KYOWA HAKKO KOGYO (JPN)	BIOKYOWA, INC.
LABSYSTEMS OY (FINLAND)	LABSYSTEMS, INC.
LUBRIZOL	AGRIGENETICS CORP.
LUCKY LTD. (S. KOREA)	LUCKY BIOTECH CORP.
MEDICARB (SWE)	CARBOHYDRATES INT'L, INC.
MICROBIOLOGICAL SCIENCES	SCOTT LABORATORIES
NATIONAL PATENTS DEVEL.	INTERFERON SCIENCES, INC.
NEW BRUNSWICK SCIENTIFIC	BIOSEARCH, INC.
ORGANON TEKNIKA (NETH)	ORGANON TEKNIKA CORP.
ORGANON TEKNIKA (NETH)	BIONETICS RESEARCH
PEDCO, INC.	BIOASSAY SYSTEMS CORP.
PETROLEUM FERM., NV	PETROFERM, INC.

Corporation	Biotechnology Subsidiary
PHARMACIA AB (SWE)	PHARMACIA LKB BIOTECHNOLOGY
PHILLIPS PETROLEUM	PROVESTA CORP.
QUEST BIOTECHNOLOGY	POLYCELL, INC.
REPAP ENTERPRISES (CANADA)	REPAP TECHNOLOGIES
RORER GROUP, INC.	MELOY LABORATORIES, INC.
SANDOZ (SWI)	ZOECON CORP.
SCHERING-PLOUGH	DNAX RESEARCH INSTITUTE
SHELL OIL CO.	TRITON BIOSCIENCES, INC.
SMITHKLINE BECKMAN	NORDEN LABS
SYNTEX	SYVA CO.
TECHAMERICA GROUP	SYNGENE PRODUCTS
3M CORP.	3M DIAGNOSTIC SYSTEMS
UNIVERSITY GENETICS	AN-CON GENETICS

Table 11-1 Large Corporations and Biotechnology Companies

TABLE 11-1 LARGE CORPORATIONS VS. THE BIOTECHNOLOGY COMPANIES

A comparison of the large corporations and the biotechnology companies in this Guide is made by looking at the average figures for the top 15 U.S. pharmaceutical companies and the 15 largest biotechnology companies.

Category	Biotechnology Company	Pharmaceutical Company
Year Founded	1979	1906
Employees	494	35,400
Revenues (1)	$73 million	$4 billion
R&D Budget (2)	$26 million	$370 million

Notes:

1. Revenues are total company sales in 1987 for the top 15 companies in each category. Pharmaceutical company data are supplemented by information from *Forbes*. Pharmaceutical company revenues are total revenues and are considerably higher than their sales of therapeutics.

2. R&D budget for top companies for 1987. Pharmaceutical company data supplemented with information from *Standard and Poors*.

3. For a more detailed ananlysis of biotechnology's impacts on the pharmaceutical industry, see: Dibner, M.D. and Osterhaus, J.T.: Biotechnology and pharmaceuticals: Merging together. *Biopharm Manufacturing*, September 1987, 56-60.

SECTION 12 PARTNERSHIPS IN BIOTECHNOLOGY

INTRODUCTION

Partnerships formed between two companies to bring together complementary strengths in biotechnology have taken many forms, from licensing agreements to outright acquisition. We have used our **Actions** database to analyze almost 500 examples of partnering over the last five years by type of partnership, industry involved, product involved and origin of the partners. Although examples of U.S/U.S. partnering are the greatest in number, perhaps due to the large number of biotechnology-related companies in the United States, we have noted a large number of U.S./Foreign partnerships as well.

One important feature of the growth of the U.S. biotechnology industry is the need that individual companies have for nurturing the new technology and creating a return on their investors' equity. To meet their needs, companies turn to strategic alliances (or partnering, partnerships, joint ventures, etc.) to add complementary strength to internal strength. By examining these strategic alliances, one can gain an overview of trends and associations within the industry.

Business activity in biotechnology (or any industry) is a continuum from research to product development to manufacturing to marketing. It is clear that the new firms in Listing 2-2 are mostly skilled in the basic research and product development with little expertise in manufacturing or marketing. The large corporations in Listing 11-1 often have marketing expertise and established channels, and have had experience with production. The corporations also have available capital, which the small firms greatly need. What the corporations lacked early on is in-house research expertise in the new science and development expertise using the new technology. Thus, it should be clear that the complementary needs have led to the formation of partnerships of a variety of types and for a variety of purposes.

Table 12-1 lists industry classification codes once again, as well as a key for the codes of external actions (partnerships) taken by the companies. The subsequent listings show the partnerships formed in biotechnology with large U.S. corporations, first listed by U.S. biotechnology firm (Listing 12-1), then by corporation (Listing 12-2). The subsequent two listings detail partnerships of U.S. biotechnology firms with foreign corporate partners, first by U.S. firm (Listing 12-3) and then by country of origin of the partner (Listing 12-4). Finally, in Listing 12-5 all partnerships are presented, sorted by the industry classification codes.

Table 12-2 shows the frequency of partnerships by industry classification, with therapeutics accounting for more than one-third of the partnerships followed by diagnostics with almost 20 percent. Figure 12-1 details the U.S./U.S. and U.S./Foreign partnerships by type of partnership. Figure 12-2 demonstrates the same categories but for health care-related partnerships. Lastly, Listing 12-6 gives an overview of acquisition of biotechnology-related companies each year, first by U.S. companies and then by non-U.S. companies.

The majority of large and small U.S. companies listed can be found in the directories in Listings 11-1 and 2-2, respectively. Since the source of the information in this section, the **Actions** database, is a different database with a somewhat broader definition of biotechnology activities than the **Companies** database, some companies mentioned in this section will therefore not be found earlier in this Guide.

Table 12-1 Classification Codes and Codes for Actions

TABLE 12-1 CLASSIFICATION CODES AND CODES FOR ACTIONS

Code	Industry Classification
A	Agriculture, Animal
B	Agriculture, Plant
C	Biomass Conversion
D	Biosensors/Bioelectronics
E	Bioseparations
F	Biotechnology Equipment
G	Biotechnology Reagents
H	Cell Culture, General
I	Chemicals, Commodity
J	Chemicals, Specialty (includes proteins and enzymes)
K	Diagnostics, Clinical Human
L	Energy
M	Food Production/Processing
N	Mining
O	Production/Fermentation
P	Therapeutics
Q	Vaccines
R	Waste Disposal/Treatment
S	Aquaculture
T	Marine Natural Products (includes algae)
U	Consulting
V	Veterinary (all animal health care)
W	Research
X	Immunological Products (non-pharmaceutical)
Y	Toxicology
Z	Venture Capital/Financing
1	Biomaterials
2	Fungi
3	Drug Delivery Systems
4	Medical Devices
5	Testing/Analytical Services

Codes For External Actions

Code	Action
A	Acquisition
E	Equity
J	Joint venture (usually unspecified in nature)
L	Licensing agreement
M	Marketing agreement
O	Other (not in other categories)
R	Research contract
V	Joint venture company formed

LISTING 12-1 PARTNERSHIPS, LISTED BY BIOTECHNOLOGY COMPANY

Company 1	Company 2	Yr	Cd.	Act.	Product
ADVANCED BIOTECH.	GENTRONIX LABS	87	K	A	MICROANALYSIS
ADVANCED GENETIC SCI.	DNA PLANT TECHNOL.	87	B	O	PLANTS
ADVANCED GENETIC SCI.	DUPONT	86	I	R	
ADVANCED GENETIC SCI.	EASTMAN KODAK	87	1	M	SNOW MAKING
AGEN USA	AMERICAN HOSP. SUPPLY	86	K	M	BLOOD SCREEN KIT
AGOURON PHARMACEUT.	CAMBRIDGE BIOSCIENCE	87	KP	J	AIDS DX/RX/VX
AGRI-DIAGNOSTICS	O.M. SCOTT & SONS	86	B	M	PLANT DX
AGRIC. GENETIC SYSTEMS	EPITOPE	88	BV	A	VET PROD/PLANTS
AGRIGENETICS	AGRIGENETICS RESEARCH	86	W	A	TECHNOLOGIES
AGRIGENETICS	SUNGENE	87	M	M	HYB. CORN SEEDS
AGRITECH SYSTEMS	HYBRITECH	86	V	L	VETERINARY DX
AGS	AGRITOPE	88	BV	A	PLANTS/VET DX
ALPHA 1 BIOMEDICALS	INTERLEUKIN-2	86	Q	V	AIDS VX
ALZA CORP.	QUEST BIOTECHNOLOGY	87	P	L	HEMOGLOBINS
AMERICAN MONITOR	QUEST BIOTECHNOLOGY	87	F	A	BT INSTRUMENTS
AMGEN	EASTMAN KODAK	86	JO	JM	CHEMICALS
AMGEN	JOHNSON & JOHNSON	86	P	J	IL-2 ANALOG
AMGEN	SMITHKLINE	86	P	EJ	SOMATOTROPIN
ANALYTAB PRODUCTS	CAMBRIDGE BIOSCIENCE	87	K	M	VIRAL DX
ANALYTICHEM	VARIAN ASSOCIATES	87		A	
APPLIED BIOSYSTEMS	SPECTROS INT'L,PLC	86	F	A	CHROMATOGRAPHY
APPLIED BIOTECHNOLOGY	DUPONT	86	PK	J	CANCER DX,THERA
APPLIED BIOTECHNOLOGY	ONCOGENETICS PART.	87	Q	J	AIDS/TB VX
APPLIED MICROBIOLOGY	AMERICAN CYANAMID	87	J	E	PROTEIN ENZYMES
APPLIED MICROBIOLOGY	AMERICAN CYANAMID	87	V	M	STAPHYLOCIDE
ARCO SEED CO.	UF GENETICS	87	B	A	
ATLANTIC ANTIBODIES	INCSTAR	87	K	A	AB'S, DX KITS
AUTOMATED DIAGNOS.	UNIVERSITY GENETICS	87	K	E	AIDS DX
BARROWS RESEARCH GRP.	PRAGMA BIO-TECH	87	P	L	CHOLESTEROL RX
BIO-RESPONSE	COOPER	86	KP	M	DX;ANTIVIRALS
BIO-RESPONSE	RETROPERFUSION SYS.	87	P	J	TPA
BIO-RESPONSE	VENTREX	86	H	J	MAB'S
BIOGEN	BAXTER TRAVENOL	86	P	L	IF-G
BIOGEN	MERCK	87	P	J	ANTICANCER RX
BIOGEN	SCHERING-PLOUGH	86	P	L	IF-A
BIOGEN	VERAX	87	H	L	CELL CULTURES
BIOMET	INGENE	87	P	J	PROTEINS
BIOMET	LIFECORE BIOMEDICAL	87	P	J	ORTHOPEDICS
BIOPROBE	AMERICAN CYANAMID	87	K	O	CHROMATOGRAPHY
BIOS	ICF	87	EF	E	BIOMATERIALS
BIOSEARCH	HP GENENCHEM	86	G	L	DNA SYNTH CHEM
BIOTECH RESEARCH LABS	DUPONT	87	K	LM	AIDS DX
BIOTECHNICA AGRICULT.	L. HERRIED SEED	87	B	A	ALFALFA
BIOTECHNICA AGRICULT.	MCALLISTER SEED CO.	87	B	A	SEEDS
BIOTECHNICA DIAGNOS.	FORSYTH DENTAL CTR.	87	K	J	PERIDONTAL DX
BIOTECHNICA INT'L	NABISCO	87	B	J	AGRONOMIC PRODS.
BIOTECHNICA INT'L	STATE FARM MUTUAL	88	B	E	AGRICULT. PRODS.
BIOTECHNICA INT'L	W.R. GRACE	87	J	J	ENZYMES
BIOTECH. DEVELOPMENT	IMMUNE RESPONSE	87	Q	M	AIDS VX
BIOTECH. GENERAL	BAXTER TRAVENOL	86	P	M	GROWTH HORM.

Listing 12-1 Partnerships, by Biotechnology Company

Company 1	Company 2	Yr	Cd.	Act.	Product
BIOTECH. GENERAL	BRISTOL-MYERS	86	P	LM	SOD
BIOTECH. GENERAL	NEUROGENETIC	87	KP	E	NEUROBIOL. DX, RX
BIOTHERAPEUTICS	ANALYTIC BIOSYSTEMS	86	P	J	DRUGS
BIOTHERAPEUTICS	BAXTER HEALTHCARE	88	P	J	CANCER RX
BIOTHERAPEUTICS	INGENE	87	KP	V	CANCER RESEARCH
BIOTHERAPEUTICS	SYNCOR	86	P	J	CANCER RX
BIOTHERAPEUTICS	XOMA	88	P	J	CANCER RX
CALGENE	PEDIGREED SEED CO.	87	B	A	COTTON
CALGON CORP.	GENENTECH	88		A	
CALIF. BIOTECHNOLOGY	AMERICAN HOME PRODS.	86	P	M	NAZDEL
CALIF. BIOTECHNOLOGY	FT. DODGE LABS	86	V	J	PET VX
CALIF. BIOTECHNOLOGY	LYPHOMED	87	3	MJ	DRUG DELIVERY
CALIF. BIOTECHNOLOGY	ORTHO	86	P	R	NAZDEL
CALIF. BIOTECHNOLOGY	WYETH	86	P	M	RENIN INHIBITOR
CALIF. INTEGRATED DIAG.	INFERGENE	87	MK	A	DX
CAMBRIDGE BIOSCIENCE	BAXTER HEALTHCARE	87	P	M	AIDS DX
CAMBRIDGE BIOSCIENCE	SMITHKLINE	86	V	M	FELINE LEUKEMIA DX
CAMBRIDGE BIOSCIENCE	SYVA	87	K	J	AIDS DX
CAMBRIDGE RES. BIOCHEM.	MILLIGEN	86	F	E	
CAREMARK	BAXTER TRAVENOL	87	P	A	HEALTH CARE
CELGENE	RITZ CHEMICALS	87	J	J	BIOPOLYMERS
CELLULAR PRODUCTS	EASTMAN KODAK	87	K	M	AIDS/T-CELL DX
CENTOCOR	ERBAMONT	86	K	L	DX
CENTOCOR	FMC	86	P	E	
CENTOCOR	IMMUNOMEDICS	87	K	L	CANCER DX
CENTOCOR	POLYCELL	86	P	L	MAB RX
CENTOCOR	REPLIGEN	87	Q	M	AIDS VX
CETUS	BEN-VENUE LABS	87	P	V	ANTICANCER RX
CETUS	BEN-VENUE LABS	86	P	M	METHOTREXATE
CETUS	CELLULAR PRODUCTS	86	G	L	IL-2
CETUS	CETUS-BEN VENUE	87	P	MJ	CANCER RX
CETUS	EASTMAN KODAK	86	K	J	DX
CETUS	ENZO BIOCHEM	86	K	J	
CETUS	EUROCETUS	87		E	ANTICANCER
CETUS	INTELLIGENT MEDICINE	86	3	J	DRUG DELIVERY
CETUS	PERKIN ELMER	86	F	V	INSTRUMENTS
CETUS	POLYCELL	88	K	J	CANCER DX
CETUS	SQUIBB	87	P	JE	VARIOUS RX
CETUS	SQUIBB	87		O	
CETUS	SQUIBB	87	P	L	ANTICANCER RX
CETUS	TRITON BIOSCIENCES	86	P	V	IF
CHARLES RIVER BIOTECH	CORNING GLASS	86	F	A	
CHEMEX	SQUIBB	87	P	L	SKIN RX
CHEMEX	UPJOHN	88	P	L	SKIN RX
CHEMICON INT'L	MELOY LABORATORIES	87		A	
CHIRON	ETHICON	86	P	J	GROWTH FACTORS
CILCO, INC.	DIAGNOSTIC, INC.	86	P	M	HYALURONIC ACID
CISTRON BIOTECH.	BAKER INSTRUMENTS	87	PG	A	
CISTRON BIOTECH.	DUPONT	86	P	L	IL-2
CLINICAL SCIENCES	DUPONT	87	K	J	HEP-B AG'S
CODON	CELANESE	87		E	
CODON	MERRILL LYNCH	86	Q	E	SWINE VX
COLLABORATIVE RES.	DOW CHEMICAL	86	M	M	RENNIN
COLLABORATIVE RES.	DOW CHEMICAL	88	JM	L	CHYMOSIN

Company 1	Company 2	Yr	Cd.	Act.	Product
COLLABORATIVE RES.	DOW CHEMICAL	88	M	J	RENNIN
COLLABORATIVE RES.	DOW CHEMICAL	86	M	P	RENNIN
COLLABORATIVE RES.	PFIZER	88	M	L	RENNIN
COLLAGEN	TARGET THERAPEUTICS	88	P3	A	DRUG DELIVERY SYS.
COOPER LASERSONICS	PFIZER	88	F	JA	BT EQUIP.
COOPER TECHNICON	MICROGENICS	87	K	R	MEDICAL DX
CYANOTECH	AMERICAN CYANAMID	87	P	JM	CHOLESTEROL RX
CYANOTECH	PHARMACHEM	86	M	C	BETA-CAROTENE
CYTOGEN	EASTMAN KODAK	86	KP	E	
CYTOGEN	EASTMAN KODAK	86	KP	R	
DAMON	BIOTHERAPY SYSTEMS	86	P	A	
DAMON	IDEC	86	P	A	B-CELL LYMPH RX
DAMON	SMITHKLINE	87	P	M	TPA
DEKALB-PFIZER GENETICS	NATIVE PLANTS	88	B	J	CORN HYBRIDS
DESERT COTTON R&D CO.	STONEVILLE SEED	88	B	A	COTTON
DIAGNON	MELOY LABORATORIES	86	Q	A	
DIAGNOSTIC PRODUCTS	MONOCLONAL AB'S	87		E	
DNA PLANT TECHNOLOGY	CAMPBELL SOUP CO.	86	B	O	DNAP-9 TOMATO
DNA PLANT TECHNOLOGY	DUPONT	87	B	J	PLANT OILS
DNA PLANT TECHNOLOGY	FARMS OF TEXAS CO.	87	M	J	HYBRID RICE
DNA PLANT TECHNOLOGY	KRAFT	87	M	O	SNACK FOODS
DNA PLANT TECHNOLOGY	MONSANTO	86	M	J	SWEETENERS
DNA PLANT TECHNOLOGY	NUTRASWEET CO.	86	M	R	SWEETENERS
DNASTAR	FIELD INVERSION TECH.	87	K	V	DX
DORR-OLIVER	W. R. GRACE	87	HF	A	BT EQUIP
ECOGEN	AMERICAN CYANAMID	86	B	RE	INSECTICIDE
ECOGEN	CILCORP VENTURES	88	B	E	BIOPESTICIDES
ECOGEN	GENETICS INSTITUTE	88	B	L	PESTICIDE
ECOGEN	ML TECHNOL. VENTURES	87	B	JR	PSEUDOMONAS
ECOGEN	MONSANTO	87	B	L	PESTICIDES
ELECTRO-NUCLEONICS	PHILLIPS PETROLEUM	87	K	R	AIDS DX
ELECTRO-NUCLEONICS	PRUTECH R&D PART. II	86	K	R	AIDS TEST
EMBRYOGEN	UPJOHN	87	A	J	LAB ANIMALS
ENDOTRONICS	CELANESE	86	WP	R	IMMUNE CELLS
ENDOTRONICS	CELANESE	86	P	E	HEPATITIS B VIRUS
ENDOTRONICS	CELANESE	86	H	E	T-CELL THERAPY
ENDOTRONICS	CELANESE	86	P	R	AIDS TREATMENT
ENDOTRONICS	SUMMA MEDICAL	86	HP	E	
ENDOTRONICS	VENTURE FUNDING LTD	87		E	
ENGENICS	EASTMAN KODAK	86	M	J	BIOCHEMICALS
ENZON	EASTMAN KODAK	87	P	E	O2 TOXICITY RX
EPITOPE	EASTMAN KODAK	88	PK	J	CANCER DX/RX
EPITOPE	INGENE	88	P	J	AIDS RX
EPITOPE	SYVA	87	K	R	AIDS DX
ETHICON	POSSIS MEDICAL	87	1	O	VASCULAR GRAFTS
EXOVIR	GENENTECH	86	P	R	EXOVIR-HZ GEL
FARNAM COMPANIES	QUIDEL	87	V	M	BOVINE DX
GEN-PROBE	FISHER SCIENTIFIC	86	Y	M	MYCOPLASMA KITS
GENE-TRAK SYSTEMS	INTEGRATED GENETICS	86	Y	M	DNA PROBE
GENELABS	SRI INTERNATIONAL	88	P	JR	AIDS RX
GENELABS	SRI INTERNATIONAL	88	P	JR	IMMUNOCONJUGATE
GENENCOR	EASTMAN KODAK	87		E	
GENENTECH	EXOVIR	86	P	JL	VIRUS TREATMENT
GENENTECH	GENZYME	87	P	JR	PROTEIN

Listing 12-1 Partnerships, by Biotechnology Company

Company 1	Company 2	Yr	Cd.	Act.	Product
GENENTECH	HP GENENCHEM	87	F	E	BT INSTRUMENTS
GENENTECH	NEORX	87	K	J	BLOOD CLOT DX
GENENTECH	NEW BRUNSWICK SCI.	86	F	L	
GENESIS LABS	MONOCLONAL ANTIBOD.	87		A	
GENETIC DESIGN	MILLIPORE	86	F	A	DNA SYNTHESIZER
GENETICS INSTITUTE	NEORX	87	P	EJ	ANTICANCER MAB
GENETICS INSTITUTE	SERAGEN	87	EP	L	BIOSEPARATION
GENETICS INSTITUTE	UNITED AGRISEEDS	87	B	A	SEEDS
GENETICS INSTITUTE	UNITED AGRISEEDS	87	B	O	HYBRID CORN
GENEX	SCHERING-PLOUGH	86	P	R	PLASMA PROTEIN
GENEX	XYDEX	87	J	A	PROTEINS
GENZYME	ALCON LABORATORIES	86	P	RL	EYE SURG. PROD.
GENZYME	HOWMEDICA	87	P	J	ORTHOPEDICS
GENZYME	NEW BRUNSWICK SCI.	87	I	A	BIOCHEMICALS
GENZYME	PFIZER	87	P	L	ORTHOPEDIC SURG
GENZYME	ZYMOGENETICS	87	PK	M	SKIN RX, DX
H.B. FULLER & CO.	GENEX	87	J	A	SPECIALTY CHEM.
HEM RESEARCH	DUPONT	87	P	E	AIDS RX
HEPAR INDUSTRIES	AMERICAN HOME PRODS.	86	P	L	HEPARIN
HILLMAN & CO.	SUTTER HILL VENTURES	87		V	
HOUSTON BIOTECH.	HOUSTON BIOTECH PART.	87	P	R	EYE DRUGS
HUNT RESEARCH CORP.	QUEST BIOTECHNOLOGY	87	P	A	HEMOGLOBIN
HYBRIDOMA SCIENCES	COVALENT TECHNOLOGY	86	K	A	
HYBRITECH	MONOCLONAL AB'S	87	K	L	VARIOUS DX
HYGEIA SCIENCES	TAMBRANDS	87	K	A	
IGEN	DUPONT	87	B	J	PLANT DISEASE DX
IGEN	GRYPHON VENTURES	87	K	E	DX ASSAYS
IMCERA BIOPRODUCTS	INT'L MINERALS & CHEM.	87	P	OM	SOMATOMEDIN
IMCLONE	FISHER SCIENTIFIC	87	K	J	HEP-B, AIDS DX
IMCLONE	LEDERLE	88	Q	J	VX
IMMUNEX	GENZYME	86	P	M	IL-2
IMMUNOLOGY VENTURES	EASTMAN KODAK	87	P	M	IL-4
IMMUNOMEDICS	JOHNSON & JOHNSON	86	KP	R	
IMMUNOSYSTEMS	DUPONT	88	B	J	PESTICIDES
IMMUNOSYSTEMS	MONSANTO AGRIC.	88	B	J	IMMUNOASSAYS
IMMUNOSYSTEMS	WESTINGHOUSE	88	B	M	HERBICIDE DX
INGENE	BEATRICE	87	M	R	SWEETENER
INGENE	BIOTHERAPEUTICS	86	P	R	CANCER THERAPY
INGENE	EASTMAN PHARMACEUT.	88	P	J	CANCER RX
INGENE	ONCOGEN	87	P	J	MAB CANCER RX
INTEGRATED GENETICS	AMOCO	86	MK	J	FOOD DX
INTEGRATED GENETICS	AMOCO	87	K	V	AIDS DX
INTEGRATED GENETICS	ORTHO	87	P	J	MSF
INTELLICORP	AMOCO	86	F	J	SOFTWARE
INTELLIGENETICS	AMOCO	86	F	A	SOFTWARE
INTERFERON SCIENCES	ANHEUSER-BUSCH	86	P	E	IF
INT'L BIOTECHNOLOGIES	EASTMAN KODAK	87		A	
INVITRON	RORER	88	P	J	FACTOR VIII:C
INVITRON	SEARLE	87	P	J	PROTEINS
INVITRON	SEARLE	86	P	J	TPA
KC BIOLOGICAL	CHARLES RIVER BIOTECH.	86	FH	A	OPTICELL
KLEINER PERKINS	ATHENA	86	KP	E	NEURAL DX
KLEINER PERKINS	GENSIA	86	P	E	HEART RX

Company 1	Company 2	Yr	Cd.	Act.	Product
KLEINER PERKINS	INSITE VISION	86	3	E	DRUG DELIVERY SYS.
LIFE TECHNOLOGIES	BECTON DICKINSON	87	K	A	DX
LIFECORE BIOMEDICAL	MEDICONTROL	88	P	J	SKIN MOISTURIZ.
LIFECORE BIOMEDICAL	ORTHOMATRIX	87	4	A	MEDICAL DEVICES
MEDI-CONTROL CORP	SEARLE	86	3	E	DRUG DELIVERY
MELOY LABORATORIES	RORER GROUP	86	P	A	
MICROBIO RESOURCES	EASTMAN KODAK	87	M	M	BETA-CAROTENE
ML TECHNOLOGY VENT.	UNITED AGRISEEDS	87	M	R	CORN
MOLECULAR GENETICS	EASTMAN KODAK	86	V	J	VET MASTITIS RX
MOLECULAR GENETICS	MOORMAN MANUF.	86	QV	RC	ANIMAL VX
MOLECULAR GENETICS	SCHERING-PLOUGH	87	K	M	TEST KITS
MOLECULAR GENETICS	TERRA INTERNATIONAL	86	B	V	CORN SEED
MOLECULON	BRISTOL-MYERS	86	3	J	DRUG DELIVERY
MONOCLONAL AB'S	ALCON LABORATORIES	86	K	J	
MONOCLONAL AB'S	ALPHA LABORATORIES	86	K	L	ELISA
MONOCLONAL AB'S	EASTMAN KODAK	88	K	L	IN VITRO DX
MONOCLONAL AB'S	MOBAY-HAVER	86	K	M	EQUINE DX
MONOCLONAL AB'S	MONOCLONAL TECHNOL.	87		A	
MONOCLONAL AB'S	ORTHO	87	K	L	OVULATION DX
MYCOGEN	LUBRIZOL	87	B	R	INSECTICIDES
MYCOGEN	MONSANTO	87	B	J	BIOPESTICIDES
NEORX	EASTMAN KODAK	86	KP	EJ	
NOVA PHARMACEUTICAL	CELANESE	86	P	E	
NOVA PHARMACEUTICAL	EASTMAN KODAK	88	F	J	SCREEN TECHNOL
ONCOGENE SCIENCE	PFIZER	86	WP	EJ	RX
ORGANOGENESIS	ELI LILLY	87	1	J	SYNTH. ARTERY
PANLABS	EASTMAN KODAK	87	P	R	RX
PLANT GENETICS	LOVELOCK SEED CO.	87	B	A	ALFALFA
PLANT GENETICS	MERRILL LYNCH	87	B	J	IMPROVED POTATOES
PLANT GENETICS	MONSANTO	87	B	J	POTATOES
PLANT GENETICS	WEYERHAEUSER	87	B	LR	PLANT ENCAPSUL.
POLYORGANIX	SYNTHATECH	87	J	A	ENZYMES
PRAXIS BIOLOGICS	BRISTOL-MYERS	87	ZQ	E	VX
PROTEIN DATABASES	MILLIPORE	87		E	
PRUTECH	ECOGEN	86	B	J	INSECTICIDES
R&D FUNDING CORP	SYNBIOTICS	86	V	J	VET RX
REPLIGEN	GILLETTE	87	P	J	DENTAL ENZYME
REPLIGEN	MERCK	87	Q	R	AIDS VX
SEPRACOR	AMERICAN CYANAMID	86	E	J	MEMBRANE TECH.
STERLING DRUG	EASTMAN KODAK	88	P	A	RX
SYNBIOTICS	INT'L MINERALS & CHEM.	87	K	E	MAB
SYNBIOTICS	INT'L MINERALS & CHEM.	87	V	RM	VET MAB'S
SYNBIOTICS	NORDEN LABORATORIES	87	V	M	LEUKEMIA DX
SYNBIOTICS	PRUTECH R&D PART. II	86	VA	R	VX
SYNERGEN	COORS BIOTECH	88	J	J	RIBOFLAVIN
SYNERGEN	DUPONT	87	P	J	AF
SYNERGEN	PROCTER & GAMBLE	86	P	J	COLLAGENASE INHIB
T CELL SCIENCES	PFIZER	86	PK	R	ARTHRITIS RX
TECHNICLONE	DAMON	86	P	R	ANTI LYMPHOMA
TECHNICON	COOPER DEVEL. CO.	86	K	A	DX
TRITON BIOSCIENCES	CETUS	86	P	J	IF-B
TRITUS	RICHARDSON-VICKS	87	P	L	IF-B
UNIVERSITY GENETICS	WORLDWIDE FARM SVC.	86	A	J	ANIMALS
VEGA BIOTECHNOLOGIES	DUPONT	86	F	EM	SYNTHESIZERS

Listing 12-1 Partnerships, by Biotechnology Company

Company 1	Company 2	Yr	Cd.	Act.	Product
VENTREX	BAXTER TRAVENOL DX	86	K	R	DX REAGENTS
XOMA	PFIZER	87	P	L	TOXIC SHOCK RX
XYTRONYX	COLGATE PALMOLIVE	87	K	L	PERIDONTAL DX

Notes:

In addition to the two companies involved, the year of partnership, industry class (Cd.; see Table 12-1), type of external action (Act.; see Table 12-1) and product involved are given, where available. Data were obtained from the public literature and company information and are compiled and stored in the **Actions** database. Most companies in the first column are detailed in Section 2, Listing 2-2, and many corporations in the second column are detailed in Section 11, Listing 11-1.

LISTING 12-2 PARTNERSHIPS, LISTED BY LARGE PARTNER

Company 1	Company 2	Yr.	Cd.	Act.	Product
ALCON LABORATORIES	GENZYME	86	P	RL	EYE SURGERY PRODS.
ALCON LABORATORIES	MONOCLONAL AB'S	86	K	J	
AMERICAN CYANAMID	APPLIED MICROBIOLOGY	87	J	E	PROTEIN ENZYMES
AMERICAN CYANAMID	APPLIED MICROBIOLOGY	87	V	M	STAPHYLOCIDE
AMERICAN CYANAMID	BIOPROBE	87	K	O	CHROMATOGRAPHY
AMERICAN CYANAMID	CYANOTECH	87	P	JM	CHOLESTEROL RX
AMERICAN CYANAMID	ECOGEN	86	B	RE	INSECTICIDE
AMERICAN CYANAMID	SEPRACOR	86	E	J	MEMBRANE TECHNOL.
AMERICAN HOME PRODS.	CALIF. BIOTECHNOLOGY	86	P	M	NAZDEL
AMERICAN HOME PRODS.	HEPAR INDUSTRIES	86	P	L	HEPARIN
AMERICAN HOSP. SUPPLY	AGEN USA	86	K	M	BLOOD SCREEN KIT
AMOCO	INTEGRATED GENETICS	86	MK	J	FOOD DX
AMOCO	INTEGRATED GENETICS	87	K	V	AIDS DX
AMOCO	INTELLICORP	86	F	J	SOFTWARE
AMOCO	INTELLIGENETICS	86	F	A	SOFTWARE
ANHEUSER-BUSCH	INTERFERON SCIENCES	86	P	E	IF
BAKER INSTRUMENTS	CISTRON BIOTECH.	87	PG	A	
BAXTER HEALTHCARE	BIOTHERAPEUTICS	88	P	J	CANCER RX
BAXTER HEALTHCARE	CAMBRIDGE BIOSCIENCE	87	P	M	AIDS DX
BAXTER TRAVENOL	BIOGEN	86	P	L	IF-G
BAXTER TRAVENOL	BIOTECH. GENERAL	86	P	M	GROWTH HORMONE
BAXTER TRAVENOL	CAREMARK	87	P	A	HEALTH CARE
BAXTER TRAVENOL DX	VENTREX	86	K	R	DX REAGENTS
BEATRICE	INGENE	87	M	R	SWEETENER
BECTON DICKINSON	LIFE TECHNOLOGIES	87	K	A	DX
BRISTOL-MYERS	BIOTECH. GENERAL	86	P	LM	SOD
BRISTOL-MYERS	MOLECULON	86	3	J	DRUG DELIVERY
BRISTOL-MYERS	PRAXIS BIOLOGICS	87	ZQ	E	VX
CAMPBELL SOUP CO.	DNA PLANT TECHNOL.	86	B	O	DNAP-9 TOMATO
CELANESE	CODON	87		E	
CELANESE	ENDOTRONICS	86	WP	R	IMMUNE CELLS
CELANESE	ENDOTRONICS	86	P	E	HEPATITIS B VIR
CELANESE	ENDOTRONICS	86	H	E	T-CELL THERAPY
CELANESE	ENDOTRONICS	86	P	R	AIDS TREATMENT
COLGATE PALMOLIVE	XYTRONYX	87	K	L	PERIDONTAL DX
COOPER	BIO-RESPONSE	86	KP	M	BLOOD DX; ANTIVIRALS
COORS BIOTECH	SYNERGEN	88	J	J	RIBOFLAVIN
CORNING GLASS	CHARLES RIVER BIOTECH	86	F	A	
DOW CHEMICAL	COLLABORATIVE RES.	86	M	M	RENNIN
DOW CHEMICAL	COLLABORATIVE RES.	88	JM	L	CHYMOSIN
DOW CHEMICAL	COLLABORATIVE RES.	88	M	J	RENNIN
DUPONT	ADVANCED GENETIC SCI.	86	I	R	
DUPONT	APPLIED BIOTECH.	86	PK	J	CANCER DX, RX
DUPONT	BIOTECH RESEARCH LABS	87	K	LM	AIDS DX
DUPONT	CISTRON BIOTECH.	86	P	L	IL-2
DUPONT	CLINICAL SCIENCES	87	K	J	HEP-B ANTIGENS
DUPONT	DNA PLANT TECHNOL.	87	B	J	PLANT OILS
DUPONT	HEM RESEARCH	87	P	E	AIDS RX
DUPONT	IGEN	87	B	J	PLANT PATHOL. DX
DUPONT	IMMUNOSYSTEMS	88	B	J	PESTICIDES

Listing 12-2 Partnerships, by Large Partner

Company 1	Company 2	Yr.	Cd.	Act.	Product
DUPONT	SYNERGEN	87	P	J	MAF
DUPONT	VEGA BIOTECHNOL.	86	F	EM	SYNTHESIZERS
EASTMAN KODAK	ADVANCED GENETIC SCI.	87	1	M	SNOW MAKING
EASTMAN KODAK	AMGEN	86	JO	JM	CHEMICALS
EASTMAN KODAK	CELLULAR PRODUCTS	87	K	M	AIDS/T-CELL DX
EASTMAN KODAK	CETUS	86	K	J	DX
EASTMAN KODAK	CYTOGEN	86	KP	E	
EASTMAN KODAK	CYTOGEN	86	KP	R	
EASTMAN KODAK	ENGENICS	86	M	J	BIOCHEMICALS
EASTMAN KODAK	ENZON	87	P	E	O2 TOXICITY RX
EASTMAN KODAK	EPITOPE	88	PK	J	CANCER DX/RX
EASTMAN KODAK	GENENCOR	87		E	
EASTMAN KODAK	IMMUNOLOGY VENTURES	87	P	M	IL-4
EASTMAN KODAK	INT'L BIOTECHNOLOGIES	87		A	
EASTMAN KODAK	MICROBIO RESOURCES	87	M	M	BETA-CAROTENE
EASTMAN KODAK	MOLECULAR GENETICS	86	V	J	VET MASTITIS RX
EASTMAN KODAK	MONOCLONAL AB'S	88	K	L	IN VITRO DX
EASTMAN KODAK	NEORX	86	KP	EJ	
EASTMAN KODAK	PANLABS	87	P	R	RX
EASTMAN KODAK	STERLING DRUG	88	P	A	RX
EASTMAN PHARMACEUT.	INGENE	88	P	J	CANCER RX
ELI LILLY	ORGANOGENESIS	87	1	J	SYNTHETIC ARTERY
ERBAMONT	CENTOCOR	86	K	L	DX
FISHER SCIENTIFIC	GEN-PROBE	86	Y	M	MYCOPLASMA DX
FISHER SCIENTIFIC	IMCLONE	87	K	J	HEP-B, AIDS DX
FMC	CENTOCOR	86	P	E	
GILLETTE	REPLIGEN	87	P	J	DENTAL ENZYME
INT'L MINERALS & CHEM.	IMCERA BIOPRODUCTS	87	P	OM	SOMATOMEDIN
INT'L MINERALS & CHEM.	SYNBIOTICS	87	K	E	MAB
INT'L MINERALS & CHEM.	SYNBIOTICS	87	V	RM	VET MAB'S
JOHNSON & JOHNSON	AMGEN	86	P	J	IL-2 ANALOG
JOHNSON & JOHNSON	IMMUNOMEDICS	86	KP	R	
KRAFT	DNA PLANT TECHNOL.	87	M	O	SNACK FOODS
LEDERLE	IMCLONE	88	Q	J	VX
LUBRIZOL	MYCOGEN	87	B	R	INSECTICIDES
MERCK	BIOGEN	87	P	J	ANTICANCER RX
MERCK	REPLIGEN	87	Q	R	AIDS VX
MERRILL LYNCH	CODON	86	Q	E	SWINE VX
MERRILL LYNCH	PLANT GENETICS	87	B	J	IMPROVED POTATOES
MILLIGEN	CAMBRIDGE RES. BIOCH.	86	F	E	
MILLIPORE	GENETIC DESIGN	86	F	A	DNA SYNTHESIZER
MILLIPORE	PROTEIN DATABASES	87		E	
MONSANTO	ALAFI PERMEA	87	F	V	BIOREACTORS
MONSANTO	DNA PLANT TECHNOL.	86	M	J	SWEETENERS
MONSANTO	ECOGEN	87	B	L	PESTICIDES
MONSANTO	MYCOGEN	87	B	J	BIOPESTICIDES
MONSANTO	PLANT GENETICS	87	B	J	POTATOES
MONSANTO AGRIC.	IMMUNOSYSTEMS	88	B	J	IMMUNOASSAYS
NABISCO	BIOTECHNICA INT'L	87	B	J	AGRONOMIC PRODS.
NEW BRUNSWICK SCI.	GENENTECH	86	F	L	
NEW BRUNSWICK SCI.	GENZYME	87	I	A	BIOCHEMICALS
NORDEN LABORATORIES	SYNBIOTICS	87	V	M	LEUKEMIA DX
NUTRASWEET CO.	DNA PLANT TECHNOL.	86	M	R	SWEETENERS
O.M. SCOTT & SONS	AGRI-DIAGNOSTICS	86	B	M	PLANT DX

Company 1	Company 2	Yr.	Cd.	Act.	Product
ORTHO	CALIF. BIOTECHNOLOGY	86	P	R	NAZDEL
ORTHO	INTEGRATED GENETICS	87	P	J	MSF
ORTHO	MONOCLONAL AB'S	87	K	L	OVULATION DX
PERKIN ELMER	CETUS	86	F	V	INSTRUMENT SYSTEMS
PFIZER	COLLABORATIVE RES.	88	M	L	RENNIN
PFIZER	COOPER LASERSONICS	88	F	JA	BT EQUIP.
PFIZER	GENZYME	87	P	L	ORTHOPEDICS
PFIZER	ONCOGENE SCIENCE	86	WP	EJ	ONCOGENES, RX
PFIZER	T CELL SCIENCES	86	PK	R	ARTHRITIS RX
PFIZER	XOMA	87	P	L	TOXIC SHOCK RX
PHILLIPS PETROLEUM	ELECTRO-NUCLEONICS	87	K	R	AIDS DX
PROCTER & GAMBLE	SYNERGEN	86	P	J	COLLAGENASE INHIB.
PRUTECH R&D PART. II	ELECTRO-NUCLEONICS	86	K	R	AIDS TEST
PRUTECH R&D PART. II	SYNBIOTICS	86	VA	R	VX
RICHARDSON-VICKS	TRITUS	87	P	L	IF-B
RITZ CHEMICALS	CELGENE	87	J	J	BIOPOLYMERS
RORER	INVITRON	88	P	J	FACTOR VIII:C
RORER	MELOY LABORATORIES	86	P	A	
SCHERING-PLOUGH	BIOGEN	86	P	L	IF-A
SCHERING-PLOUGH	GENEX	86	P	R	PLASMA PROTEIN
SCHERING-PLOUGH	MOLECULAR GENETICS	87	K	M	TEST KITS
SEARLE	INVITRON	87	P	J	PROTEINS
SEARLE	INVITRON	86	P	J	TPA
SEARLE	MEDI-CONTROL CORP.	86	3	E	DRUG DELIVERY
SMITHKLINE	AMGEN	86	P	EJ	SOMATOTROPIN
SMITHKLINE	CAMBRIDGE BIOSCIENCE	86	V	M	FELINE LEUKEMIA DX
SMITHKLINE	DAMON	87	P	M	TPA
SQUIBB	CETUS	87	P	JE	RX
SQUIBB	CETUS	87		O	
SQUIBB	CETUS	87	P	L	ANTICANCER RX
SQUIBB	CHEMEX	87	P	L	SKIN RX
SRI INTERNATIONAL	GENELABS	88	P	JR	AIDS RX
SRI INTERNATIONAL	GENELABS	88	P	JR	IMMUNOCONJUGATE
STATE FARM MUTUAL	BIOTECHNICA INT'L	88	B	E	AGRICULTURAL PRODS.
SYVA	CAMBRIDGE BIOSCIENCE	87	K	J	AIDS DX
SYVA	EPITOPE	87	K	R	AIDS DX
UNITED AGRISEEDS	GENETICS INSTITUTE	87	B	A	SEEDS
UNITED AGRISEEDS	GENETICS INSTITUTE	87	B	O	HYBRID CORN
UNITED AGRISEEDS	ML TECHNOLOGY VENT.	87	M	R	CORN
UPJOHN	CHEMEX	88	P	L	SKIN RX
UPJOHN	EMBRYOGEN	87	A	J	LAB ANIMALS
W. R. GRACE	BIOTECHNICA INT'L	87	J	J	ENZYMES
W. R. GRACE	DORR-OLIVER	87	HF	A	EQUIPMENT
WESTINGHOUSE B-A	IMMUNOSYSTEMS	88	B	M	HERBICIDE DX
WEYERHAEUSER	PLANT GENETICS	87	B	LR	PLANT ENCAPSULATOR
WYETH	CALIF. BIOTECHNOLOGY	86	P	M	RENIN INHBITOR

Listing 12-2 Partnerships, by Large Partner

Notes:

In addition to the two companies involved, the year of partnership (Yr.), industry class (Cd.; see Table 12-1), type of external action (Act.; see Table 12-1) and product involved are given, where available. Data were obtained from the public literature and company information and are compiled and stored in the **Actions** database. Most companies in the second column are detailed in Section 2, Listing 2-2, and many corporations in the first column are detailed in Section 11, Listing 11-1.

LISTING 12-3 PARTNERSHIPS WITH FOREIGN COMPANIES

U.S. Company	Foreign Company	Country
ADVANCED GENETIC SCIENCES Year: 86 Industry:	KARLSHAMNS OLJEFABRIKER Action: E Product:	SWEDEN
AGRACETUS Year: 87 Industry: B	DAINIHON JOACHUGIKU Action: L Product: TERMITICIDE	JAPAN
AGRI-DIAGNOSTICS Year: 86 Industry: B	CIBA-GEIGY Action: J Product: DX KITS	SWITZ
AIMS RESEARCH Year: 87 Industry: K	AIMS BIOTECH Action: R Product: AIDS DX	CANADA
AMERICAN CYANAMID Year: 88 Industry: P	QUADRA LOGIC TECHNOLOGIES Action: J Product: CANCER RX	CANADA
AMERICAN CYANAMID Year: 88 Industry: P	QUADRA LOGIC TECHNOLOGIES Action: E Product: HUMAN, VET RX	CANADA
AMGEN Year: 86 Industry: P	FERMENTA Action: E Product:	SWEDEN
ARTHUR D. LITTLE Year: 87 Industry: P	TORAY RESEARCH CENTER Action: O Product: RX SCREENING	JAPAN
BAXTER-TRAVENOL Year: 88 Industry: P	TEVA PHARMACEUTICAL Action: A Product: RX	ISRAEL
BECKMAN INSTRUMENTS Year: 88 Industry: J	BRITISH BIOTECHNOLOGY, LTD. Action: L Product: DESIGNER GENES	UK
BECKMAN INSTRUMENTS Year: 87 Industry: F	CRUACHEM Action: M Product: DNA SYNTHESIZER	SCOTLAND
BETHESDA RESEARCH LABS Year: 86 Industry: G	BIOPROCESSING, LTD. Action: M Product: GROWTH FACTOR	UK
BIOASSAY SYSTEMS Year: 86 Industry: H	TORAY-FUJI BIONICS Action: L Product: VIRUS DX	JAPAN
BIOGEN Year: 86 Industry: Q	BEHRINGWERKE Action: M Product: MALARIA VX	WGER
BIOGEN Year: 87 Industry: P	GLAXO Action: A Product: IL-2, GM-CSF	UK
BIOGEN Year: 87 Industry: P	HOFFMANN-LA ROCHE Action: LA Product: B CELL GROWTH FACTOR	SWITZ
BIOGEN Year: 87 Industry: B	PLANT GENETIC SYSTEMS Action: L Product: HERBICIDE RESISTANCE	BELGIUM
BIONETICS RESEARCH Year: 87 Industry: Y	ORGANON TEKNIKA Action: LM Product: SALMONELLA DX	NETH
BIOREACTOR TECHNOLOGY Year: 86 Industry: F	C. ITOH Action: M Product: BIOREACTOR	JAPAN
BIOSTAR MEDICAL PRODUCTS Year: 87 Industry: K	C. ITOH Action: M Product: DX	JAPAN

Listing 12-3 Partnerships with Foreign Companies

U.S. Company	Foreign Company	Country
BIOSTAR MEDICAL PRODUCTS Year: 86 Industry:	SKANDIGEN AB Action: E Product:	SWEDEN
BIOSYM TECHNOLOGY Year: 87 Industry: F	MITSUBISHI Action: M Product: BT EQUIPMENT	JAPAN
BIOTECHNICA INTERNATIONAL Year: 87 Industry: B	ENICHEM AGRICOLTURA SPA Action: O Product: RHIZOBIA BACT.	ITALY
BIOTECHNICA INTERNATIONAL Year: 88 Industry: P	HOFFMANN-LA ROCHE Action: R Product: VITAMINS	SWITZ
BIOTECHNICA INTERNATIONAL Year: 86 Industry:	SEAGRAMS Action: E Product:	CANADA
CALBIOCHEM Year: 88 Industry: J	BIODOR HOLDING Action: A Product: BIOCHEMICALS	SWITZ
CALGENE Year: 87 Industry: M	L DAEHNFELDT Action: M Product: EDIBLE RAPESEED	DENMARK
CALGENE Year: 87 Industry: B	NIPPON STEEL CO. Action: J Product: PLANT OILS	JAPAN
CALGENE Year: 87 Industry: B	TOAGOSEI CHEMICAL Action: L Product: COTTON, CORN	JAPAN
CALIFORNIA BIOTECHNOLOGY Year: 86 Industry: P	BYK GULDEN LOMBERG CHEMISCH Action: L Product:	WGER
CALIFORNIA BIOTECHNOLOGY Year: 86 Industry: P	HOFFMANN-LA ROCHE Action: RC Product:	SWITZ
CALIFORNIA BIOTECHNOLOGY Year: 87 Industry: P	ORGANON INTERNATIONAL Action: J Product: EPO	NETH
CALIFORNIA BIOTECHNOLOGY Year: 87 Industry: P	ORGANON INTERNATIONAL Action: J Product: ALZHEIMER'S RX	NETH
CAMBRIDGE BIOSCIENCE Year: 87 Industry: Q	INSTITUT MERIEUX Action: JR Product: AIDS VX	FRANCE
CAMBRIDGE BIOSCIENCE Year: 86 Industry: V	VIRBAC Action: J Product: VX	FRANCE
CELANESE Year: 86 Industry:	HOECHST Action: A Product:	WGER
CENTOCOR Year: 87 Industry: K	AMERSHAM CANADA Action: M Product: CARDIOVASC. DX	CANADA
CENTOCOR Year: 87 Industry: P	TAKEDA Action: J Product: TOXIC SHOCK RX	JAPAN
CETUS Year: 87 Industry:	EUROCETUS Action: E Product:	NETH
CETUS Year: 87 Industry: F	OLYMPUS OPTICAL Action: M Product: BLOOD TYPE EQUIPMENT	JAPAN
CHARLES RIVER BIOTECH Year: 88 Industry: K	AJINOMOTO Action: J Product: MABS	JAPAN

U.S. Company	Foreign Company	Country
CHEMEX Year: 88 Industry: P	TAKEDA Action: L Product: ANTI-ALLERGY RX	JAPAN
CHIRON Year: 87 Industry: P	BEHRINGWERKE Action: L Product: TPA	WGER
CHIRON Year: 87 Industry: Q	CIBA-GEIGY Action: V Product: VX	SWITZ
CHIRON Year: 86 Industry: Q	CIBA-GEIGY Action: J Product: VX	SWITZ
CHIRON Year: 87 Industry: P	HITACHI Action: M Product: ULCER RX	JAPAN
CHIRON Year: 87 Industry: K	LUCKY, LTD. Action: J Product: DX	S KOREA
CHIRON Year: 86 Industry: J	PETROFINA GROUP Action: J Product: ENZYMES	BELGIUM
CODON Year: 86 Industry: Q	NORDEN LABORATORIES Action: M Product: SWINE VX	CANADA
COLLABORATIVE RESEARCH Year: 87 Industry: K	SPECIAL REFERENCE LABS Action: L Product: GENETIC, CA DXS	JAPAN
COOPER Year: 88 Industry: P	CIBA-GEIGY Action: A Product: CONTACT LENS	SWITZ
COOPER TECHNICON Year: 87 Industry:	ALFA-LAVAL Action: A Product: INDUSTRIAL SYSTEMS	SWEDEN
COORS Year: 87 Industry: M	MEIJI SEIKA Action: J Product: FOOD ADDITIVES	JAPAN
CPC INTERNATIONAL Year: 87 Industry: M	AJINOMOTO Action: A Product: FOOD PRODUCTS	JAPAN
CPC INTERNATIONAL Year: 87 Industry: B	FERUZZI Action: A Product: CORN	ITALY
CREATIVE BIOMOLECULES Year: 87 Industry: P	BANEXI Action: R Product: TPA	FRANCE
CYANOTECH Year: 86 Industry: BP	DAIKYO OIL CO. Action: E Product:	JAPAN
CYTOCULTURE INTERNATIONAL Year: 86 Industry: H	MCA DEVELOPMENT Action: P Product: MAB'S	NETH
DAMON Year: 86 Industry: K	ARES-SERONO Action: O Product: DX MABS	NETH
DAMON Year: 86 Industry: P	CONNAUGHT LABS Action: V Product: INSULIN TECHNOLOGY	CANADA
DAMON Year: 87 Industry: P	YAMANOUCHI PHARMACEUTICALS Action: L Product: TPA	JAPAN
DDI PHARMACEUTICALS Year: 87 Industry: P	ALLELIX Action: J Product: HUMAN SOD	CANADA

Listing 12-3 Partnerships with Foreign Companies

U.S. Company	Foreign Company	Country
DIAGNOSTIC PRODUCTS Year: 86 Industry: K	DAINIPPON Action: M Product: IMMUNO-DX	JAPAN
DUPONT Year: 87 Industry: F	ALOKA Action: M Product: DNA SEQUENCER	JAPAN
DUPONT Year: 86 Industry: K	CONNAUGHT LABS Action: J Product: DX	CANADA
DUPONT Year: 88 Industry:	ENICHEM AGRICOLTURA Action: E Product:	ITALY
DUPONT Year: 88 Industry: P	HOFFMANN-LA ROCHE Action: J Product: CANCER RX	SWITZ
ECOGEN Year: 87 Industry: M	PLANT GENETIC SYSTEMS Action: J Product: LETTUCE	BELGIUM
EDWARD J. FUNK & SONS Year: 86 Industry: B	BP NUTRITION Action: A Product: SEEDS	UK
ELECTRO-NUCLEONICS Year: 86 Industry: K	ORGANON TEKNIKA Action: M Product:	NETH
ELECTRO-NUCLEONICS Year: 86 Industry: K	PHARMACIA Action: M Product: DX	SWEDEN
ELECTRO-NUCLEONICS Year: 87 Industry: K	PHARMACIA Action: A Product: DX	SWEDEN
ELECTRO-NUCLEONICS Year: 86 Industry:	PHARMACIA Action: E Product:	SWEDEN
ELI LILLY Year: 87 Industry: P	HOFFMANN-LA ROCHE Action: L Product: HGH	SWITZ
ENDOTRONICS Year: 86 Industry: V	CELIAS Action: A Product: MAMMALIAN CELLS	UK
ENDOTRONICS Year: 86 Industry: P	NIPPON CHEMICAL INDUSTRIAL CO. Action: J Product: HGH	JAPAN
ENDOTRONICS Year: 86 Industry: K	SINO AMERICAN BIOTECHNOLOGY Action: J Product: MAB DX	CHINA
ENVIRONMENTAL DIAGNOSTICS Year: 88 Industry: BM	TRANSIA Action: M Product: AG/FOOD TEST KIT	FRANCE
EPITOPE Year: 87 Industry: K	ORGANON TEKNIKA Action: M Product: AIDS DX	NETH
FLOW GENERAL Year: 88 Industry: HK	LABSYSTEMS Action: M Product: DX,CELL CULTURE	FINLAND
GENELABS Year: 87 Industry: P	SANDOZ Action: JM Product: RX	SWITZ
GENENTECH Year: 87 Industry: P	BOEHRINGER INGELHEIM Action: M Product: TPA	WGER
GENENTECH Year: 87 Industry: X	DAINIPPON Action: L Product: TNF	JAPAN

U.S. Company	Foreign Company	Country

GENENTECH
Year: 87 Industry: P
MITSUBISHI   JAPAN
Action: LM   Product: RX

GENENTECH
Year: 87 Industry: P
QUADRA LOGIC TECHNOLOGIES   CANADA
Action: J   Product: IMMUNOSUPPRESSANT

GENETICS INSTITUTE
Year: 87 Industry: P
BOEHRINGER MANNHEIM   WGER
Action: L   Product: EPO

GENETICS INSTITUTE
Year: 86 Industry: P
WELLCOME   UK
Action: V   Product: DRUGS

GENETICS INSTITUTE
Year: 87 Industry: P
CHUGAI PHARMACEUTICAL   JAPAN
Action: M   Product: EPO

GENETICS INSTITUTE
Year: 87 Industry: P
SANDOZ   SWITZ
Action: L   Product: CA/AIDS RX, CSF

GENETICS INSTITUTE
Year: 86 Industry: P
WELLCOME   UK
Action: JL   Product: TPA

GENETICS INSTITUTE
Year: 87 Industry: PK
WELLCOME   UK
Action: V   Product: IF/TPA/FAC VIII

GENEX
Year: 86 Industry: P
BEHRINGWERKE   WGER
Action: R   Product: RX PROTEIN

GENEX
Year: 87 Industry: P
NIPPON OIL CO   JAPAN
Action: L   Product: VITAMIN B12

GENZYME
Year: 87 Industry: P
HOFFMANN-LA ROCHE   SWITZ
Action: J   Product: HYALURONIC ACID

GENZYME
Year: 86 Industry: J
NAGASE & CO.   JAPAN
Action: J   Product: AMYLASE

HOUSTON BIOTECHNOLOGY
Year: 87 Industry: P
SANTEN PHARMACEUTICAL   JAPAN
Action: L   Product: CATARACT RX

HYBRITECH
Year: 86 Industry: K
CELLTECH   UK
Action: R   Product: DX KITS

HYBRITECH
Year: 86 Industry: P
SANOFI RECHERCHE   FRANCE
Action: J   Product: T CELL MARKER

ICN
Year: 87 Industry:
HOFFMANN-LA ROCHE   SWITZ
Action: E   Product:

IDETEK
Year: 87 Industry: Y
NOVO INDUSTRI   DENMARK
Action: E   Product:

IGENE
Year: 86 Industry: M
BIOSOPH LABORATORIES   UK
Action: V   Product: FLAVORS

IGI, INC.
Year: 87 Industry: Q
COOPERS ANIMAL HEALTH   UK
Action: L   Product: ENCAPSULATED VX

IMMUNONUCLEAR
Year: 86 Industry: K
BOOTS-CELLTECH   UK
Action: M   Product: IMMUNORADIOMETER

IMMUNOMEDICS
Year: 87 Industry: K
KURARAY   JAPAN
Action: M   Product: IN VITRO DX

Listing 12-3 Partnerships with Foreign Companies

U.S. Company	Foreign Company	Country

IMMUNOSYSTEMS
Year: 87 Industry: Y
TRANSIA LABS FRANCE
Action: M Product: ENVIRONMENT DX

INCSTAR
Year: 87 Industry: P
SANDOZ SWITZ
Action: J Product: ANTI-REJECTION RX

INGENE
Year: 86 Industry: M
MITSUBISHI JAPAN
Action: J Product: SWEETENERS

INTEGRATED GENETICS
Year: 87 Industry: P
BASF WGER
Action: O Product: TPA

INTEGRATED GENETICS
Year: 87 Industry: P
BEHRINGWERKE WGER
Action: J Product: EPO

INTEGRATED GENETICS
Year: 86 Industry: Q
CONNAUGHT LABS CANADA
Action: J Product: HEP B VX

INTEGRATED GENETICS
Year: 86 Industry: A W
GRANADA LABS UK
Action: RC Product: HORMONES

INTEGRATED GENETICS
Year: 87 Industry: K
SPECIAL REFERENCE LABS JAPAN
Action: J Product: DX SERVICES

INTEGRATED GENETICS
Year: 87 Industry: P
TOYOBO JAPAN
Action: J Product: TPA

JOHNSON & JOHNSON
Year: 88 Industry: Q
BRITISH BIOTECHNOLOGY UK
Action: E Product: AIDS VX

JOHNSON & JOHNSON
Year: 87 Industry:
CELLTECH UK
Action: E Product:

LIFE TECHNOLOGIES
Year: 87 Industry: K
TORAY INDUSTRIES JAPAN
Action: M Product: DNA DETECTOR

LIPOSOME COMPANY
Year: 87 Industry: Q
CONNAUGHT LABS CANADA
Action: J Product: INFLUENZA VX

LIPOSOME COMPANY
Year: 87 Industry: P
SANTEN JAPAN
Action: J Product: ANTI-ALLERGY RX

LIPOSOME TECHNOLOGY
Year: 86 Industry: P
TAKEDA JAPAN
Action: R Product:

MOLECULAR GENETICS
Year: 86 Industry: V
NORDEN LABORATORIES CANADA
Action: M Product: ANIMAL HEALTH CARE

MOLECULAR GENETICS
Year: 86 Industry: B
RHONE-POULENC AGROCHIMIE FRANCE
Action: R Product: CORN DISEASES

MOLECULON
Year: 87 Industry: 3
F.H. FAULDING & CO. AUSTRALIA
Action: E Product: RX DELIVERY

MUREX
Year: 87 Industry: K
KYOWA HAKKO KOGYO JAPAN
Action: M Product: AIDS DX

MUREX
Year: 86 Industry: KZ
PILOT LABORATORIES CANADA
Action: E Product: AIDS DX

MYCOGEN
Year: 87 Industry: B
KUBOTA JAPAN
Action: J Product: BIOPESTICIDES

U.S. Company			Foreign Company		Country
NEOGEN			SOLVAY		BELGIUM
	Year: 86	Industry: V	Action: J	Product: DX	
NOVO BIOLABS USA			NOVO INDUSTRI		DENMARK
	Year: 86	Industry: K	Action: M	Product: DX	
NPI			KYOWA HAKKO KOGYO		JAPAN
	Year: 87	Industry: B	Action: J	Product: PLANTS	
OCCIDENTAL PETROLEUM			MITSUI TOATSU CHEMICALS		JAPAN
	Year: 86	Industry: M	Action: V	Product: HYBRID RICE	
ORTHO			CELLTECH		UK
	Year: 87	Industry: P	Action: V	Product: EPO	
PANLABS			FINNISH SUGAR CO		FINLAND
	Year: 87	Industry: M	Action: L	Product: ENZYMES	
PFIZER			CELLTECH		UK
	Year: 88	Industry: JM	Action: L	Product: CHYMOSIN	
PHILLIPS PETROLEUM			BISSENDORF BIOSCIENCES GMBF	WGER	
	Year: 86	Industry: P	Action: J	Product: GH RELEASNG FACTOR	
PHILLIPS PETROLEUM			KYOWA HAKKO KOGYO		JAPAN
	Year: 87	Industry: M	Action: J	Product: SALMON HORMONE	
PLANT CELL RESEARCH INSTITUTE			MONTEDISON		ITALY
	Year: 87	Industry: B	Action: A	Product:	
PLANT CELL RESEARCH INSTITUTE			MONTEDISON		ITALY
	Year: 87	Industry: B	Action: E	Product: PLANT PRODUCTS	
PLANT GENETICS			CALBEE POTATO		JAPAN
	Year: 87	Industry: M	Action: L	Product: POTATO TUBERS	
PLANT GENETICS			FRANZANI		ARGENTINA
	Year: 87	Industry: B	Action: L	Product: ALFALFA SEEDS	
PLANT GENETICS			NITTOH SANYO		JAPAN
	Year: 87	Industry: B	Action: L	Product: ALFALFA SEEDS	
PLANT GENETICS			TWYFORD SEEDS		UK
	Year: 87	Industry: B	Action: V	Product: AGRICULT. PRODS.	
PRAGMA BIO-TECH			EIKEN CHEMICAL		JAPAN
	Year: 87	Industry: K	Action: J	Product: CANCER DX	
PROTEIN TECHNOLOGIES			CJB DEVELOPMENTS LTD		UK
	Year: 86	Industry: F	Action: M	Product: ELECTROPHORESIS	
PRUTEC			DJ VAN DER HAVE BV		NETH
	Year: 86	Industry: B	Action: J	Product: SEEDS	
QUIDEL			CLONATEC		FRANCE
	Year: 86	Industry: K	Action: JM	Product: DX	
QUIDEL			INVESGEN		SPAIN
	Year: 86	Industry: K	Action: M	Product: TEST KITS	
QUIDEL			TOSHIN CHEMICAL		JAPAN
	Year: 87	Industry: K	Action: L	Product: PREGNANCY DX	

Listing 12-3 Partnerships with Foreign Companies

U.S. Company				Foreign Company		Country
REPLIGEN				E. MERCK		WGER
	Year: 88	Industry: P		Action: J	Product: PLATELET FACTOR IV	
REPLIGEN				SANDOZ		SWITZ
	Year: 87	Industry: BR		Action: V	Product: WASTE/PESTICIDE	
REPLIGEN				TIEDEMANNS		NORWAY
	Year: 87	Industry: SB		Action: R	Product: AQUACULT/TOBACCO	
RICHARDSON-VICKS				ALLELIX		CANADA
	Year: 87	Industry: K3		Action: J	Product: HOME DX, DRUG DELIV.	
ROHM & HAAS				SUMITOMO		JAPAN
	Year: 87	Industry: 1		Action: JM	Product: ACRYLIC PLASTIC	
ROHM & HAAS				TOSOH		JAPAN
	Year: 87	Industry: F		Action: V	Product: BT EQUIPMENT	
SCHERING-PLOUGH				SHIONOGI		JAPAN
	Year: 87	Industry: P		Action: L	Product: ANTIBIOTIC	
SENETEK				VHJ TESLA		CZECH
	Year: 87	Industry: F		Action: V	Product: BT EQUIPMENT	
SERONO LABS				DIAGNOSTICS PASTEUR		FRANCE
	Year: 86	Industry: K		Action: M	Product: TEST KITS	
SMITHKLINE				BOEHRINGER MANNHEIM		WGER
	Year: 86	Industry: P		Action: J	Product: HEART RX	
SMITHKLINE				BRITISH BIOTECHNOLOGY, LTD.		UK
	Year: 88	Industry: Q		Action: E	Product: AIDS VX	
SMITHKLINE				BRITISH BIOTECHNOLOGY, LTD.		UK
	Year: 87	Industry: P		Action: J	Product: TPA	
SMITHKLINE				SUMITOMO		JAPAN
	Year: 87	Industry: P		Action: V	Product: NON-PRESCRIP. RX	
SQUIBB				NOVO INDUSTRI		DENMARK
	Year: 86	Industry: T		Action: M	Product: HUMAN INSULIN	
SRI INTERNATIONAL				MONTEDISON		ITALY
	Year: 87	Industry: B		Action: OE	Product:	
STEARNS CATALYTIC CO.				CHIYODA CHEMICAL ENGINEERING		JAPAN
	Year: 86	Industry: OP		Action: O	Product: GENERAL	
SUMMA MEDICAL				COSMO BIO		JAPAN
	Year: 87	Industry: KY		Action: L	Product: MAB	
SUMMA MEDICAL				HOFFMANN-LA ROCHE		SWITZ
	Year: 87	Industry: K		Action: L	Product: BLOOD CLOT DX	
SYNBIOTICS				NORDEN-EUROPE		UK
	Year: 86	Industry: V		Action: M	Product: VET DX	
SYNERGEN				CIBA-GEIGY		SWITZ
	Year: 86	Industry: P		Action: J	Product: ELASTASE INHIBITOR	
T CELL SCIENCES				BOEHRINGER MANNHEIM		WGER
	Year: 86	Industry: G		Action: M	Product: MABS	

U.S. Company	Foreign Company	Country
T CELL SCIENCES Year: 87 Industry: P	YAMANOUCHI PHARMACEUTICAL Action: M Product: IL-2	JAPAN
T CELL SCIENCES Year: 87 Industry: K	YAMANOUCHI PHARMACEUTICAL Action: R Product: CA DX	JAPAN
TECHAMERICA Year: 87 Industry: P	SDS BIOTECH Action: A Product: RX	SWEDEN
UNIGENE LABORATORIES Year: 87 Industry: P	FARMITALIA CARLO ERBA Action: R Product: HORMONES	ITALY
UNION CARBIDE Year: 88 Industry: MO	GIST-BROCADES Action: A Product: YEAST/ENZYMES	NETH
UNION CARBIDE Year: 86 Industry: B	RHONE-POULENC Action: A Product: AGRICHEMICALS	FRANCE
UNIVERSITY GENETICS Year: 86 Industry: A	INTERNATIONAL EMBRYOS Action: V Product: ANIMAL DX	UK
UNIVERSITY GENETICS Year: 86 Industry: P	L'OREAL Action: L Product: TCF	FRANCE
UPJOHN Year: 87 Industry: B	BRUINSMA SEEDS Action: A Product: SEEDS	NETH
UPJOHN Year: 87 Industry: P	CHUGAI PHARMACEUTICALS Action: M Product: ARTHRITIS RX	JAPAN
UPJOHN Year: 87 Industry: P	SANKYO Action: M Product: CEPHALOSPORIN RX	JAPAN
VESTAR Year: 88 Industry: P	MALLINCKRODT DIAGNOSTICA Action: M Product: CANCER IMAGING	HOLLAND
VESTAR Year: 88 Industry: P	NOBEL MEDICA Action: M Product: CANCER IMAGING	SWEDEN
VIRAL TECHNOLOGIES Year: 88 Industry: Q	NIPPON ZEON Action: L Product: AIDS VX	JAPAN
VIVOTECH Year: 86 Industry: P	CANADIAN ASTRONAUTICS Action: J Product: INSULIN	CANADA
XOMA Year: 87 Industry: P K	HOFFMANN-LA ROCHE Action: J Product: IL-2, MAB'S	SWITZ
ZYMOGENETICS Year: 87 Industry: P	TEIJIN Action: L Product: PROTEIN C	JAPAN

Notes: Actions are from 1986 to early 1988 and are from the companies or the public literature. The first company listed is the U.S. Company, the second the foreign partner. In some cases, the industry codes and products were not available. See Table 12-1 for industry and action codes. The same actions, listed by country, appear in Listing 12-4.

Listing 12-4 Countries of Foreign Partners

LISTING 12-4 COUNTRIES OF FOREIGN PARTNERS

Foreign Company	U.S. Partner	Product
Argentina		
FRANZANI	PLANT GENETICS	ALFALFA SEEDS
Australia		
F.H. FAULDING & CO.	MOLECULON	RX DELIVERY
Belgium		
PETROFINA GROUP	CHIRON	ENZYMES
PLANT GENETIC SYSTEMS	BIOGEN	HERBICIDE RESISTANCE
PLANT GENETIC SYSTEMS	ECOGEN	LETTUCE
SOLVAY	NEOGEN	DX
Canada		
AIMS BIOTECH	AIMS RESEARCH	AIDS DX
ALLELIX	DDI PHARMACEUTICALS	HUMAN SOD
ALLELIX	RICHARDSON-VICKS	HOME DX, DRUG DELIV.
AMERSHAM CANADA	CENTOCOR	CARDIOVASCULAR DX
CANADIAN ASTRONAUTICS	VIVOTECH	INSULIN
CONNAUGHT LABS	DAMON	INSULIN TECHNOLOGY
CONNAUGHT LABS	DUPONT	DX
CONNAUGHT LABS	INTEGRATED GENETICS	HEP B VX
CONNAUGHT LABS	LIPOSOME COMPANY	INFLUENZA VX
NORDEN LABORATORIES	CODON	SWINE VX
NORDEN LABORATORIES	MOLECULAR GENETICS	ANIMAL HEALTH CARE
PILOT LABORATORIES	MUREX	AIDS DX
QUADRA LOGIC TECHNOL.	AMERICAN CYANAMID	CANCER RX
QUADRA LOGIC TECHNOL.	GENENTECH	IMMUNOSUPPRESSANT
SEAGRAMS	BIOTECHNICA INT'L	
China		
SINO AMERICAN BIOTECH.	ENDOTRONICS	MAB DX
Czechoslovakia		
VHJ TESLA	SENETEK	BT EQUIPMENT
Denmark		
L. DAEHNFELDT	CALGENE	EDIBLE RAPESEED
NOVO INDUSTRI	IDETEK	
NOVO INDUSTRI	NOVO BIOLABS USA	DX
NOVO INDUSTRI	SQUIBB	HUMAN INSULIN
Finland		
FINNISH SUGAR CO.	PANLABS	ENZYMES
LABSYSTEMS	FLOW GENERAL	DX, CELL CULTURE
France		
BANEXI	CREATIVE BIOMOLECULES	TPA
CLONATEC	QUIDEL	DX
DIAGNOSTICS PASTEUR	SERONO LABS	DX KITS

Foreign Company	U.S. Partner	Product
France (Cont.)		
INSTITUT MERIEUX	CAMBRIDGE BIOSCIENCE	AIDS VX
L'OREAL	UNIVERSITY GENETICS	TCF
OHF SANTE ANIMALE	ALCIDE	GERMICIDE DIP
RHONE-POULENC AGROCHIMIE	UNION CARBIDE	AGCHEM
RHONE-POULENC AGROCHIMIE	MOLECULAR GENETICS	CORN DISEASES
SANOFI RECHERCHE	HYBRITECH	T CELL MARKER
TRANSIA	ENVIRONMENTAL DIAGN.	AG/FOOD TEST KIT
TRANSIA	IMMUNOSYSTEMS	ENVIRONMENT DX
VIRBAC	CAMBRIDGE BIOSCIENCE	VX
Israel		
TEVA PHARMACEUTICAL	BAXTER-TRAVENOL	RX
Italy		
ENICHEM AGRICOLTURA	DUPONT	
ENICHEM AGRICOLTURA	BIOTECHNICA INT'L	RHIZOBIA BACTERIA
FARMITALIA CARLO ERBA	UNIGENE LABORATORIES	HORMONES
FERUZZI	CPC INTERNATIONAL	CORN
MONTEDISON	PLANT CELL RESEARCH INST.	
MONTEDISON	SRI INTERNATIONAL	
Japan		
AJINOMOTO	CHARLES RIVER BIOTECH	MABS
AJINOMOTO	CPC INTERNATIONAL	FOOD PRODUCTS
ALOKA	DUPONT	DNA SEQUENCER
C. ITOH	BIOREACTOR TECHNOLOGY	BIOREACTOR
C. ITOH	BIOSTAR MEDICAL PRODS.	DX
CALBEE POTATO	PLANT GENETICS	POTATO TUBERS
CHIYODA CHEMICAL ENG.	STEARNS CATALYTIC CO.	GENERAL
CHUGAI PHARMACEUTICALS	GENETICS INSTITUTE	EPO
CHUGAI PHARMACEUTICALS	UPJOHN	ARTHRITIS RX
COSMO BIO	SUMMA MEDICAL	MAB
DAIKYO OIL CO.	CYANOTECH	
DAINIHON JOACHUGIKU	AGRACETUS	TERMITICIDE
DAINIPPON	DIAGNOSTIC PRODUCTS	IMMUNODIAGNOSTICS
DAINIPPON	GENENTECH	TNF
EIKEN CHEMICAL	PRAGMA BIO-TECH	CANCER DX
HITACHI	CHIRON	ULCER RX
KUBOTA	MYCOGEN	BIOPESTICIDES
KURARAY	IMMUNOMEDICS	IN VITRO DX
KYOWA HAKKO KOGYO	MUREX	AIDS DX
KYOWA HAKKO KOGYO	NPI	PLANTS
KYOWA HAKKO KOGYO	PHILLIPS PETROLEUM	SALMON HORMONE
MEIJI SEIKA	COORS	FOOD ADDITIVES
MITSUBISHI	BIOSYM TECHNOLOGY	BT EQUIPMENT
MITSUBISHI	GENENTECH	RX
MITSUBISHI	INGENE	SWEETENERS
MITSUI TOATSU CHEMICALS	OCCIDENTAL PETROLEUM	HYBRID RICE
NAGASE & CO.	GENZYME	AMYLASE
NIPPON CHEMICAL INDUS.	ENDOTRONICS	HGH
NIPPON OIL CO.	GENEX	VITAMIN B12
NIPPON STEEL CO.	CALGENE	PLANT OILS
NIPPON ZEON	VIRAL TECHNOLOGIES	AIDS VX

Listing 12-4 Countries of Foreign Partners

Foreign Company	U.S. Partner	Product
Japan (Cont.)		
NITTOH SANYO	PLANT GENETICS	ALFALFA SEEDS
INSTITUT MERIEUX	CAMBRIDGE BIOSCIENCE	AIDS VX
OLYMPUS OPTICAL	CETUS	BLOOD TYPE INST.
OTSUKA	SALUTAR	IMAGING AGENTS
SANKYO	UPJOHN	CEPHALOSPORIN RX
SANTEN PHARMACEUTICAL	LIPOSOME COMPANY	ANTI-ALLERGY RX
SANTEN PHARMACEUTICAL	HOUSTON BIOTECHNOLOGY	CATARACT RX
SHIONOGI	SCHERING-PLOUGH	ANTIBIOTIC
SPECIAL REFERENCE LABS	COLLABORATIVE RESEARCH	GENETIC/CA DX
SPECIAL REFERENCE LABS	INTEGRATED GENETICS	DX SERVICES
SUMITOMO	ROHM & HAAS	ACRYLIC PLASTIC
SUMITOMO	SMITHKLINE	NON-PRESCRIP RX
TAKEDA	CENTOCOR	TOXIC SHOCK RX
TAKEDA	CHEMEX	ANTI-ALLERGY RX
TAKEDA	LIPOSOME TECHNOLOGY	
TEIJIN	ZYMOGENETICS	PROTEIN C
TOAGOSEI CHEMICAL	CALGENE	COTTON, CORN
TORAY INDUSTRIES	LIFE TECHNOLOGIES	DNA DETECTOR
TORAY RESEARCH CENTER	ARTHUR D. LITTLE	RX SCREENING
TORAY-FUJI BIONICS	BIOASSAY SYSTEMS	VIRUS DX
TOSHIN CHEMICAL	QUIDEL	PREGNANCY DX
TOSOH	ROHM & HAAS	BT EQUIPMENT
TOYOBO	INTEGRATED GENETICS	TPA
YAMANOUCHI PHARMACEUT.	DAMON	TPA
YAMANOUCHI PHARMACEUT.	T CELL SCIENCES	IL-2
YAMANOUCHI PHARMACEUT.	T CELL SCIENCES	CA DX
Netherlands		
ARES-SERONO	DAMON	DX MABS
BRUINSMA SEEDS	UPJOHN	SEEDS
DJ VAN DER HAVE BV	PRUTEC	SEEDS
EUROCETUS	CETUS	
GIST-BROCADES	UNION CARBIDE	YEAST/ENZYMES
MALLINCKRODT DIAGNOSTICA	VESTAR	CANCER IMAGING
MCA DEVELOPMENT	CYTOCULTURE INT'L	MAB'S
ORGANON INTERNATIONAL	CALIFORNIA BIOTECH.	EPO
ORGANON INTERNATIONAL	CALIFORNIA BIOTECH.	ALZHEIMER'S RX
ORGANON TEKNIKA	BIONETICS RESEARCH	SALMONELLA DX
ORGANON TEKNIKA	ELECTRO-NUCLEONICS	
ORGANON TEKNIKA	EPITOPE	AIDS DX
Norway		
TIEDEMANNS	REPLIGEN	AQUACULT., TOBACCO
South Korea		
LUCKY, LTD.	CHIRON	DX
Spain		
INVESGEN	QUIDEL	TEST KITS

Foreign Company	U.S. Partner	Product
Sweden		
AB FERMENTA	AMGEN	
ALFA-LAVAL	COOPER TECHNICON	INDUSTRIAL SYST.
KARLSHAMNS OLJEFABRIKER	ADVANCED GENETIC SCI.	
NOBEL MEDICA	VESTAR	CANCER IMAGING
PHARMACIA	ELECTRO-NUCLEONICS	DX
SDS BIOTECH	TECHAMERICA	RX
SKANDIGEN AB	BIOSTAR MEDICAL PRODS.	
Switzerland		
BIODOR HOLDING	CALBIOCHEM	BIOCHEMICALS
CIBA-GEIGY	AGRI-DIAGNOSTICS	DX KITS
CIBA-GEIGY	CHIRON	VX
CIBA-GEIGY	COOPER	CONTACT LENS
CIBA-GEIGY	SYNERGEN	ELASTASE INHIBITOR
HOFFMANN-LA ROCHE	BIOGEN	B CELL GROWTH FACT.
HOFFMANN-LA ROCHE	BIOTECHNICA INT'L	VITAMINS
HOFFMANN-LA ROCHE	CALIFORNIA BIOTECH.	
HOFFMANN-LA ROCHE	DUPONT	CANCER RX
HOFFMANN-LA ROCHE	ELI LILLY	HGH
HOFFMANN-LA ROCHE	GENZYME	HYALURONIC ACID
HOFFMANN-LA ROCHE	ICN	
HOFFMANN-LA ROCHE	SUMMA MEDICAL	BLOOD CLOT DX
HOFFMANN-LA ROCHE	XOMA	IL-2, MAB'S
SANDOZ	GENELABS	THERAPEUTICS
SANDOZ	GENETICS INSTITUTE	CA/AIDS RX, CSF
SANDOZ	INCSTAR	ANTI-REJECTION RX
SANDOZ	REPLIGEN	WASTE/PESTICIDE
United Kingdom		
BIOPROCESSING, LTD.	BETHESDA RESEARCH LABS	GROWTH FACTOR
BIOSOPH LABORATORIES	IGENE	FLAVORS
BOOTS-CELLTECH	IMMUNO NUCLEAR	IMMUNORADIOMETER
BP NUTRITION	EDWARD J. FUNK & SONS	SEEDS
BRITISH BIOTECHNOLOGY	JOHNSON & JOHNSON	AIDS VX
BRITISH BIOTECHNOLOGY	BECKMAN INSTRUMENTS	GENES
BRITISH BIOTECHNOLOGY	SMITHKLINE	AIDS VX
BRITISH BIOTECHNOLOGY	SMITHKLINE	TPA
CELIAS	ENDOTRONICS	MAMMALIAN CELLS
CELLTECH	HYBRITECH	DX KITS
CELLTECH	JOHNSON & JOHNSON	
CELLTECH	ORTHO	EPO
CELLTECH	PFIZER	CHYMOSIN
CJB DEVELOPMENTS LTD	PROTEIN TECHNOLOGIES	ELECTROPHORESIS
COOPERS ANIMAL HEALTH	IGI, INC.	ENCAPSULATED VX
CRUACHEM	BECKMAN INSTRUMENTS	DNA SYNTHESIZER
GLAXO	BIOGEN	IL-2, GM-CSF
GRANADA LABS	INTEGRATED GENETICS	HORMONES
INTERNATIONAL EMBRYOS	UNIVERSITY GENETICS	ANIMAL DX
NORDEN-EUROPE	SYNBIOTICS	VETERINARY DX
TWYFORD SEEDS	PLANT GENETICS	AGRICULT. PRODS.
WELLCOME	GENETICS INSTITUTE	RX; IF, TPA, FAC VIII

Listing 12-4 Countries of Foreign Partners

Foreign Company	U.S. Partner	Product
West Germany		
BASF	INTEGRATED GENETICS	TPA
BEHRINGWERKE	INTEGRATED GENETICS	EPO
BEHRINGWERKE	BIOGEN	MALARIA VX
BEHRINGWERKE	CHIRON	TPA
BEHRINGWERKE	GENEX	RX PROTEIN
BISSENDORF BIOSCIENCES	PHILLIPS PETROLEUM	GH RELEASING FACTOR
BOEHRINGER INGELHEIM	GENENTECH	TPA
BOEHRINGER MANNHEIM	GENETICS INSTITUTE	EPO
BOEHRINGER MANNHEIM	SMITHKLINE	HEART RX
BOEHRINGER MANNHEIM	T CELL SCIENCES	MABS
BYK GULDEN LOMBERG CHEM.	CALIFORNIA BIOTECH.	
E. MERCK	REPLIGEN	PLATELET FACTOR IV
HOECHST	CELANESE	

Notes:

All partnerships were between 1986 and early 1988. The product is shown, where information was available. For more details on these partnerships, see Listing 12-3.

LISTING 12-5 ALL PARTNERSHIPS BY INDUSTRY CLASSIFICATION

Company 1	Company 2	Ctry.	Yr.	Cd.	Product
Animal Agriculture (A)					
EMBRYOGEN	UPJOHN		87	J	LAB ANIMALS
INTEGRATED GENETICS	GRANADA LABS	UK	86	R	HORMONES
UNIVERSITY GENETICS	INT'L EMBRYOS	UK	86	V	ANIMAL DX
UNIVERSITY GENETICS	WORLDWIDE FARM SVC.		86	J	ANIMALS
Plant Agriculture (B)					
ADVANCED GENETIC SCI.	DNA PLANT TECHNOLOGY		87	O	PLANTS
ADVANCED GENETIC SCI.	AGRITOPE		88	A	PLANTS, VET DX
AGRACETUS	DAINIHON JOACHUGIKU	JAPAN	87	L	TERMITICIDE
AGRI-DIAGNOSTICS	CIBA-GEIGY	SWITZ	86	J	DX KITS
AGRI-DIAGNOSTICS	O.M. SCOTT & SONS		86	M	PLANT DX
AGRIC. GENETIC SYSTEMS	EPITOPE		88	A	VET PROD, PLANTS
ARCO SEED CO.	UF GENETICS		87	A	
BIOGEN	PLANT GENETIC SYSTEMS	BELG	87	L	HERBICIDE RESIS.
BIOTECHNICA AGRIC.	L. HERRIED SEED		87	A	ALFALFA
BIOTECHNICA AGRIC.	MCALLISTER SEED CO.		87	A	SEEDS
BIOTECHNICA INT'L	NABISCO		87	J	AG PRODS.
BIOTECHNICA INT'L	STATE FARM MUTUAL		88	E	AG PRODS.
BIOTECHNICA INT'L	ENICHEM AGRICOLTURA	ITALY	87	O	RHIZOBIA BACT.
CALGENE	NIPPON STEEL CO.	JAPAN	87	J	PLANT OILS
CALGENE	PEDIGREED SEED CO.		87	A	COTTON
CALGENE	TOAGOSEI CHEMICAL	JAPAN	87	L	COTTON, CORN
CALLAHAN ENTERPRISES	RHONE-POULENC AG	FRANCE	86	J	CORN, SOYA
CPC INT'L	FERUZZI	ITALY	87	A	CORN
DEKALB-PFIZER GENETICS	NATIVE PLANTS		88	J	CORN HYBRIDS
DESERT COTTON R&D CO.	STONEVILLE SEED		88	A	COTTON
DNA PLANT TECHNOLOGY	CAMPBELL SOUP CO.		86	J	DNAP-9 TOMATO
DNA PLANT TECHNOLOGY	DUPONT		87	J	PLANT OILS
ECOGEN	AMERICAN CYANAMID		86	RE	INSECTICIDE
ECOGEN	CILCORP. VENTURES		88	E	BIOPESTICIDES
ECOGEN	GENETICS INSTITUTE		88	L	PESTICIDE
ECOGEN	ML TECHNOL. VENTURES		87	JR	PSEUDOMONAS
ECOGEN	MONSANTO		87	L	PESTICIDES
EDWARD J. FUNK & SONS	BP NUTRITION	UK	86	A	SEEDS
ENVIRONMENTAL DIAG.	TRANSIA	FRANCE	88	M	AG, FOOD TEST KIT
GENETICS INSTITUTE	UNITED AGRISEEDS		87	A	SEEDS
GENETICS INSTITUTE	UNITED AGRISEEDS		87	O	HYBRID CORN
IGEN	DUPONT		87	J	PLANT PATH. DX
IMMUNOSYSTEMS	DUPONT		88	J	PESTICIDES
IMMUNOSYSTEMS	MONSANTO AGRIC.		88	J	IMMUNOASSAYS
IMMUNOSYSTEMS	WESTINGHOUSE		88	M	HERBICIDE DX
MOLECULAR GENETICS	RHONE-POULENC AG	FRANCE	86	R	CORN DISEASES
MOLECULAR GENETICS	TERRA INT'L		86	V	CORN SEED
MYCOGEN	KUBOTA	JAPAN	87	J	BIOPESTICIDES
MYCOGEN	LUBRIZOL		87	R	INSECTICIDES
MYCOGEN	MONSANTO		87	J	BIOPESTICIDES
NPI	KYOWA HAKKO KOGYO	JAPAN	87	J	PLANTS
O.M. SCOTT & SONS	AGRI-DIAGNOSTICS		86	M	TURFGRASS RX
PLANT CELL RES. INST.	MONTEDISON	ITALY	87	A	
PLANT GENETICS	FRANZANI	ARG.	87	L	ALFALFA SEEDS

Listing 12-5 All Partnerships

Company 1	Company 2	Ctry.	Yr.	Cd.	Product
Plant Agriculture (Cont.)					
PLANT GENETICS	LOVELOCK SEED CO.		87	A	ALFALFA
PLANT GENETICS	MERRILL LYNCH		87	J	POTATOES
PLANT GENETICS	MONSANTO		87	J	POTATOES
PLANT GENETICS	NITTOH SANYO	JAPAN	87	L	ALFALFA SEEDS
PLANT GENETICS	TWYFORD SEEDS	UK	87	V	AG PRODS.
PLANT GENETICS	WEYERHAEUSER		87	LR	PLANT ENCAPSUL.
PRUTECH	DJ VAN DER HAVE BV	NETH	86	J	SEEDS
PRUTECH	ECOGEN		86	J	INSECTICIDES
REPLIGEN	SANDOZ	SWITZ	87	V	WASTE, PESTICIDE
SRI INT'L	MONTEDISON	ITALY	87	OE	
UNION CARBIDE	RHONE-POULENC	FRANCE	86	A	AGCHEM
UPJOHN	BRUINSMA SEEDS	NETH	87	A	SEEDS
Separations (E)					
BIOS	ICF		87	E	BIOMATERIALS
GENETICS INSTITUTE	SERAGEN		87	L	BIOSEPARATION
SEPRACOR	AMERICAN CYANAMID		86	J	MEMBRANE TECH
Equipment (F)					
ALAFI PERMEA	MONSANTO		87	V	BIOREACTORS
AMERICAN MONITOR	QUEST BIOTECHNOLOGY		87	A	INSTRUMENTS
APPLIED BIOSYSTEMS	SPECTROS INT'L, PLC		86	A	CHROMATOG.
BECKMAN INSTRUMENTS	CRUACHEM	UK	87	M	DNA SYNTH.
BIOREACTOR TECHNOL.	C. ITOH	JAPAN	86	M	BIOREACTOR
BIOSYM TECHNOLOGY	MITSUBISHI	JAPAN	87	M	BT EQUIPMENT
CAMBRIDGE RES. BIOCHEM.	MILLIGEN		86	E	
CETUS	OLYMPUS OPTICAL	JAPAN	87	M	BLOOD TYP EQUIP
CETUS	PERKIN ELMER		86	V	INST. SYSTEMS
CHARLES RIVER BIOTECH	CORNING GLASS		86	A	
COOPER LASERSONICS	PFIZER		88	JA	EQUIPMENT
DUPONT	ALOKA	JAPAN	87	M	DNA SEQUENCER
GENENTECH	HP GENENCHEM		87	E	BT INSTRUMENTS
GENENTECH	NEW BRUNSWICK SCI.		86	L	
GENETIC DESIGN	MILLIPORE		86	A	DNA SYNTH.
INTELLICORP	AMOCO		86	J	SOFTWARE
INTELLIGENETICS	AMOCO		86	A	SOFTWARE
KC BIOLOGICAL	CHARLES RIVER BIOTECH.		86	A	OPTICELL
PROTEIN TECHNOLOGIES	CJB DEVELOPMENTS, LTD.	UK	86	M	INSTRUMENTS
ROHM & HAAS	TOSOH	JAPAN	87	V	EQUIPMENT
SENETEK	VHJ TESLA	CZECH	87	V	EQUIPMENT
VEGA BIOTECHNOLOGIES	DUPONT		86	E	SYNTHESIZERS
Reagents (G)					
BETHESDA RES. LABS	BIOPROCESSING, LTD.	UK	86	M	GROWTH FACTOR
BIOSEARCH	HP GENENCHEM		86	L	DNA SYNTH.CHEM
CETUS	CELLULAR PRODUCTS		86	L	IL-2 REAGENTS
T CELL SCIENCES	BOEHRINGER MANNHEIM	WGER	86	M	MABS
Cell Culture (H)					
BIO-RESPONSE	VENTREX		86	J	MAB'S
BIOASSAY SYSTEMS	TORAY-FUJI BIONICS	JAPAN	86	L	VIRUS DX
BIOGEN	VERAX		87	L	CELL CULTURES
CYTOCULTURE INT'L	MCA DEVELOPMENT	NETH	86	P	MAB'S

Company 1	Company 2	Ctry.	Yr.	Cd.	Product
Cell Culture (Cont.)					
ENDOTRONICS	CELANESE		86	E	T-CELL THERAPY
FLOW GENERAL	LABSYSTEMS	FIN	88	M	DX, CELL CULTURE
ENDOTRONICS	SUMMA MEDICAL		86	E	
Commodity Chemicals (I)					
ADVANCED GENETIC SCI.	DUPONT		86	R	
GENZYME	NEW BRUNSWICK SCI.		87	A	BIOCHEMICALS
Specialty Chemicals (J)					
AMGEN	EASTMAN KODAK		86	JM	CHEMICALS
APPLIED MICROBIOLOGY	AMERICAN CYANAMID		87	E	ENZYMES
BECKMAN INSTRUMENTS	BRITISH BIOTECH., LTD	UK	88	L	DESIGNER GENES
BIOTECHNICA INT'L	W.R. GRACE		87	J	ENZYMES
CALBIOCHEM	BIODOR HOLDING	SWITZ	88	A	BIOCHEMICALS
CELGENE	RITZ CHEMICALS		87	J	BIOPOLYMERS
CHIRON	PETROFINA GROUP	BELG	86	J	ENZYMES
COLLABORATIVE RES.	DOW CHEMICAL		88	L	CHYMOSIN
GENEX	XYDEX		87	A	PROTEINS
GENZYME	NAGASE & CO.	JAPAN	86	J	AMYLASE
H.B. FULLER & CO.	GENEX		87	A	SPECIALTY CHEM.
PFIZER	CELLTECH	UK	88	L	CHYMOSIN
POLYORGANIX	SYNTHATECH		87	A	ENZYMES
SYNERGEN	COORS BIOTECH		88	J	RIBOFLAVIN
Clinical Diagnostics (K)					
ADVANCED BIOTECH.	GENTRONIX LABS		87	A	MICROANALYSIS
AGEN USA	AMERICAN HOSP. SUPPLY		86	M	BLOOD SCREEN
AGOURON PHARMACEUT.	CAMBRIDGE BIOSCIENCE		87	J	AIDS DX, RX, VX
AIMS RESEARCH	AIMS BIOTECH	CAN	87	R	AIDS DX
ANALYTAB PRODUCTS	CAMBRIDGE BIOSCIENCE		87	M	VIRAL DX
ATLANTIC ANTIBODIES	INCSTAR		87	A	AB'S, DX KITS
AUTOMATED DIAGNOS.	UNIVERSITY GENETICS		87	E	AIDS DX
BIO-RESPONSE	COOPER		86	M	DX; ANTIVIRALS
BIOPROBE	AMERICAN CYANAMID		87	O	CHROMATOG.
BIOSTAR MEDICAL PRODS	C. ITOH	JAPAN	87	M	DX
BIOTECH. GENERAL	NEUROGENETIC		87	E	DX; RX
BIOTECH RESEARCH LABS	DUPONT		87	LM	AIDS DX
BIOTECHNICA DIAGN.	FORSYTH DENTAL CENTER		87	J	PERIDONTAL DX
BIOTHERAPEUTICS	INGENE		87	V	CANCER RESCH.
CAMBRIDGE BIOSCIENCE	SYVA		87	J	AIDS DX
CELLULAR PRODUCTS	EASTMAN KODAK		87	M	AIDS, T-CELL DX
CENTOCOR	AMERSHAM CANADA	CAN	87	M	CARDIOVASC. DX
CENTOCOR	ERBAMONT		86	L	DX
CENTOCOR	IMMUNOMEDICS		87	L	CANCER DX
CETUS	EASTMAN KODAK		86	J	DX
CETUS	ENZO BIOCHEM		86	J	
CETUS	POLYCELL		88	J	CANCER DX
CHARLES RIVER BIOTECH	AJINOMOTO	JAPAN	88	J	MABS
CHIRON	LUCKY, LTD.	S KOR	87	J	DX
CLINICAL SCIENCES	DUPONT		87	J	HEP-B AG'S
COLLABORATIVE RES.	SPECIAL REFERENCE LABS	JAPAN	87	L	GENETIC, CA DX
COOPER TECHNICON	MICROGENICS		87	R	MEDICAL DX
DAMON	ARES-SERONO	NETH	86	O	DX MABS

Listing 12-5 All Partnerships

Company 1	Company 2	Ctry.	Yr.	Cd.	Product
Clinical Diagnostics (Cont.)					
DIAGNOSTIC PRODUCTS	DAINIPPON	JAPAN	86	M	IMMUNODIAG.
DNASTAR	FIELD INVERSION TECHNOL.		87	V	DX
DUPONT	CONNAUGHT LABS	CAN	86	J	DX
ELECTRO-NUCLEONICS	ORGANON TEKNIKA	NETH	86	M	
ELECTRO-NUCLEONICS	PHARMACIA	SWE	86	M	DX
ELECTRO-NUCLEONICS	PHARMACIA	SWE	87	A	DX
ELECTRO-NUCLEONICS	PHILLIPS PETROLEUM		87	R	AIDS DX
ELECTRO-NUCLEONICS	PRUTECH R & D PART. II		86	R	AIDS TEST
ENDOTRONICS	SINO-AMER. BIOTECH.	CHINA	86	J	MAB DX
EPITOPE	ORGANON TEKNIKA	NETH	87	M	AIDS DX
EPITOPE	SYVA		87	R	AIDS DX
GENENTECH	NEORX		87	J	BLOOD CLOT DX
HYBRIDOMA SCIENCES	COVALENT TECHNOLOGY		86	A	
HYBRITECH	CELLTECH	UK	86	R	DX KITS
HYBRITECH	MONOCLONAL AB'S		87	L	VARIOUS DX
HYGEIA SCIENCES	TAMBRANDS		87	A	
IGEN	GRYPHON VENTURES		87	E	DX ASSAYS
IMCLONE	FISHER SCIENTIFIC		87	J	HEP-B, AIDS DX
IMMUNO NUCLEAR	BOOTS-CELLTECH	UK	86	M	
IMMUNOMEDICS	JOHNSON & JOHNSON		86	R	
IMMUNOMEDICS	KURARAY	JAPAN	87	M	IN VITRO DX
INTEGRATED GENETICS	AMOCO		87	V	AIDS DX
INTEGRATED GENETICS	SPECIAL REFERENCE LABS	JAPAN	87	J	DX SERVICES
KLEINER PERKINS	ATHENA		86	E	NEURAL DX
LIFE TECHNOLOGIES	BECTON DICKINSON		87	A	DX
LIFE TECHNOLOGIES	TORAY INDUSTRIES	JAPAN	87	M	DNA DETECTOR
MOLECULAR GENETICS	SCHERING-PLOUGH		87	M	TEST KITS
MONOCLONAL AB'S	ALCON LABS		86	J	
MONOCLONAL AB'S	ALPHA LABS		86	L	ELISA
MONOCLONAL AB'S	EASTMAN KODAK		88	L	IN VITRO DX
MONOCLONAL AB'S	MOBAY-HAVER		86	M	HORSE PREG. TEST
MONOCLONAL AB'S	ORTHO		87	L	OVULATION DX
MUREX	KYOWA HAKKO KOGYO	JAPAN	87	M	AIDS DX
MUREX	PILOT LABS	CAN	86	E	AIDS DX
NEORX	EASTMAN KODAK		86	EJ	
NOVO BIOLABS USA	NOVO INDUSTRI	DEN	86	M	DX
PRAGMA BIO-TECH	EIKEN CHEMICAL	JAPAN	87	J	CANCER DX
QUIDEL	CLONATEC	FRANCE	86	JM	DX
QUIDEL	INVESGEN	SPAIN	86	M	TEST KITS
QUIDEL	TOSHIN CHEMICAL	JAPAN	87	L	PREGNANCY DX
RICHARDSON-VICKS	ALLELIX	CAN	87	J	DX, DRUG DELIV.
SALUTAR	OTSUKA	JAPAN	87	L	IMAGING AGENTS
SERONO LABS	DIAGNOSTICS PASTEUR	FRANCE	86	M	TEST KITS
SUMMA MEDICAL	COSMO BIO	JAPAN	87	L	MAB
SUMMA MEDICAL	HOFFMAN-LA ROCHE	SWITZ	87	L	BLOOD CLOT DX
SYNBIOTICS	INT'L MINERALS & CHEM.		87	E	MAB
T CELL SCIENCES	YAMANOUCHI PHARMA.	JAPAN	87	R	CA DX
TECHNICON	COOPER DEVELOPMENT CO.		86	A	DX
VENTREX	BAXTER-TRAVENOL DX		86	R	DX REAGENTS
XYTRONYX	COLGATE PALMOLIVE		87	L	PERIDONTAL DX

Company 1	Company 2	Ctry.	Yr.	Cd.	Product
Food Production/Processing (M)					
AGRIGENETICS	SUNGENE		87	M	HYB. CORN SEEDS
CALGENE	L DAEHNFELDT	DEN	87	M	RAPESEED
CALIF. INTEGRATED DIAG.	INFERGENE		87	A	DX
COLLABORATIVE RES.	DOW CHEMICAL		86	M	RENNIN
COLLABORATIVE RES.	DOW CHEMICAL		86	P	RENNIN
COLLABORATIVE RES.	DOW CHEMICAL		88	J	RENNIN
COLLABORATIVE RES.	PFIZER		88	L	RENNIN
COORS	MEIJI SEIKA	JAPAN	87	J	FOOD ADDITIVES
CPC INT'L	AJINOMOTO	JAPAN	87	A	FOOD PRODUCTS
CYANOTECH	PHARMACHEM		86	R	BETA-CAROTENE
DNA PLANT TECHNOLOGY	FARMS OF TEXAS CO.		87	J	HYBRID RICE
DNA PLANT TECHNOLOGY	KRAFT		87	O	SNACK FOODS
DNA PLANT TECHNOLOGY	MONSANTO		86	J	SWEETENERS
DNA PLANT TECHNOLOGY	NUTRASWEET CO.		86	R	SWEETENERS
ECOGEN	PLANT GENETIC SYSTEMS	BELG	87	J	LETTUCE
ENGENICS	EASTMAN KODAK		86	J	BIOCHEMICALS
IGENE	BIOSOPH LABS	UK	86	V	FLAVORS
INGENE	BEATRICE		87	R	SWEETENER
INGENE	MITSUBISHI	JAPAN	86	J	SWEETENERS
INTEGRATED GENETICS	AMOCO		86	J	FOOD DX
MICROBIO RESOURCES	EASTMAN KODAK		87	M	BETA-CAROTENE
ML TECHNOLOGY VENT.	UNITED AGRISEEDS		87	R	CORN
OCCIDENTAL PETROLEUM	MITSUI TOATSU CHEM.	JAPAN	86	V	HYBRID RICE
PANLABS	FINNISH SUGAR CO.	FIN	87	L	ENZYMES
PHILLIPS PETROLEUM	KYOWA HAKKO KOGYO	JAPAN	87	J	SALMON HORM.
PLANT GENETICS	CALBEE POTATO	JAPAN	87	L	POTATO TUBERS
UNION CARBIDE	GIST-BROCADES	NETH	88	A	YEAST, ENZYMES
Production/Fermentation (O)					
STEARNS CATALYTIC CO.	CHIYODA CHEMICAL ENG.	JAPAN	86	O	GENERAL
Therapeutics (P)					
ALZA CORP.	QUEST BIOTECHNOLOGY		87	L	HEMOGLOBINS
AMERICAN CYANAMID	QUADRA LOGIC TECHNOL.	CAN	88	JE	CANCER RX
AMGEN	AB FERMENTA	SWE	86	E	
AMGEN	JOHNSON & JOHNSON		86	J	IL-2 ANALOG
AMGEN	SMITHKLINE		86	EJ	SOMATOTROPIN
APPLIED BIOTECHNOLOGY	DUPONT		86	J	CANCER DX, RX
ARTHUR D. LITTLE	TORAY RESEARCH CENTER	JAPAN	87	O	RX SCREENING
BARROWS RESEARCH GRP.	PRAGMA BIO-TECH		87	L	CHOLESTEROL RX
BAXTER TRAVENOL	TEVA PHARM.	ISRAEL	88	A	RX
BIO-RESPONSE	RETROPERFUSION SYS.		87	J	TPA
BIOGEN	BAXTER TRAVENOL		86	L	IF-G
BIOGEN	GLAXO	UK	87	A	IL-2, GM-CSF
BIOGEN	HOFFMANN-LA ROCHE	SWITZ	87	LA	B CELL GROWTH
BIOGEN	MERCK		87	J	ANTICANCER RX
BIOGEN	SCHERING-PLOUGH		86	L	ALPHA-IF
BIOMET	INGENE		87	J	PROTEINS
BIOMET	LIFECORE BIOMEDICAL		87	J	ORTHOPEDICS
BIOTECH. GENERAL	BAXTER TRAVENOL		86	M	GROWTH HORM.
BIOTECH. GENERAL	BRISTOL-MYERS		86	LM	SOD
BIOTECHNICA INT'L	HOFFMANN-LA ROCHE	SWITZ	88	R	VITAMINS
BIOTHERAPEUTICS	ANALYTIC BIOSYSTEMS		86	J	DRUGS

Listing 12-5 All Partnerships

Company 1	Company 2	Ctry.	Yr.	Cd.	Product
Therapeutics (Cont.)					
BIOTHERAPEUTICS	BAXTER HEALTH CARE		88	J	CANCER RX
BIOTHERAPEUTICS	SYNCOR		86	J	CANCER RX
BIOTHERAPEUTICS	XOMA		88	J	CANCER RX
CALIFORNIA BIOTECH.	AMERICAN HOME PRODS.		86	M	NAZDEL
CALIFORNIA BIOTECH.	ORTHO		86	R	NAZDEL
CALIFORNIA BIOTECH.	WYETH		86	M	RENIN INHIB.
CALIFORNIA BIOTECH.	GULDEN LOMBERG CHEM.	WGER	86	L	
CALIFORNIA BIOTECH.	HOFFMANN-LA ROCHE	SWITZ	86	R	
CALIFORNIA BIOTECH.	ORGANON INT'L	NETH	87	J	EPO
CALIFORNIA BIOTECH.	ORGANON INT'L	NETH	87	J	ALZHEIMER'S RX
CAMBRIDGE BIOSCIENCE	BAXTER HEALTHCARE		87	M	AIDS DX
CAREMARK	BAXTER TRAVENOL		87	A	HEALTH CARE
CENTOCOR	FMC		86	E	
CENTOCOR	POLYCELL		86	L	MAB THERAPS
CENTOCOR	TAKEDA	JAPAN	87	J	TOXIC SHOCK RX
CETUS	BEN-VENUE LABS		86	M	METHOTREXATE
CETUS	BEN-VENUE LABS		87	V	ANTICANCER RX
CETUS	CETUS-BEN VENUE		87	MJ	CANCER RX
CETUS	SQUIBB		87	JE	VARIOUS RXS
CETUS	SQUIBB		87	L	ANTICANCER RX
CETUS	TRITON BIOSCIENCES		86	V	IF
CHEMEX	SQUIBB		87	L	SKIN RX
CHEMEX	TAKEDA	JAPAN	88	L	ALLERGY RX
CHEMEX	UPJOHN		88	L	SKIN RX
CHIRON	BEHRINGWERKE	WGER	87	L	TPA
CHIRON	ETHICON		86	J	GROWTH FACT.
CHIRON	HITACHI	JAPAN	87	M	ULCER RX
CILCO, INC.	DIAGNOSTIC INC		86	M	HYAL. ACID
CISTRON BIOTECH.	BAKER INSTRUMENTS		87	A	
CISTRON BIOTECH.	DUPONT		86	L	IL-2
COOPER	CIBA-GEIGY	SWITZ	88	A	CONTACT LENS
CREATIVE BIOMOLECULES	BANEXI	FRANCE	87	R	TPA
CYANOTECH	AMERICAN CYANAMID		87	JM	CHOLESTEROL RX
DAMON	BIOTHERAPY SYSTEMS		86	A	
DAMON	CONNAUGHT LABS	CAN	86	V	INSULIN
DAMON	IDEC		86	A	B-CELL LYMPH RX
DAMON	SMITHKLINE		87	M	TPA
DAMON	YAMANOUCHI PHARM.	JAPAN	87	L	TPA
DUPONT	HOFFMANN-LA ROCHE	SWITZ	88	J	CANCER RX
ELI LILLY	HOFFMANN-LA ROCHE	SWITZ	87	L	HGH
ENDOTRONICS	CELANESE		86	E	HEPATITIS B RX
ENDOTRONICS	CELANESE		86	R	AIDS RX
ENDOTRONICS	NIPPON CHEMICAL IND.	JAPAN	86	J	HGH
ENZON	EASTMAN KODAK		87	E	O2 TOXICITY RX
EPITOPE	EASTMAN KODAK		88	J	CANCER DX, RX
EPITOPE	INGENE		88	J	AIDS RX
EXOVIR	GENENTECH		86	R	EXOVIR-HZ GEL
GENELABS	SANDOZ	SWITZ	87	JM	RX
GENELABS	SRI INT'L		88	JR	AIDS RX
GENELABS	SRI INT'L		88	JR	CONJUGATE
GENENTECH	BOEHRINGER INGELHEIM	WGER	87	M	TPA
GENENTECH	DAINIPPON	JAPAN	87	L	TNF
GENENTECH	EXOVIR		86	JL	VIRUS RX

Company 1	Company 2	Ctry.	Yr.	Cd.	Product
Therapeutics (Cont.)					
GENENTECH	GENZYME		87	JR	PROTEIN
GENENTECH	MITSUBISHI	JAPAN	87	LM	RX
GENENTECH	QUADRA LOGIC TECHNOL.	CAN	87	J	SUPPRESSANT
GENETICS INSTITUTE	BOEHRINGER MANNHEIM	WGER	87	L	EPO
GENETICS INSTITUTE	WELLCOME	UK	86	V	DRUGS
GENETICS INSTITUTE	WELLCOME	UK	87	V	IF, TPA, FACT. VIII
GENETICS INSTITUTE	CHUGAI PHARM.	JAPAN	87	M	EPO
GENETICS INSTITUTE	NEORX		87	EJ	ANTICANCER MAB
GENETICS INSTITUTE	SANDOZ	SWITZ	87	L	CA, AIDS RX, CSF
GENETICS INSTITUTE	WELLCOME	UK	86	JL	TPA
GENEX	BEHRINGWERKE	WGER	86	R	RX PROTEIN
GENEX	NIPPON OIL CO	JAPAN	87	L	VITAMIN B12
GENEX	SCHERING-PLOUGH		86	R	PLASMA PROTEIN
GENZYME	ALCON LABS		86	RL	EYE SURG. PROD.
GENZYME	HOFFMANN-LA ROCHE	SWITZ	87	J	HYAL. ACID
GENZYME	HOWMEDICA		87	J	ORTHO SURG PROD
GENZYME	PFIZER		87	L	ORTHOPEDICS
GENZYME	ZYMOGENETICS		87	M	SKIN RX, DX
HEM RESEARCH	DUPONT		87	E	AIDS RX
HOUSTON BIOTECH.	HOUSTON BIOTECH PART.		87	R	EYE RX
HOUSTON BIOTECH.	SANTEN PHARM.	JAPAN	87	L	CATARACT RX
HUNT RESEARCH CORP.	QUEST BIOTECHNOLOGY		87	A	HEMOGLOBIN
HYBRITECH	SANOFI RECHERCHE	FRANCE	86	J	T CELL MARKER
IMCERA BIOPRODUCTS	INT'L MINERALS & CHEM.		87	OM	SOMATOMEDIN
IMMUNEX	GENZYME		86	M	IL-2
IMMUNOLOGY VENTURES	EASTMAN KODAK		87	M	IL-4
INCSTAR	SANDOZ	SWITZ	87	J	ANTI-REJECT. RX
INGENE	BIOTHERAPEUTICS		86	R	CANCER RX
INGENE	EASTMAN PHARMACEUT.		88	J	CANCER RX
INGENE	ONCOGEN		87	J	ANTI-CA MAB
INTEGRATED GENETICS	BASF	WGER	87	O	TPA
INTEGRATED GENETICS	BEHRINGWERKE	WGER	87	J	EPO
INTEGRATED GENETICS	ORTHO		87	J	MSF
INTEGRATED GENETICS	TOYOBO	JAPAN	87	J	TPA
INTERFERON SCIENCES	ANHEUSER-BUSCH		86	E	IF
INVITRON	RORER		88	J	FACTOR VIII:C
INVITRON	SEARLE		86	J	TPA
INVITRON	SEARLE		87	J	PROTEINS
KLEINER PERKINS	GENSIA		86	E	HEART RX
LIFECORE BIOMEDICAL	MEDICONTROL		88	J	SKIN LOTION
LIPOSOME COMPANY	SANTEN	JAPAN	87	J	ALLERGY RX
LIPOSOME TECHNOLOGY	TAKEDA	JAPAN	86	R	
MELOY LABS	RORER GROUP		86	A	
NOVA PHARM.	CELANESE		86	E	
ORTHO	CELLTECH	UK	87	V	EPO
PANLABS	EASTMAN KODAK		87	R	RX
PHILLIPS PETROLEUM	BISSENDORF BIOSCIENCES	WGER	86	J	GH REL. FACTOR
REPLIGEN	E. MERCK	WGER	88	J	PLATELET FACTOR
REPLIGEN	GILLETTE		87	J	DENTAL ENZYME
SCHERING-PLOUGH	SHIONOGI	JAPAN	87	L	ANTIBIOTIC
SMITHKLINE	BOEHRINGER MANNHEIM	WGER	86	J	HEART RX
SMITHKLINE	BRITISH BIOTECH. LTD	UK	87	J	TPA
SMITHKLINE	SUMITOMO	JAPAN	87	V	RX

Listing 12-5 All Partnerships

Company 1	Company 2	Ctry.	Yr.	Cd.	Product
Therapeutics (Cont.)					
SQUIBB	NOVO INDUSTRI	DEN	86	M	HUMAN INSULIN
STERLING DRUG	EASTMAN KODAK		88	A	RX
SYNERGEN	CIBA-GEIGY CORP.	SWITZ	86	J	ELASTASE INHIB.
SYNERGEN	DUPONT		87	J	AF
SYNERGEN	PROCTER & GAMBLE		86	J	COLLAGENASE
T CELL SCIENCES	PFIZER		86	R	ARTHRITIS DRUG
T CELL SCIENCES	YAMANOUCHI PHARM.	JAPAN	87	M	IL-2
TECHAMERICA	SDS BIOTECH	SWE	87	A	RX
TECHNICLONE	DAMON		86	R	ANTI LYMPHOMA
TRITON BIOSCIENCES	CETUS		86	J	IF-B
TRITUS	RICHARDSON-VICKS		87	L	IF-B
UNIGENE LABS	FARMITALIA CARLO ERBA	ITALY	87	R	HORMONES
UNIVERSITY GENETICS	L'OREAL	FRANCE	86	L	TCF
UPJOHN	CHUGAI PHARMA	JAPAN	87	M	ARTHRITIS RX
UPJOHN	SANKYO	JAPAN	87	M	RX
VESTAR	MALLINCKRODT DIAG.	NETH	88	M	CANCER IMAGING
VESTAR	NOBEL MEDICA	SWE	88	M	CANCER IMAGING
XOMA	HOFFMANN-LA ROCHE	SWITZ	87	J	IL-2, MAB'S
XOMA	PFIZER		87	L	TOXIC SHOCK RX
ZYMOGENETICS	TEIJIN	JAPAN	87	L	PROTEIN C
Vaccines (Q)					
ALPHA 1 BIOMEDICALS	INTERLEUKIN-2		86	V	AIDS VX
APPLIED BIOTECHNOLOGY	ONCOGENETICS PARTNERS		87	J	AIDS, TB VX
BIOGEN	BEHRINGWERKE	WGER	86	M	MALARIA VX
BIOTECH. DEVELOPMENT	IMMUNE RESPONSE		87	M	AIDS VX
CAMBRIDGE BIOSCIENCE	INSTITUT MERIEUX	FRANCE	87	JR	AIDS VX
CENTOCOR	REPLIGEN		87	M	AIDS VX
CHIRON	CIBA-GEIGY CORP.	SWITZ	86	J	VX
CHIRON	CIBA-GEIGY CORP.	SWITZ	87	V	VX
CODON	MERRILL LYNCH		86	E	SWINE VX
CODON	NORDEN LABS	CAN	86	M	SWINE VX
DIAGNON	MELOY LABS		86	A	
IMCLONE	LEDERLE		88	J	VX
INTEGRATED GENETICS	CONNAUGHT LABS	CAN	86	J	HEP B VX
JOHNSON & JOHNSON	BRITISH BIOTECHNOLOGY	UK	88	E	AIDS VX
LIPOSOME COMPANY	CONNAUGHT LABS	CAN	87	J	INFLUENZA VX
MOLECULAR GENETICS	MOORMAN MANUF.		86	R	ANIMAL VX
PRAXIS BIOLOGICS	BRISTOL-MYERS		87	E	VX
REPLIGEN	MERCK		87	R	AIDS VX
SMITHKLINE	BRITISH BIOTECH., LTD.	UK	88	E	AIDS VX
VIRAL TECHNOLOGIES	NIPPON ZEON	JAPAN	88	L	AIDS VX
Aquaculture (S)					
REPLIGEN	TIEDEMANNS	NOR	87	R	AQUACULTURE
Veterinary Areas (V)					
AGRITECH SYSTEMS	HYBRITECH		86	L	VETERINARY DX
ALCIDE	OHF SANTE ANIMALE	FRANCE	88	M	GERMICIDE DIP
APPLIED MICROBIOLOGY	AMERICAN CYANAMID		87	M	STAPHYLOCIDE
CALIF. BIOTECHNOLOGY	FT. DODGE LABS		86	J	PET VX
CAMBRIDGE BIOSCIENCE	SMITHKLINE		86	M	FELINE LEUK. KIT

Company 1	Company 2	Ctry.	Yr.	Cd.	Product
Veterinary Areas (Cont.)					
CAMBRIDGE BIOSCIENCE	VIRBAC	FRANCE	86	J	VX
ENDOTRONICS	CELIAS	UK	86	A	MAMMAL. CELLS
FARNAM COMPANIES	QUIDEL		87	M	BOVINE DX
MOLECULAR GENETICS	EASTMAN KODAK		86	J	VET MASTITIS RX
NEOGEN	SOLVAY	BELG	86	J	DX
R&D FUNDING CORP.	SYNBIOTICS		86	J	VET RX
SYNBIOTICS	INT'L MINERALS & CHEM.		87	RM	VET MAB'S
SYNBIOTICS	NORDEN LABS		87	M	LEUKEMIA DX
SYNBIOTICS	NORDEN-EUROPE	UK	86	M	VETERNARY DX
SYNBIOTICS	PRUTECH R&D PART. II		86	R	VX
Research (W)					
AGRIGENETICS	AGRIGENETICS RES. ASSOC.		86	A	TECHNOLOGIES
Immunological Products (X)					
ENDOTRONICS	CELANESE		86	R	IMMUNE CELLS
ONCOGENE SCIENCE	PFIZER		86	EJ	ONCOGENES, GF'S
Toxicology (Y)					
BIONETICS RESEARCH	ORGANON TEKNIKA	NETH	87	LM	SALMONELLA DX
GEN-PROBE	FISHER SCIENTIFIC		86	M	MYCOPLASMA DX
GENE-TRAK SYSTEMS	INTEGRATED GENETICS		86	M	DNA PROBE
IDETEK	NOVO INDUSTRI	DEN	87	E	
IMMUNOSYSTEMS	TRANSIA LABS	FRANCE	87	M	ENVIRONMENT DX
Biomaterials (1)					
ADVANCED GENETIC SCI.	EASTMAN KODAK		87	M	SNOW MAKING
ETHICON	POSSIS MEDICAL		87	O	VASC. GRAFTS
ORGANOGENESIS	ELI LILLY		87	J	SYNTH. ARTERY
ROHM & HAAS	SUMITOMO	JAPAN	87	JM	ACRYLIC PLASTIC
Drug Delivery (3)					
CALIF. BIOTECHNOLOGY	LYPHOMED		87	MJ	DRUG DELIV.
CETUS	INTELLIGENT MEDICINE		86	J	DRUG DELIV.
COLLAGEN	TARGET THERAPEUTICS		88	A	DRUG DELIV.
KLEINER PERKINS	INSITE VISION		86	E	DRUG DELIV.
MEDI-CONTROL CORP.	SEARLE		86	E	DRUG DELIV.
MOLECULON	BRISTOL-MYERS		86	J	DRUG DELIV.
MOLECULON	F.H. FAULDING & CO.	AUSTRL	87	E	DRUG DELIV.
Medical Devices (4)					
LIFECORE BIOMEDICAL	ORTHOMATRIX		87	A	MEDICAL DEVICES

Notes:

These actions are combined from Listings 12-1 and 12-3 and are from the 26 month period of January 1986 through February 1988. The code refers to the type of action, as described in Table 12-1. The country for Company 1 is U.S. and the country for Company 2 is U.S., except where otherwise marked. There are no actions listed for codes C, D, L, N, R, T, U, Z, 2 or 5.

Table 12-2 Partnerships by Industry Classification

TABLE 12-2 PARTNERSHIPS BY INDUSTRY CLASSIFICATION

Code	Classification	Number (1)	Percent of Total
A	Agriculture, Animal	5	1.1
B	Agriculture, Plant	52	11.3
C	Biomass Conversion		
D	Biosensors/Bioelectronics		
E	Bioseparations	3	0.7
F	Biotechnology Equipment	24	5.2
G	Biotechnology Reagents	5	1.1
H	Cell Culture, General	9	2.0
I	Chemicals, Commodity	2	0.4
J	Chemicals, Specialty	13	2.8
K	Diagnostics, Clinical Human	89	19.3
L	Energy		
M	Food Production/Processing	28	6.1
N	Mining		
O	Production/Fermentation	3	0.7
P	Therapeutics	163	35.4
Q	Vaccines	20	4.3
R	Waste Disposal/Treatment	1	0.2
S	Aquaculture	1	0.2
T	Marine Natural Products	1	0.2
U	Consulting		
V	Veterinary	19	4.1
W	Research	3	0.7
X	Immunological Products	1	0.2
Y	Toxicology	5	1.1
Z	Venture Capital/Financing	2	0.4
1	Biomaterials	4	0.9
2	Fungi		
3	Drug Delivery Systems	7	1.5
4	Medical Devices	1	0.2
5	Testing/Analytical Services		

Notes:
1. Number of actions in each classification. Includes all actions in Listings 12-1 and 12-3.
2. Percent of total actions, 461 over 26 month period of January 1986 through February 1988.

FIGURE 12-1 PARTNERSHIPS WITH U.S. AND FOREIGN COMPANIES

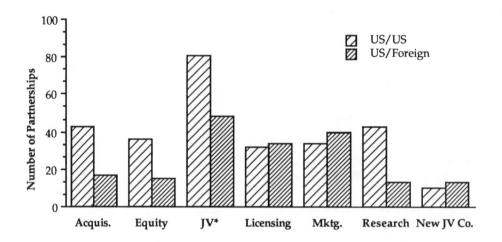

Partnerships during the 26-month period of January 1986 through February 1988, were sorted between U.S. company/U.S. company partnerships or U.S. company/foreign company partnerships. Partnerships involved acquisition, equity purchased, unspecified joint actions (JV*), Licensing agreements, marketing agreements, research contracts, or the formation of new joint venture companies. Details of these partnerships appear in Listings 12-1 through 12-4. A total of 461 actions are represented. Data are from public literature and from the companies.

Figure 12-2 Partnerships in Health Care

FIGURE 12-2 PARTNERSHIPS IN HEALTH CARE

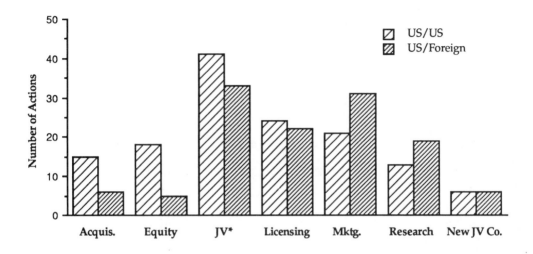

Partnerships involving health care (Codes K, P and Q; Diagnostics, Therapeutics and Vaccines) during the 26-month period of January 1986 through February 1988, were sorted by type and as occurring between U.S. companies and other U.S. companies or foreign companies. Partnerships involved acquisition, equity purchased, unspecified joint actions (JV*), licensing agreements, marketing agreements, research contracts, or the formation of new joint venture companies. Details of these partnerships appear in Listings 12-1 through 12-5. A total of 251 actions are represented. Data are from public literature and the companies.

LISTING 12-6 ACQUISITION OF BIOTECHNOLOGY COMPANIES

Acquired Company	Purchaser	Yr.	Cd.	Products
Acquisitions by U.S. Companies				
BENTLY LABS	AMERICAN HOSPITAL SUPPLY	81	O	
NEW ENGLAND NUCLEAR	DUPONT	81	G	REAGENTS
ACUGENICS	COOPER	82	K	DX
BECKMAN	SMITHKLINE	82	FG	PEPTIDES & AB'S
BIO-SCIENCES ENTERPRISES	AMERICAN HOSPITAL SUPPLY	82	K	DX TEST KITS
CALIFORNIA IMMUNO DIAG.	TECHAMERICA	82	KP	ANIMAL DX
DNAX	SCHERING-PLOUGH	82	P	
HYLAND DIAGNOSTICS	COOPER	82	K	DX
MORTON-NORWICH PHARM.	PROCTER & GAMBLE	82	P	RX
AMICON	W. R. GRACE	83	F	FILTRATION
CALIFORNIA MEDICAL CHEM.	CREATIVE BIOMOLECULES	83	K	KITS
ELECTRO MOLECULAR VENT.	GENTRONIX	83	D	
IMMULOK*	JOHNSON & JOHNSON	83	K	DX
MERIDIAN BIOMED	VENTREX	83	P	
TEXAS BIOSCIENCE	VENTREX	83		
WORTHINGTON DIAGNOSTICS	COOPER	83	K	DX
AGRIGENETICS	LUBRIZOL	84		
AMERICAN VETERINARY PROD.	IMMUNOGENETICS	84	V	
ATLANTIC ANTIBODIES	CHARLES RIVER LABS	84	G	AB'S
BIOCLINICAL SYSTEMS	VENTREX	84	K	
ICL SCIENTIFIC DIAGNOS.**	HYBRIDOMA SCIENCES	84	K	DX
IMMUNO NUCLEAR	IMMUNO BIOTECH	84	K	RIA'S
UNIVERSITY GENETICS	TRANS WORLD GENETICS	84		
VIRAGEN	AUTOMATED MEDICAL LABS	84	K	
AMERICAN BIONUCLEAR*	BIO RECOVERY TECHNOLOGY	85	K	
BARTELS IMMUNODIAGNOST.	AMERICAN HOSPITAL SUPPLY	85	K	DX
GENETIC SYSTEMS	BRISTOL-MYERS	85	K	MABS
HYBRITECH	ELI LILLY	85	KP	MABS
LATTINGTOWN CYTOLOGY CTR.	ENZO	85		
LIFE TECHNOLOGIES**	LYPHOMED	85	S	
ONCOGEN	BRISTOL-MYERS	85		
SEARLE	MONSANTO	85	P	RX
AGRIGENETICS	AGRIGENETICS RES. ASSOCS.	86	W	
CORNING GLASS**	CHARLES RIVER BIOTECH	86	F	
DAMON	IDEC	86	P	B-CELL LYMPH RX
DAMON*	BIOTHERAPY SYSTEMS	86	P	
DIAGNON*	MELOY LABORATORIES	86	Q	
GENETIC DESIGN	MILLIPORE	86	F	DNA SYNTH.
HYBRIDOMA SCIENCES	COVALENT TECHNOLOGY	86		
INTELLIGENETICS	AMOCO	86	F	SOFTWARE
KC BIOLOGICAL	CHARLES RIVER BIOTECH.	86	FH	OPTICELL
KEY PHARMACEUTICALS	SCHERING-PLOUGH	86	P	
KRATOS ANALYTICAL**	APPLIED BIOSYSTEMS	86	F	
MELOY LABORATORIES	RORER GROUP	86	P	
SPECTROS INTERNATIONAL**	APPLIED BIOSYSTEMS	86	F	
TECHNICON	COOPER DEVELOPMENT CO.	86	K	DX
AMERICAN MONITOR	QUEST BIOTECHNOLOGY	87	F	INSTRUMENTS
ANALYTICHEM	VARIAN ASSOCIATES	87		
ARCO SEED CO.**	UF GENETICS	87	B	

Listing 12-6 Acquisition

Acquired Company	Purchaser	Yr.	Cd.	Products
ATLANTIC ANTIBODIES	INCSTAR	87	K	AB'S, DX KITS
BAKER INSTRUMENTS	CISTRON BIOTECHNOLOGY	87	PG	
CALIF. INTEGRATED DIAGNOST.	INFERGENE	87	MK	DX
CAREMARK	BAXTER TRAVENOL	87	P	HEALTH CARE
DORR-OLIVER	W.R. GRACE	87	HF	EQUIPMENT
GENESIS LABS	MONOCLONAL ANTIBODIES	87		
GENEX**	H.B. FULLER & CO.	87	J	SPECIALTY CHEM.
GENTRONIX LABORATORIES**	ADVANCED BIOTECH.	87	K	MICROANALYSIS
GENZYME**	NEW BRUNSWICK SCIENTIFIC	87	I	BIOCHEMICALS
HUNT RESEARCH CORP.	QUEST BIOTECHNOLOGY	87	P	HEMOGLOBIN
HYGEIA SCIENCES	TAMBRANDS	87	K	
INTERNATIONAL BIOTECH.	EASTMAN KODAK	87		
LIFE TECHNOLOGIES**	BECTON DICKINSON	87	K	DX
LOVELOCK SEED CO.	PLANT GENETICS	87	B	ALFALFA
MCALLISTER SEED CO.	BIOTECHNICA INT'L	87	B	SEEDS
MELOY LABORATORIES**	CHEMICON INTERNATIONAL	87		
MONOCLONAL ANTIBODIES	MONOCLONAL TECHNOL.	87	X	
ORTHOMATRIX	LIFECORE BIOMEDICAL	87	F	MEDICAL DEVICES
PEDIGREED SEED CO.	CALGENE	87	B	COTTON
XYDEX	GENEX	87	J	PROTEINS
COOPER LASERSONICS	PFIZER	88	F	EQUIPMENT

Acquired Company	Purchaser	Yr.	Cd.	Products
Acquisitions by Non-U.S. Companies				
ALPHA THERAPEUTICS	GREEN CROSS (JPN)	78	P	IF
TEXASGULF	ELF-AQUITAINE (FR)	81		
BIOMEDICAL RESEARCH LABS	HOFFMANN-LA ROCHE (SWI)	82	K	LAB TESTS
P-L BIOCHEMICALS	PHARMACIA (SWE)	82	G	BIOCHEMICALS
ZOECON	SANDOZ (SWI)	83	B	SEEDS
AMERICAN DIAGNOSTICS	HOFFMANN-LA ROCHE (SWI)	84	K	DX
PURIFICATION ENGINEERING	RHONE-POULENC AG. (FR)	85	HJ	CHEMICALS
CELANESE	HOECHST (WGER)	86		
CELIAS** (UK)	ENDOTRONICS	86	V	MAMMAL. CELLS
BIOGEN	GLAXO (UK)	87	P	IL-2, GM-CSF
BIOGEN	HOFFMANN-LA ROCHE (SWI)	87	P	BCGF
BRUINSMA SEEDS (NETH)	UPJOHN	87	B	SEEDS
COOPER TECHNICON**	ALFA-LAVAL (SWE)	87		INDUSTRIAL SYST.
CPC INTERNATIONAL	AJINOMOTO (JPN)	87	M	FOOD PRODUCTS
PHARMACIA** (SWE)	ELECTRO-NUCLEONICS	87	K	DX
PLANT CELL RESEARCH INST**	MONTEDISON (IT)	87	B	
TECHAMERICA	SDS BIOTECH (SWE)	87	P	RX
COOPER**	CIBA-GEIGY (SWI)	88	P	CONTACT LENS

Notes:

Biotechnology-related acquisitions listed in the **Actions** database are presented here, sorted by year. Where available, company industry codes (Cd.; see Table 12-1) and products are given. Acquisitions are divided between U.S./U.S. acquisitions and those between U.S. and foreign companies. Most of the small and large U.S. companies listed here are detailed in the directories (Listings 2-2 and 11-1, respectively).

* A merger between two companies

** Acquisition of a division only of the company listed in the first column

SECTION 13 STATE BIOTECHNOLOGY CENTERS

INTRODUCTION

One of the most recent events in the growth of biotechnology in the United States has been the advent of state support for biotechnology. Beginning in 1981 with the creation of the North Carolina Biotechnology Center, more than 40 of these centers have evolved (see Figure 13-1). Each of these centers has a different focus, source of support and set of programs. This section provides a description of and data on these biotechnology centers.

Data Collection

To determine the range and scope of state support for biotechnology, centers from four other states provided support to the Biotechnology Information Program of the North Carolina Biotechnology Center to conduct a study of all state centers in the United States. Data were collected between September 1987 and February 1988. A questionnaire was sent to state-supported biotechnology centers in 30 states. We asked questions about center structure, programs, sources of funding, facilities, personnel and goals. A few centers, originally contacted, did not fit our definition of a biotechnology center (i.e., a clear focus on biotechnology and some or all direct state funding specifically for its support). Included in the study were responses from 41 centers representing 25 states, as well as one federally funded center, allowing an overview of the centers. Although great care was taken to compile an exhaustive list of centers, it is possible that additional state centers remain unidentified and are thus not included in this study. An overview of the centers appears in Table 13-1.

Mission

The basic mission of these centers falls into five main categories. The most common mission, seen in half of the centers, is to support research projects. These centers are generally associated with a single university. Another 25 percent of the centers focus primarily on economic development within their state. In general, centers focusing on economic development are older and larger (in budget, space and personnel) than those in any other category. The remaining 25 percent of the state centers have a primary focus in industry, service facilities, or training of staff and students. These centers are considerably smaller in space and personnel than those in the above two categories, and represent an average of only $500,000 in annual budget.

Affiliation

Seventy percent of the centers were affiliated with a single university or university system. Only 20 percent of the biotechnology centers are state agencies. The remaining centers are private, non-profit centers. These are larger on average than most and are not affiliated with a university. Budgets of these centers are above average and their programs are varied.

Funding

The primary funding source (providing more than half of the funding) was state funds in 70 percent of the centers. Federal funding was a primary source in an additional 25 percent, with the remainder having primarily industry funding, private foundation funding, or numerous sources of funding. Only one center, at the Massachusetts Institute of Technology, was initiated by federal funds and not by a state.

Common Themes

Despite the major differences in affiliation, funding and focus, our study demonstrates some common threads that underlie most of the centers. Most report supporting state and local biotechnology functions, such as conferences, seminars and workshops. Most have newsletters and other communications vehicles. Many are interested in biotechnology education and training -- from high school through the post-graduate level. Additionally, most are concerned with legislation and regulation. To date, only one center reported legislation related to biotechnology in its state.

Notes

1.   As a result of this study, the book <u>Directory of States' Biotechnology Centers</u> is available from the North Carolina Biotechnology Center (Information Program, PO Box 13547, Research Triangle Park, NC 27709-3547). This directory contains additional personnel and project data and is updated periodically. Write for details.

2.   See: Dibner, Mark: An analysis of state-sponsored biotech centers in the United States. Genetic Engineering News 8(1): 21, January 1988.

3.   This project was originally supported by state centers from Maryland, Michigan, New York (Cornell) and Wisconsin in conjunction with the North Carolina Biotechnology Center.

TABLE 13-1 CHARACTERISTICS OF STATE BIOTECHNOLOGY CENTERS

Category	Mean	Minimum	Maximum	Sum*
Year Founded	1985	1981	1987	N/A
1987/1988 Annual Budget	$2,188,322	$50,000	$7,985,000	$83,156,249
Full-Time Employees	16	1	140	672
Faculty Members	30	0	250	934
Facilities (square feet)	13,951	190	60,000	585,942

Notes:
Data are from 42 biotechnology centers.
* Sum represents mean data extrapolated for all centers. (N/A= not applicable)

Figure 13-1 State Biotechnology Centers Started per Year

FIGURE 13-1 STATE BIOTECHNOLOGY CENTERS STARTED PER YEAR

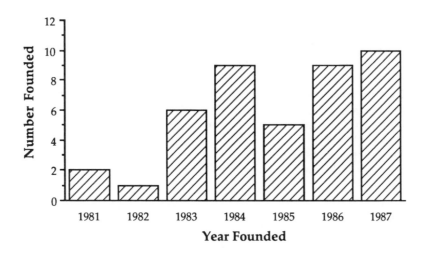

The number of state biotechnology centers founded per year is shown. The first state centers, in North Carolina and Michigan, were founded in 1981 and this figure shows centers founded through 1987. For more details on these centers, see Listing 13-1.

TABLE 13-2 QUICK LIST OF STATE BIOTECHNOLOGY CENTERS

State Center	Contact Person	Telephone
AR Arkansas Biotechnology Center	Collis R. Geren	501-575-2651
AZ Arizona State Univ., Dept. of Zoology	Dr. Ann Kammer	602-965-3156
AZ Univ. of Arizona Div. of Biotechnology	Dr. John H. Law	602-621-5769
CA CSU Prog. for Educ. & Res. in Biotech.	Dr. A. Stephen Dahms	619-265-5578
CA UCI Gene Research & Biotech. Prog.	Dr. G. Wesley Hatfield	714-856-5344
CA UC Biotech. Research & Educ. Prog.	Susanne Huttner	213-206-6814
CO Colorado Inst. for Research in Biotech.	Dr. Robert H. Davis	303-492-7314
CT Univ. of Connecticut Biotech. Center	Dr. Todd M. Schuster	203-486-4333
FL Biotechnology Research & Devel. Inst.	Lenie Breeze	904-462-3904
FL Interdisciplinary Ctr. for Biotech. Research	Lenie Breeze	904-392-7283
FL Prog. in Medical Technology	Dr. Robert Reeves	904-644-1855
GA Health Science Technology Center	Virginia Orndorff	404-721-2812
GA Univ. of Georgia Biotechnology Center	Dr. Leonard Mortenson	404-542-1693
IA Office of Biotechnology	Walter Fehr	515-294-9818
IL Biotechnology Center	Janet H. Glaser	217-333-1695
IL Plant Molecular Biology Center	Michelle Sabarini	815-753-7841
IN Center for Plant Biotechnology	Thomas K. Hodges	317-494-4657
IN Institute for Molecular and Cellular Biol.	Rhea Percival	812-335-4183
MA Biotech. Process Engineering Center	Ruth Ayers	617-253-2504
MD Engineering Research Ctr. Biotech. Prog.	Edward Sybert	301-454-1479
MD Maryland Biotechnology Inst.	Jane McGlade	301-454-1628
MI Michigan Biotechnology Inst.	Dr. Jack H. Pincus	517-337-3181
NC North Carolina Biotechnology Center	W. Steven Burke	919-541-9366
NJ NJ Ctr. for Advanced Biotech. & Medicine	Robert Namovicz	201-463-4665
NM Plant Genetic Engineering Laboratory	Mindy McAbee	505-646-5794
NY Center for Biotechnology	Ginny Llobell	516-632-8521
NY Cornell University Biotechnology Prog.	Raymond B. Snyder	607-255-2300
OH Edison Animal Biotechnology Center	Robert Mallot	614-593-4713
OH Edison Biotechnology Center	Dorothy C. Baunach	216-229-9445
OH Ohio State Univ. Biotechnology Center	Linda J. Carter	614-292-5670
OR Center for Gene Research and Biotech.	Dr. Christopher Matthew	503-754-4511
PA Center for Molecular Bioscience/Biotech.	Dr. Cinda Herndon-King	215-758-3645
PA Penn State Biotechnology Institute	Mary Ann Bjalme	814-863-3650
PA Pittsburgh Center for Biotech. & Bioeng.	Dr. Jerome S. Schultz	412-648-7956
RI RI Partnership for Science & Technology	Bruce Lang	401-277-2601
TN Biotechnology Prog.	Dr. D. K. Dougall	615-974-6841
TN Center for Environmental Biotechnology	Dr. James Blackburn	615-974-5219
UT Center for Biopolymers at Interfaces	Arline Allen	801-581-5455
UT Center for Excellence in Biotechnology	Steven D. Aust	801-750-2753
VA Center for Bioprocess/Product Devel.	Dr. Elmer Gaden, Jr.	804-924-6276
VA Center for Biotechnology	Lloyd Wolfinbarger, Jr.	804-440-3617
VA Institute of Biotechnology	Dr. Terry Woodworth	804-786-8565
WI Univ. of Wisconsin Biotechnology Center	Dr. Margaret vanBoldrik	608-262-5077

Listing 13-1 Directory of State Centers

LISTING 13-1 DIRECTORY OF STATE BIOTECHNOLOGY CENTERS

Centers are listed in order of state code.

Arkansas Biotechnology Center
Biomass Center Farm
University of Arkansas
Fayetteville, **AR** 72701

Telephone: 501-575-2651

Director:	Collis R. Geren, Coordinator
Started:	1986
Mission:	To provide biotechnical expertise and equipment for public and private sector scientists
Personnel:	Full-Time: 6 Part-Time: 1
Funding:	86-$200,000 87-$193,500

Concentration Areas: MAB production, biomass conversion, custom fermentation.

Department of Zoology
Arizona State University
Tempe, **AZ** 85287

Telephone: 602-965-3156

Started:	1987
Contact:	Dr. Ann Kammer
Mission:	In start-up stages. Not specified yet.

University of Arizona, Division of Biotechnology
ARL, Biosciences West 364
Tucson, **AZ** 85721

Telephone: 602-621-5769

Director:	Dr. John H. Law, Director of Biotechnology
Started:	1986
Mission:	To obtain and operate major equipment in core user facilities; to coordinate biotechnology research and education; to interface with industry.
Personnel:	Full-Time: 7
Funding:	86-$1,000,000 87-$2,000,000

Concentration Areas: Macromolecular structure, molecular design, EM, MABS, biomaterial processing.

California State University Program for Education and Research in Biotechnology (CSUPERB)
Molecular Biology Institute
San Diego State University
San Diego, CA 92101

Telephone: 619-229-2822

Director: Dr. A. Stephen Dahms
Started: 1987
Mission: Coordination of biotechnology education and research development/
             funding in the 9-campus California State University System.
Personnel: Full-Time: 3
Funding: 87-$150,000

Concentration Areas: Being determined.

UCI Gene Research and Biotechnology Program
University of California
Department of Microbiology and Molecular Genetics
Irvine, CA 92717

Telephone: 714-856-5344

Director: Dr. G. Wesley Hatfield, Professor
Started: 1984
Mission: A campus-wide, comprehensive project to identify, clarify
             and unify the efforts of the many UCI researchers working
             in biotechnology and to interface with the surrounding
             business, industry, governmental and lay communities.
Personnel: Full-Time: 2   Part-Time: 1
Funding: 87-$955,000

Concentration Areas: Molecular genetics, biochemical & protein engineering,
             medical devices, diagnostics.

University of California Biotechnology Research & Education Program
Molecular Biology Institute
UCLA, 405 Hilgard Avenue
Los Angeles, CA 90024

Telephone: 213-206-6814

Director: Dr. Paul D. Boyer
Started: 1985

340

Listing 13-1 Directory of State Centers

University of California Biotechnology Research & Education Program (Cont.)
Mission: To: 1. Foster and support research in biotechnology and
   related issues, 2. Promote training at undergraduate, graduate
   and postdoctoral levels, 3. Inform the public, government and
   industry on biotechnology development and public implications.
Personnel: Full-Time: 2 Part-Time: 1
Funding: 85-$1,500,000 86-$1,500,000 87-$1,500,000

Concentration Areas: Broad research, training, many scientific and social areas.

Colorado Institute for Research in Biotechnology (CIRB)

Department of Chemical Engineering
University of Colorado
Boulder, CO 80309

Telephone: 303-492-7314

Director: Drs. Davis and Murphy, Co-Directors
Started: 1987
Mission: To integrate various academic and research programs
   within the state which will provide basic technical
   support for industrial development and the promotion of
   biotechnology industry within Colorado.
Personnel: Full-Time: 0 Part-Time: 2 Contract: 0 Faculty: 8
Funding: 87- $50,000

Concentration Areas: Microbial cells, cultured plant/animal cells, cell components.

University of Connecticut Biotechnology Center

University of Connecticut
U125-75 North Eagleville Rd.
Storrs, CT 06268

Telephone: 203-486-4333

Director: Dr. Todd M. Schuster, Director
Started: 1986
Mission: Program started in July, 1986 with funds from the State
   Center for Excellence. Other information not available.

Biotechnology Research and Development Institute
1 Progress Blvd.
PO Box 26
Alachua, FL 32615

Telephone: 904-462-3904

Director: Dr. John F. Gerber
Started: 1987
Mission: To nurture biotechnology as a science and a business by turning
             biotech discoveries at University of Florida into useful products.
Personnel: Full-Time: 3   Contract: 1
Funding: 87-$237,000

Interdisciplinary Center for Biotechnology Research
University of Florida
Bldg. 106
Gainesville, FL 32611

Telephone: 904-392-7283

Director: T.W. O'Brien and L.C. Hannah, Co-Directors
Started: 1987
Mission: To coordinate biotechnology research activities at the campus level.
Personnel: Full-Time: 28   Faculty: 157
Funding: 87-$3,000,000

Program in Medical Technology
Florida State University
Tallahassee, FL 32306

Telephone: 904-644-1855

Director: Dr. John Hozier
Started: 1987
Mission: Training and research in biomedical technology with
             emphasis on human genetics and molecular diagnostics.
Personnel: Full-Time: 3   Part-Time: 1 Faculty: 45
Funding: 87-$100,000

Concentration Areas: To include human diagnostics, molecular genetics.

Listing 13-1 Directory of State Centers

Health Science Technology Center
Bldg. CN, Rm 128
Medical College of Georgia
Augusta, **GA** 30912

Telephone: 404-721-2812

Director: Virginia Orndorff, Program Manager
Started: 1987
Mission: To create jobs, businesses, products, revenues and state
taxes through start-up and relocated health science-
related technology enterprises within Georgia. A branch of
Georgia's Advanced Technology Development Center (ATDC).
Personnel: Full-Time: 1 Part-Time: 1 Faculty: 0
Funding: 87-$220,000

Concentration Areas: Health sciences.

University of Georgia Biotechnology Center
University of Georgia
Athens, **GA** 30602

Telephone: 404-542-1693

Director: Dr. Leonard E. Mortenson
Started: 1982
Mission: To facilitate University research in biotechnology. Award
start-up grants for new projects. Integrate activities
between departments, schools and colleges. Liaison between
industry and the University of Georgia faculty.
Personnel: Full-Time: 62 Part-Time: 3 Faculty: 61
Funding: 84-$853,297 85-$451,380 86-$2,730,529 87-$3,263,862

Concentration Areas: Plant molecular biology, complex carbohydrates,
fermentation.

Office of Biotechnology
Iowa State University
1391 Agronomy
Ames, **IA** 50011

Telephone: 515-294-9818

Director: Walter Fehr, Biotechnology Coordinator
Started: 1984

Office of Biotechnology, Iowa State University (Cont.)

Mission: To coordinate development of University-wide biotechnology
             program in Colleges of Agriculture, Engineering, Family and
             Consumer Science, Science and Humanities, and Veterinary Medicine.
             Coordinate with activities of the Biotechnology Consortium.
Personnel: Full-Time: 3   Faculty: 0
Funding: 84-$7,500 85-$7,500 86-$4,250,000 87-$4,500,000 88-$4,500,000

Concentration Areas: Primarily agricultural biotechnology.

Biotechnology Center
901 S. Mathews St.
Urbana, IL 61801

Telephone: 217-333-1695

Director: Lowell P. Hager
Started: 1983
Mission: Support of centralized laboratories with analytical and
             scientific services.
Personnel: Full-Time: 10   Part-Time: 1   Faculty: 85
Funding: 85-$250,000 86-$473,000 87-$758,000 88-$1,015,000

Plant Molecular Biology Center
Northern Illinois University
319 Montgomery Hall
DeKalb, IL 60115

Telephone: 815-753-7841

Director: Arnold Hampel
Started: 1986
Mission: Dedicated to high-quality research in plant molecular biology.
             University/industry interactions, industrial projects are undertaken.
Personnel: Full-Time: 17   Part-Time: 1   Faculty: 25
Funding: 86-$240,000 87-$600,000 88-$858,000

Concentration Areas: Host/parasite interactions, plant virology, transposable
                      elements, etc.

Listing 13-1 Directory of State Centers

Center for Plant Biotechnology
Agricultural Research Building
Purdue University
West Lafayette, IN 47907

Telephone: 317-494-4657

Director: Thomas K. Hodges
Started: 1985
Mission: To coordinate a grant from the Indiana Corporation for Science and Technology for research on agricultural plant biotechnology.
Personnel: Full-Time: 10   Part-Time: 2   Faculty: 10
Funding: 87-$5,200,000

Concentration Areas: Plant biotechnology.

Institute for Molecular and Cellular Biology
Indiana University
Jordan Hall 322A
Bloomington, IN 47405

Telephone: 812-335-4183

Director: Dr. Rudolf A. Raff
Started: 1983
Mission: To perform basic research in cellular and molecular biology, both plant and animal. University/industry ties are fostered.
Personnel: Full-Time: 10   Part-Time: 2   Contract: 2   Faculty: 32
Funding: 87-$720,467

Concentration Areas: Basic research.

Biotechnology Process Engineering Center
20A-207 Massachusetts Institute of Technology
18 Vassar St.
Cambridge, MA 02139

Telephone: 617-253-0805

Director: Dr. Daniel I. C. Wang
Started: 1985
Mission: A federally funded program to provide interdisciplinary education and research in biotechnology process engineering.
Personnel: Full-Time: 140   Part-Time: 75 Contract: 0   Faculty: 17
Funding: 85-$4,500,000 86-$5,500,000 87-$6,300,000 88-$6,350,000

Engineering Research Center, Biotechnology Program
University of Maryland, Bldg 081
College Park, **MD** 20742

Telephone: 301-454-1479

Director: Edward Sybert, Manager
Started: 1984
Mission: To provide a center for work on rDNA fermentation, plant
cell growth, and mammalian cell culture.
Personnel: Full-Time: 2   Faculty: 6
Funding: 84-$140,000   85-$208,000   86-$334,700   87-$315,000

Concentration Areas: Microbial fermentation scaleup, plant/cell culture,
rDNA scaleup.

Maryland Biotechnology Institute
1326 Chemistry Building
University of Maryland
College Park, **MD** 20742

Telephone: 301-454-1628

Director: Dr. Rita R. Colwell
Started: 1985
Mission: To serve as a focus for new ventures in biotechnology at the
University.   MBI has 4 research centers and 1 public issues program:
Center for Marine Biotech, Medical Biotech Center, Agricultural
Biotech Center and the Center for Advanced Research in Biotech.
Personnel: Full-Time: 52   Part-Time: 16 Contract: 6   Faculty: 36
Funding: 86-$2,680,000 87-$5,690,000 88-$7,440,000

Concentration Areas: Marine, agricultural, medical, advanced research-protein
structure.

Michigan Biotechnology Institute
PO Box 27609
3900 Collins Rd.
Lansing, **MI** 48909

Telephone: 517-337-3181

Director: Dr. J. G. Zeikus, President
Started: 1981

346

Listing 13-1 Directory of State Centers

Michigan Biotechnology Institute (Cont.)
Mission: A private, non-profit corporation to broaden the industrial base and to create jobs by development of biotechnology innovations. Help establish new companies, expand existing businesses, attract others to state.
Personnel: Full-Time: 40   Part-Time: 39   Faculty: 14
Funding: 84-$893,000   85-$1,646,000   86-$2,691,000   87-$7,985,000

Concentration Areas: Industrial enzymes/bioelectronics, fermentation, biomaterials.

North Carolina Biotechnology Center
PO Box 13547
79 Alexander Drive, 4501 Bldg.
Research Triangle Park, NC 27709

Telephone: 919-541-9366

Director: Dr. Charles E. Hamner, Jr., President
Started: 1981
Mission: To promote economic development in North Carolina by catalyzing biotechnology in the state. Includes growth of biotechnology infrastructure in academia as well as economic development. A locus for biotechnology interactions in NC.
Personnel: Full-Time: 24   Part-Time: 2   Contract: 3   Faculty: 0
Funding: 84-$1,000,000   85-$6,500,000   86-$6,500,000   87-$6,500,000   88-$6,500,000

Concentration Areas: Lymphocyte technology, forestry, marine biotechnology, bioelectronics, plant molecular biology, public education, biotechnology industry information, economic development, many others.

NJ Center for Advanced Biotechnology and Medicine
675 Hoes Lane
Piscataway, NJ 08854

Telephone: 201-463-4665

Director: Aaron J. Shatkin
Started: 1984
Mission: Basic molecular biology including structural biology, molecular genetics, cell and developmental biology and molecular pharmacology.
Personnel: Full-Time: 32   Part-Time: 2   Faculty: 3
Funding: 84-$200,000   85-$1,300,000   86-$1,500,000   87-$3,200,000   88-$3,500,000

Concentration Areas: Molecular virology and molecular genetics research.

Plant Genetic Engineering Laboratory for Desert Adaptation
New Mexico State University, Box 3GL
Las Cruces, NM 88003

Telephone: 505-646-5453

Director: Dr. John D. Kemp
Started: 1983
Mission: R and D activities in plant biotechnology for desert
agriculture. Encourages economic development through
technology transfer and applied research.
Personnel: Full-Time: 12   Part-Time: 15 Contract: 0 Faculty: 8
Funding: 84-$650,000 85-$825,000 86-$812,000 87-$650,000 88-$650,000

Concentration Areas: Crop improvement for dry regions. Stress tolerance.

Center for Biotechnology
Room 130, Life Sciences Bldg.
SUNY at Stony Brook
Stony Brook, NY 11794

Telephone: 516-632-8521

Director: Dr. Richard K. Koehn
Started: 1983
Mission: To advance economic development through programs designed to
promote biotechnology-related research and industry collaboration.
Personnel: Full-Time: 4   Part-Time: 1 Contract: 1
Funding: 84-$1,000,000 85-$1,000,000 86-$1,000,000 87-$1,000,000 88-$1,000,000

Concentration Areas: Medical biotechnology.

Cornell University Biotechnology Program
Box 547 Baker Laboratory
Ithaca, NY 14853

Telephone: 607-255-2300

Director: Dr. Gordon Hammes
Started: 1983
Mission: Foster basic biotechnology research and applications for economic
development. Composed of Cornell Univ. Biotechnology Institute
(industry consortium), US Army Center for Excellence in
Biotechnology and the NY Center for Advanced Technology in
Biotechnology (Agriculture).

Listing 13-1 Directory of State Centers

Cornell University Biotechnology Program (Cont.)
Personnel: Full-Time: 3   Part-Time: 4
Funding: 84-$2,000,000 85-$2,250,000 86-$3,500,000 87-$5,000,000

Concentration Areas: Molecular genetics, cell biology in plants, animals and
                       cell production.

Edison Animal Biotechnology Center
Wilson Hall/West Green
Ohio University
Athens, **OH** 45701

Telephone: 614-593-4713

Director: Dr. Thomas E. Wagner
Started: 1984
Mission: ·   A working "bridge institute" between Ohio Univ., Case
             Western, and Ohio State Univ. and corporate affiliates.
             Research projects on animal gene transfer, and molecular
             farming of medical and industrial proteins.
Personnel: Full-Time: 38   Faculty: 2
Funding: 85-$800,000 86-$1,300,000 87-$1,400,000 88-$1,600,000

Concentration Areas: Transgenic animals, molecular farming

Edison Biotechnology Center
11000 Cedar Avenue
University Research Center, Building One
Cleveland, **OH** 44106

Telephone: 216-229-9445

Director: Dorothy C. Baunach, Interim Director
Started: 1987
Mission: To promote the development of biotechnology-based business
             in Northeast Ohio.
Funding: 87-88- $2,000,000

Ohio State University Biotechnology Center
Room 206, Rightmire Hall
1061 Carmack Rd.
Columbus, **OH** 43210

Telephone: 614-292-5670

Ohio State University Biotechnology Center (Cont.)

Director: Pappachan Kolattukudy
Started: 1986
Mission: To coordinate a campus-wide network of interacting
               research teams in 7 colleges. To train graduate students, postdocs
               and other professional staff in biotechnology. Center supports a
               3-way partnership between the Center, government, and industry.
Personnel: Full-Time: 23   Part-Time: 8   Contract: 0   Faculty: 6
Funding: 86-$1,472,800 87-$1,967,445

Concentration Areas: Plants/microbes, neurobiotechnology, and industrial
                       microbiology.

Center for Gene Research and Biotechnology
Oregon State University
Corvallis, OR 97331

Telephone: 503-754-4511

Contact: Dr. Christopher Matthew
Started: 1983
Mission: An inter-departmental, inter-college center for collaborative research,
               shared biotechnology research facilities. To foster new funding
               for biotechnology research and to attract key faculty members.
Personnel: Full-Time: 2   Part-Time: 3   Faculty: 1
Funding: 86-$425,000   87-$600,000

Concentration Areas: A variety of plant and health biotechnology areas.

Center for Molecular Bioscience and Biotechnology
Lehigh University
Chandler-Ullmann Hall 17
Bethlehem, PA 18015

Telephone: 215-758-3645

Director: Dr. Arthur Humphrey
Started: 1986
Mission: To coordinate and integrate existing university-wide
               programs in the biosciences. To foster interdisciplinary
               research, promote efficient use of resources, provide
               training, and to promote interaction with private sector.
Personnel: Full-Time: 3   Faculty: 26
Funding: 87-$4,000,000

350

Listing 13-1 Directory of State Centers

Penn State Biotechnology Institute
519 Biotechnology Hdqtrs Bldg.
University Park, **PA** 16802

Telephone: 814-863-3650

Director:	Dr. Jean E. Brenchley
Started:	1984
Mission:	To conduct applications-oriented research, conduct research leading to new products and processes, promote interaction with private sector, promote information exchange, and training and education for biotechnology professionals.
Personnel:	Full-Time: 5 Part-Time: 1 Faculty: 66

Concentration Areas: Numerous stated areas of scientific interest.

Pittsburgh Center for Biotechnology & Bioengineering
911 William Pitt Union
Univ. of Pittsburgh
Pittsburgh, **PA** 15260

Telephone: 412-648-7956

Director:	Dr. Jerome S. Schultz
Started:	1987
Mission:	To promote entrepreneurial activities and research in the health and life sciences with emphasis on molecular basis of biological recognition systems, immunology, neuroscience, imaging, biosensors, and molecular modeling as related to clinical practice.
Personnel:	Full-Time: 3

RI Partnership for Science & Technology
7 Jackson Walkway
Providence, **RI** 02903

Telephone: 401-277-2601

Director:	Bruce Lang, Executive Director
Started:	1985
Mission:	To fund applied research projects at a RI university, college, or hospital. To help RI businesses grow. Some biotech projects.
Funding:	$2,000,000 (Open budget, not an annual sum)

Concentration Areas: Funding of applied research. Some projects relate to biotechnology.

Biotechnology Program
M303 Walters Life Sciences Building
University of Tennessee
Knoxville, TN 37996

Telephone: 615-974-6841

Director: Dr. D. K. Dougall
Started: 1984
Mission: To provide graduate training in biotechnology to the Master's level.
Personnel: Full-Time: 0   Part-Time: 5 Contract: 0 Faculty: 25
Funding: 87-$175,000

Center for Environmental Biotechnology
583 Dabney Hall
University of Tennessee
Knoxville, TN 37996

Telephone: 615-974-5219

Director: Dr. Gary S. Sayler
Started: 1986
Mission: Basic and applied research leading to the development of
             safe and effective direct environmental application of
             microbial process and microorganisms.
Personnel: Full-Time: 2   Part-Time: 5 Contract: 2 Faculty: 20
Funding: 87-$1,650,000

Concentration Areas: Hazardous chemicals

Center for Biopolymers at Interfaces
College of Engineering
University of Utah, 2234 Merrill Bldg.
Salt Lake City, UT 84112

Telephone: 801-581-5455

Director: Dr. Karin Caldwell
Started: 1986
Mission: To study adsorption and desorption of proteins from polymeric
             surfaces. Has many industrial applications. Support from 10
             companies, especially contact lens companies. Also at Univ.
             of Utah: companion Center for Sensor Technology.
Personnel: 14 Faculty from 4 colleges in the University
Funding: 87- $225,000

Listing 13-1 Directory of State Centers

Center for Excellence in Biotechnology
Utah State University
Logan, UT 84322

Telephone: 801-750-2753

Director: Steven D. Aust
Started: 1987
Mission: To foster the development of biotechnology at the University
To provide service laboratories for MABs, macromolecular
synthesis and analysis and fermentation facilities.
Personnel: Full-Time: 17 Part-Time: 5 Faculty: 2
Funding: 87-$1,573,000

Center for Bioprocess/Product Development
University of Virginia
Thornton Hall
Charlottesville, VA 22901

Telephone: 804-924-6276

Director: Dr. Donald J. Kirwan
Started: 1987
Mission: To work on improved bioreactor design and performance
using immobilized cell technology, bioseparations and
hybridoma cell culture/scale-up.
Personnel: Full-Time: 6 Faculty: 6
Funding: 88-$900,000

Concentration Areas: Separations, bioreactors and hybridoma cell culture.

Center for Biotechnology
Old Dominion University
Norfolk, VA 23529

Telephone: 804-440-3617

Director: Lloyd Wolfinbarger, Jr.
Started: 1986
Mission: To facilitate research efforts in toxicity, tissue transplantation and
marine environment between regional industry and faculty.
Personnel: Full-Time: 0 Part-Time: 2 Faculty: 8
Funding: 86-$21,500 87-$260,000

Institute of Biotechnology (IBT)
Center for Innovative Technology (CIT)
Box 126 MCV Station, VCU
Richmond, VA 23298

Telephone: 804-786-8565

Director: Dr. Francis L. Macrina, Director IBT
Started: 1984
Mission: To enhance, mobilize and transfer Virginia's science/technology
             resources to promote economic development. The Institute of
             Biotechnology is one of 4 research-funding entities within CIT.
Personnel: Full-Time: 2   Part-Time: 1   Contract: 0   Faculty: 1
Funding: 84-$1,947,751 85-$1,575,613 86-$717,752 87-$1,482,975

Concentration Areas: Many areas, including agriculture, toxicology, diagnostics
             and engineering.

University of Wisconsin Biotechnology Center
1710 University Avenue
Madison, WI 53705

Telephone: 608-262-8606

Director: Dr. Richard R. Burgess
Started: 1984
Mission: Develop campus-wide technical resource facilities.
             Promote formation of research units working on practical
             problems. Facilitate univ/industry interactions. Attract
             companies to state. Secure research/training funds. Coordinate
             academic programs.
Personnel: Full-Time: 56   Faculty: 250
Funding: 84-$551,777 85-$1,637,838 86-$2,171,238 87-$2,170,000

Concentration Areas: Service Facilities: Biocomputing, information resources,
             protein/DNA purification sequencing/synthesis, plant cell and tissue
             culture, hybridoma, transgenic mouse. Consortia: Biopulping,
             bioprocess and metabolic engineering.

Notes:
Telephone numbers of centers may vary from telephone numbers of contact persons in Table 13-2. Not all information was made available from each center.

SECTION 14 AN ANALYSIS OF TRENDS

355

THE U.S. BIOTECHNOLOGY INDUSTRY

Our Biotechnology Information Program was set up to monitor growth and changes in the biotechnology industry. From the data contained in this Guide and our additional studies, a number of current trends in the U.S. biotechnology industry can be observed. Six of these trends are described below.

I. Industry Growth: What Is It?

There have been more than 360 small companies formed in the United States alone to use the new technologies of genetic engineering, monoclonal antibody production, and large scale cell culture. Figure 4-1 shows the number of U.S. biotechnology companies founded each year from 1971 to 1987. The "boom" years of biotechnology company founding between 1980 and 1984 have been followed by a decrease in foundings per year over the past few years. There have been some acquisitions of companies, a few have folded, and others have merged. With these companies subtracted as individual business units, there appears to be only slight growth of the industry. However, the number of business units alone may be a deceptive benchmark. The revenues generated by publicly-offered biotechnology firms has been growing over the past few years. Also, as seen in Figure 6-3, there has been an average of 10-fold growth in personnel in these companies from the end of their first year to the present, representing 145 percent per year. And the company executives predict growth to continue at an annual rate of 27 percent over the next five years. Thus, although the growth in number of companies has slowed, the industry is growing in both revenues and personnel.

II. Movement to Large Industry

The analysis in Section 9 shows that the early biotechnology companies had academic roots, with the majority of founders of companies before 1980 coming from academia. In contrast, biotechnology companies over the past three years have been founded by three times as many people from industry as from academia. This trend is paralleled by a trend of commercial biotechnology moving from academia to the small biotechnology firms to large corporations. The average biotechnology firm was founded in 1981 and the large U.S. corporations with significant programs in biotechnology listed in Section 11 announced involvement on average in 1983. A simple comparison of the average large pharmaceutical company and the average of the 12 largest biotechnology firms appears in Table 11-1 confirms the disparity between the competitors. For example, the pharmaceutical corporations have 350 times as many employees and 235 times the average annual revenues of the biotechnology firms. In contrast, the R&D budget of the pharmaceutical corporations is only 73 times that of the biotechnology firms, but this is still a sizable difference in absolute research resources.

III. Changing Focus of the Biotechnology Firms

Figure 9-2 confirms that the biotechnology firms are changing focus. Company executives rated research of highest importance in the companies' first years, but the future trend is towards a higher importance on production and marketing. A comparison of personnel between the companies' first years and today indicates that the focus on production and marketing, once virtually non-existent in the firm, now account for 21 percent of the personnel. Clearly, due to the long development time of the products of biotechnology, it is only now necessary for the companies to shift focus. This trend should continue.

IV. Symbiotic Relationships

The large corporations, many just beginning their involvement in the new biotechnologies, require a means for gaining access to the technology while building their in-house expertise. On the other hand, the much smaller biotechnology firms have had a need to sell their technology know-how and generate revenues. The result has been hundreds of strategic alliances formed. As outlined in Section 12, these alliances include joint ventures, marketing agreements, licensing agreements, and research contracts. Each of the larger biotechnology firms has formed many of these relationships and we have noted that the largest U.S. pharmaceutical firms have from one to ten partnerships. From these relationships, the biotechnology firms get multiple sources of income and the larger corporations can have access to numerous potential products. Partnering between the large and small firms (both U.S.) is still on the rise, as measured by instances per year (Figure 12-1).

V. There Is Increasing Patent Chaos for Biotechnology

When our questionnaire data is extrapolated to the 360 companies currently in our database, there are an estimated 10,000 biotechnology patent applications on file in the U.S. Patent Office. This figure does not include the patent applications by foreign companies or large U.S. corporations. With the many biotechnology patents awaiting decision, there is likely to be a backlog. Also, patent protection for individual products is still in question. Take the case of TPA, with an estimated 42 companies in 24 partnerships working on the product, and an estimated $500 million in annual sales. If one company gets strong patent protection, can the others survive? If one company does not get full protection, will the TPA pie sliced 15, 20 or 30 ways bring enough revenues for all companies to recoup their R&D investments?

VI. Markets Declining, New Markets Opening

Just a few years ago, it was predicted that, by the early 1990's, the market for pharmaceuticals made by biotechnology would reach up to $20 billion or perhaps 20 percent of the total pharmaceutical market. More recently, these figures have been downgraded considerably. Last year, the Boston Consulting Group estimated that therapeutics made by biotechnology processes would account for only about $3 billion by 1990, or only about 4.5 percent of the world market. Although it is likely that this figure will grow in subsequent years, what is less likely is that biotechnology will ever capture 20 percent of the therapeutics market.

357

However, the science that makes biotechnological processes possible is also rapidly advancing. As we learn more about normal and pathological cell biology, new drug opportunities will arise. Also, biotechnology should allow us to make natural substances in quantity, permitting discovery of new drug entities and therapeutic classes. These new drugs may be discovered and/or produced using more classical chemistry and pharmacology, but biotechnology will play a role in the new markets they will create. One example of this may be the synthesis of interleukins and their receptors to create compounds that are therapeutics for arthritis. The actual drugs may not be made by biotechnological processes, but biotechnology will play a key role. Thus, the pharmaceutical companies that have learned to work with biotechnology will benefit both directly and indirectly from the process.

As with other new technologies, new business opportunities should arise that are related to biotechnology. There will be new opportunities for companies that produce separation and fermentation equipment. One can sense that, although there are only six pharmaceuticals currently on the U.S. market made using the new technologies, many more will follow and equipment and reagents will be needed for their development and production.

Conclusions

The biotechnology industry in the United States may now account for as many as 50,000 employees and a few billion dollars in revenues. It is probable that we will see both of these figures grow substantially in the 1990's. The form that the biotechnology industry follows is yet to be established. The role of the small company may become limited as the large corporations develop their abilities with the new technologies. Small firms that continue to rely on partnerships with the large companies may have trouble as the partnerships become less necessary to the large partners. Other small firms may survive through self-reliance or having established a unique niche in product or technology.

Another potential barrier to the growth of the biotechnology industry in the U.S. is the uncertain patent situation. With the intense competition among the firms, patent protection may be the only remedy to protect investment. A possible backlog at the patent office may hurt the smaller firms.

Regulatory issues also remain unresolved and may have a costly effect on the small firms. The delay in approval for Genentech's TPA was estimated to cost the company $40 million in lost revenues each month. Newer technologies, such as genetically-altered plants and animals, also need to have regulatory issues clarified or companies working with these technologies may, too, face costly delays.

What is clear is that the impacts of biotechnology on the economy and society will be tremendous. Most of the products of biotechnology are still in the development stage and, thus, most of the impacts of biotechnology are in our future. Professionals from farmers to physicians will see tremendous benefits from the new technologies. New markets will open and new industries and sub-industries will grow. Governments and companies will realize considerable revenue generated by the new technologies. We have an exciting decade ahead of us.

Table 14-1 The Biotechnology Industry

TABLE 14-1 THE BIOTECHNOLOGY INDUSTRY

I. The U.S. biotechnology industry is composed of:

    A. Biotechnology companies:
       360 in number
       Average founded in 1981
       Account for more than 31,000 employees

    B. Large corporations working in biotechnology:
       60 in number
       Average began working with biotechnology in 1983
       Estimate that biotechnology employees number 10,000 (1)

    C. Ancillary industry (reagents, equipment, separations, services, etc.) (2):
       Unknown number, likely more than 1,000 companies
       Unknown average entry date
       Unknown number of employees

II. Of the biotechnology companies:
       56% are privately held; smallest personnel number and revenues
       28% are public; median personnel number and revenues
       16% are subsidiaries; highest personnel number and revenues

       21% are primarily in therapeutics
       17% are primarily in diagnostics
       11% are primarily in reagents

       26.8% are in California
       10.0% are in Massachusetts
       7.5% are in New Jersey

Notes:
1. The estimated numbers of biotechnology employees in the corporations are extrapolated from data from 20 representative corporations.
2. The ancillary industry is highly varied and includes many disciplines from paper filter manufacturers to lawyers. It is likely that there are more than 1,000 of these companies in the United States. We have no suitable data representing this group.

TABLE 14-2 MEAN AND MEDIAN BIOTECHNOLOGY COMPANY DATA

	Mean Value	Median Value	Total Industry (4)
Personnel	90.3	30	31,811
Funding (1)	$18.9	$7.5	$6,785
Revenues (2)	$11.0	$2.7	$3,949
R&D Budget (3)	$4.6	$1.7	$1,665

Notes:
1. Funding is total company financing (in $ millions) to date taken from data supplied by 109 companies in mid-1987.
2. Revenues are total expected revenues (in $ millions) for current fiscal year 1987-1988 taken from data supplied by 129 biotechnology companies.
3. R&D Budget is for 1987 (in $ millions) and is taken from data supplied by 97 biotechnology companies.
4. Total industry data are mean values extrapolated to the biotechnology industry composed of the 360 U.S. biotechnology firms in Listing 2-2.

APPENDICES

APPENDIX A CLASSIFICATION CODES FOR COMPANIES AND OTHER ABBREVIATIONS

Code	Industry Classification
A	Agriculture, Animal
B	Agriculture, Plant
C	Biomass Conversion
D	Biosensors/Bioelectronics
E	Bioseparations
F	Biotechnology Equipment
G	Biotechnology Reagents
H	Cell Culture, General
I	Chemicals, Commodity
J	Chemicals, Specialty (includes proteins and enzymes)
K	Diagnostics, Clinical Human
L	Energy
M	Food Production/Processing
N	Mining
O	Production/Fermentation
P	Therapeutics
Q	Vaccines
R	Waste Disposal/Treatment
S	Aquaculture
T	Marine Natural Products (includes algae)
U	Consulting
V	Veterinary (all animal health care)
W	Research
X	Immunological Products (non-pharmaceutical)
Y	Toxicology
Z	Venture Capital/Financing
1	Biomaterials
2	Fungi
3	Drug Delivery Systems
4	Medical Devices
5	Testing/Analytical Services

Appendix A

Products

The following abbreviations are those most commonly used in the listings of company products on the pages that follow:

Abbrev.	Meaning
AB	Antibody
AG	Agriculture
ANF	Atrial natriuretic factor
BGH	Bovine growth hormone (bovine somatotropin)
BSA	Bovine serum albumin
CSF	Colony stimulating factor
CV	Cardiovascular
DX	Diagnostic(s)
EGF	Epidermal growth factor
EIA	Enzyme immunoassay
ELISA	Enzyme-linked immunosorbent assay
EPO	Erythropoietin
FGF	Fibroblast growth factor
GM-CSF	Granulocyte macrophage colony stimulating factor
HCG	Human chorionic gonadotropin
HEP B	Hepatitis B
HGH	Human growth hormone
HSA	Human serum albumin
HSV	Herpes simplex virus
IF	Interferon (IF-A=interferon-alpha, etc.)
IL	Interleukin (IL-2=interleukin-2, etc.)
KPA	Kidney plasminogen activator
MAB	Monoclonal antibody
PTA	Parathyroid hormone
RE	Restriction endonuclease; restriction enzyme
RIA	Radioimmunoassay
RX	Therapeutic(s), Drug(s)
SOD	Superoxide dismutase
TNF	Tumor necrosis factor
TPA	Tissue plasminogen activator
VX	Vaccine(s)

Financing Type

The following abbreviations are used to indicate the type of company financing:

Abbrev.	Meaning
PRI	Privately held
PUB	Public stock
SUB	Subsidiary

APPENDIX B   COMPANIES NO LONGER IN BUSINESS
OR WITH NAMES CHANGED

In the process of data collection, contact was attempted with a number of companies listed in our reference sources. Those listed below are companies we believe to be no longer working with biotechnology or out of business. In all cases, multiple attempts to contact the companies or their top management were made.

Below this list is a second list of companies that have changed names.

Companies no longer in the biotechnology industry

Company Name	Location
ADVANCED GENETICS RESEARCH INSTITUTE	OAKLAND, CA
SUB OF AGRION CORP. -- OUT OF BUSINESS	
AGROTEK	WESTPORT, CT
SUB OF UNIVERSITY GENETICS	
AMERICAN BIOGENICS	DECATUR, IL
PRI -- OWNED BY BIOASSAY SYSTEMS CORP.	
ANITECH, INC.	MOUNTAIN VIEW, CA
PRI -- WAS 50-50 JV-- MONOCLONAL AB'S & RALSTON PURINA	
AUTOMATED MEDICAL LABS	HIALEAH, FL
BALL BIOTECH CO.	W. CHICAGO, IL
PRI -- OWNED BY GEORGE A BALL & CO.	
BIOASSAY SYSTEMS CORP.	WOBURN, MA
SUB OF PEDCO INC	
BIOCHEMICAL R&D CO.	HARLINGEN, TX
PRI	
BIOLOGICS CORP.-SEE FERMENTA ANIMAL	OMAHA, NE
SUB OF TECH AMERICA (BOUGHT OUT BY FERMENTA)	
BIOQUAL INC.	ROCKVILLE, MD
PUBLIC, BOUGHT BY DIAGNON	
BREIT LABORATORIES	W. SACRAMENTO, CA
PRI – IN BANKRUPTCY PROCEEDINGS 2/87	
BW BIOTEC, INC.	CHICAGO, IL
SUB -- JV WITH TOYO ENGINEERING	
CAMBRIDGE DIAGNOSTICS, INC.	CAMBRIDGE, MA
PRI -- VESTIGIAL PARENT OF TOXICON	
CATALYTICA ASSOCIATES, INC.	MOUNTAIN VIEW, CA
CELTEK, INC.	NORMAN, OK
PRI -- OUT OF BUSINESS 2/87	

Appendix B

Companies no longer in the biotechnology industry (Cont.)

Company Name	Location
CHIMERIX	GLASTONBURY, CT

PRI -- PHONE NOT IN SERVICE AS OF 2/25/88

CLINICAL BIOTECHNOLOGIES, INC. LENEXA, KS
PRI -- PHONE NO LONGER LISTED 1/88

CORDIS LABORATORIES, INC. MIAMI, FL

DIMENSION DIAGNOSTICS ST. CHARLES, MI
PRI -- NO PHONE LISTED/WORKING 1/88

ENGENICS, INC. MENLO PARK, CA
PRI -- OUT OF BUSINESS 1/88

ENZYME TECHNOLOGY, INC. LENEXA, KS
PRI -- INACTIVE SUB OF CLINICAL BIOTECHNOLOGIES, INC.

FERMENTEC CORP. LOS ALTOS, CA
PRI -- OUT OF BUSINESS IN 1987

INT'L GENETIC SCIENCES CHAPEL HILL, NC
JV BETWEEN 1ST MISSISSIPPI AND INT'L GENETIC SCI.

LITTON BIONETICS, INC. KENSINGTON, MD
PURCHASED BY ORGANON TEKNIKA

MUTECH MADISON, WI
PRI -- PHONE DISCONNECTED

NUCLEAR & GENETIC TECHNOLOGY DEER PARK, NY
PUB -- IN CHAP. 11, 1/87; STILL IN CHAP. 11, 2/88

PEPTIDE LABORATORY BERKELEY, CA
PRI -- NOT ABLE TO CONTACT

REPLICON SAN FRANCISCO, CA
PRI -- PHONE DOES NOT ANSWER

RORER BIOTECH, INC. SPRINGFIELD, VA
SUB -- RORER GROUP-ABSORBED BY PARENT CO 2/88

SDS BIOTECH PAINESVILLE, OH
FORMER SUB OF DIAMOND SHAMROCK AND JAPANESE CO.

SYNAX, INC. CAMBRIDGE, MA

ULTRAFERM, INC. WHEELING, IL
PRI -- DISCONNECTED PHONE

VERTEX CORP., BIOPROCESS DIVISION MENLO PARK, CA
PRI -- A PARTNERSHIP WITH A COMPANY IN FINLAND

Companies with changed names

Old Name	New Name
AMERICAN BIONETICS	ABN
AXONICS	3M DIAGNOSTIC SYSTEMS
BIOCLINICAL SYSTEMS, INC.	MERGED W/BINAX, INC., S. PORTLAND, ME
BIOLOGICAL ENERGY CORP.	REPAP TECHNOLOGIES
BIOTHERAPY SYSTEMS, INC.	IDEC PHARMACEUTICAL
BIOTHERM	CARDIOVASCULAR DIAGNOSTICS (4/88)
CARCINEX, INC.	INTEK DIAGNOSTICS
DIAGNOSTIC, INC.	LIFECORE, INC.
ESCAGEN CORP.	ESCAGENETICS CORP. (3/88)
HAWAII BIOTECHNOLOGIES	HAWAII BIOTECHNOLOGY GROUP, INC.
INTERLEUKIN-2	CEL-SCI CORP. (3/88)
INT'L MEDICAL DIAGN.	PORTON MEDICAL LABS
ISSAQUAH RES. INSTITUTE	SEATTLE BIOMED. RESEARCH INSTITUTE
LIFECORE, INC.	LIFECORE BIOMEDICAL, INC. (3/88)
TECHAMERICA GROUP	FERMENTA ANIMAL HEALTH

APPENDIX C   NEW COMPANIES AND LAST-MINUTE ADDITIONS

The following companies have been identified during the final preparation of this Guide. More details on these companies are being added to the Companies database and will be included in the next edition.

AC BIOTECHNICS, INC.
100 CROSSWAYS PARK W., SUITE 204
WOODBURY, NY 11797
Telephone: 516-496-3300
Products: ANAEROBIC WASTEWATER TREATMENT

President: WILLIAM BONKOWSKI
R&D Dir: JERZY KOLLAJTIS
Financing: SUB
Started: 1985
Class Code: R

AMERICAN ENZYME CORP.
6825 HOBSON VALLEY DR.
WOODBRIDGE, IL 60517
Telephone: 312-810-0272
Products: PROTEINS & ENZYMES FOR DX,
PHARMACEUTICALS & FOOD INDUSTRY

President: A.J. MIRABELLI
R&D Dir: DR. DIRK SIKKEMA
Financing: SUB
Started: 1987
Class Code: J

ANTIGENICS, INC.
700 BUSINESS CENTER DR.
HORSHAM, PA 19044
Telephone: 215-441-5400
Products: AIDS RESEARCH

CEO: DR. CHARLES BERKOFF
President: DR. CHARLES BERKOFF
R&D Dir: DR. CHARLES BERKOFF
Financing: PRI
Started: 1984
Class Code: P

BIOSOURCE GENETICS CORP.
3333 VACA VALLEY PARKWAY
VACAVILLE, CA 95688
Telephone: 707-446-5501
Products: RDNA PLANTS TO PRODUCE BIO
POLYMERS FOR USE IN FOOD, COSMETICS,
SPECIALTY CHEMICALS INDUSTRY

CEO: ROBERT L. ERWIN
President: ROBERT L. ERWIN
R&D Dir: LAURENCE GRILL
Financing: PRI
Started: 1987
Class Code: B

CELGENE CORP.
7 POWDER HORN DR.
WARREN, NJ 07060
Telephone: 201-271-1001
Products: SOIL MICROORGANISMS TO PRODUCE
SPECIALTY & FINE CHEMICALS

CEO: LOUIS FERNANDEZ
President: LOUIS FERNANDEZ
R&D Dir: PETER J. KRETSCHMER
Financing: PUB
Started: 1986
Class Code: J

CELTRIX LABORATORIES
2500 FABER PLACE
PALO ALTO, CA 94303
Telephone: 415-856-0200
Products: CERAMIC COLLAGEN FOR BONE REPAIR;
PROTEINS FOR GROWTH FACTORS

CEO: HOWARD PALESSKY
President: DR. BRUCE B. PHARRISS
R&D Dir: DR. DALE STRINGFELLOW
Financing: SUB
Started: 1987
Class Code: J

CEPHALON, INC.

145 BRANDYWINE PARKWAY
WESTCHESTER, PA 19380
Telephone: 215-344-0200
Products: RX FOR ALZHEIMER'S & PARKINSON'S
   DISEASES & OTHER DEGENERATIVE
   BRAIN DISORDERS

President: DR. FRANK BALDINO, JR.
R&D Dir: DR. FRANK BALDINO, JR.
Financing: PRI
Started: 1987
Class Code: P

ENZYTECH

763-D CONCORD AVE.
CAMBRIDGE, MA 02138
Telephone: 617-661-0940
Products: PROTEIN ENHANCEMENT TECHNOLOGY;
   COMMERCIAL ENZYMES; DRUG DELIVERY

CEO: MARK B. SKALETSKY
President: DR. LEONARD STARK
R&D Dir: DR. AKIVA GROSS
Financing: PRI
Started: 1987
Class Code: P

GENSIA PHARMACEUTICALS, INC.

11075 ROSELLE ST.
SAN DIEGO, CA 92121
Telephone: 619-546-8300
Products: RX FOR CARDIOVASCULAR/CEREBRO-
   VASCULAR/ NEUROLOGICAL DISEASES

CEO: DAVID F. HALE
President: DAVID F. HALE
R&D Dir: DR. ROSS DIXON
Financing: PRI
Started: 1986
Class Code: P

HELIX BIOCOR

3905 ANNAPOLIS LANE
MINNEAPOLIS, MN 55447
Telephone: 612-553-7736
Products: PROTEINS FOR USE IN
   THERAPEUTICS

CEO: MANUEL VILLAFANA
President: RICHARD KRAMP
R&D Dir: JOHN SALSTROM
Financing: PRI
Started: 1987
Class Code: J

HUMAGEN

1500 AVEON ST. EXT.
CHARLOTTESVILLE, VA 22901
Telephone: 804-979-4000
Products: SPERM FERTILITY DX; CONTRACEPTIVE VX;
   RX DELIVERY SYSTEM FOR IMPOTENCY

CEO: JAMES WHIDDEN
President: JAMES WHIDDEN
R&D Dir: DR. JOHN HERR
Financing: PRI
Started: 1984
Class Code: K

IMMUCOR, INC.

3130 GATEWAY DR.
NORCROSS, GA 30071
Telephone: 404-441-2051
Products: DX; MABS TO HUMAN BLOOD GROUPS,
   PLATELET AB TESTS, HIV TESTS, BLOOD
   BANK REAGENTS

CEO: EDWARD L. GALLUP
President: EDWARD L. GALLUP
R&D Dir: DR. LYLE SINOR
Financing: PUB
Started: 1982
Class Code: K

IMMUNE RESPONSE, INC.

333 W. WAKDER DR., SUITE 700
CHICAGO, IL 60606
Telephone: 312-444-2078
Products: CANCER VX

President: DR. LAWRENCE EIDLEN
R&D Dir: DR. LAWRENCE EIDLEN
Financing: PRI
Started: 1986
Class Code: Q

MOLECULON

230 ALBANY ST.
CAMBRIDGE, MA 02139
Telephone: 617-577-9900
Products: DRUG DELIVERY SYSTEMS FOR
HUMAN AND VET USES

CEO: ARTHUR S. OBERMAYER
President: GEORGE GOLDENBERG
R&D Dir: LARRY D. NICHOLS
Financing: PRI
Started: 1982
Class Code: P

NOVAGENE, INC.

55 BRIARHOLLOW LANE
HOUSTON, TX 77027
Telephone: 713-621-8440
Products: VET VX

CEO: MALON KIT
R&D Dir: MALON KIT
Financing: PRI
Started: 1987
Class Code: V

OROS SYSTEMS, INC.

222 THIRD ST., SUITE 3220
CAMBRIDGE, MA 02142
Telephone: 617-868-6767
Products: AUTOMATED SYSTEM TO CHARACTERIZE
& PURIFY MABS; PROTEIN PURIFICATION

CEO: GLYN EDWARDS
President: MICHAEL BOSS
R&D Dir: DR. ANDREW KENNEY
Financing: SUB
Started: 1987
Class Code: K

PHARMATECH

P.O. BOX 730
ALACHUA, FL 32615
Telephone: 904-462-1210
Products: DRUG DELIVERY OF SYNTHETIC PEPTIDES
THROUGH BLOOD-BRAIN BARRIER

CEO: DR. WARREN STERN
President: DR. WARREN STERN
Financing: PRI
Started: 1982
Class Code: 3

POLYBAC CORP.

954 MARCON BLVD.
ALLENTOWN, PA 18013
Telephone: 215-264-8740
Products: POLLUTION CONTROL MICROBES; CROP
DISEASE CONTROL MICROBES

CEO: TERRY HERZOG
President: DR. RALPH GUTTMAN
R&D Dir: DR. RALPH GUTTMAN
Financing: PRI
Started: 1975
Class Code: R

SYNTHATECH

1290 INDUSTRIAL WAY
ALBANY, OR 97321
Telephone: 503-967-6575
Products: ENZYME SYNTHESIS

CEO: MICHAEL MITTON
President: MICHAEL MITTON
R&D Dir: PAUL AHRENS
Financing: PUB
Started: 1980
Class Code: J

THERMASCAN, INC.

500 5TH AVE., SUITE 2610
NEW YORK, NY 10110
Telephone: 212-944-9344
Products: AIDS DX; SYNTHETIC PROTEINS

CEO: DETLEV BAURS-KREY
President: DETLEV BAURS-KREY
R&D Dir: DR. NANIK D. GYAN
Financing: PUB
Started: 1983
Class Code: K

VIRAL TECHNOLOGIES, INC.

777 14TH ST. N.W., SUITE 410
WASHINGTON, DC 20005
Telephone: 202-628-0348
Products: AIDS VX

CEO: MAXIMILIAN DE CLARA
President: J.J. FINKELSTEIN
R&D Dir: DR. ALLAN GOLDSTEIN
Financing: PRI
Started: 1986
Class Code: Q

APPENDIX D REFERENCES AND INFORMATION SOURCES

The following information sources are those most frequently used in the gathering of data for our databases and this Guide.

Books and Directories
Bioscan. Oryx Press, Phoenix, AZ, 1987-1988.

The Biotechnology Directory 1988. J. Coombs and Y.R. Alston. Stockton Press, New York, NY; Macmillan Press Ltd., London, UK, 1987.

Commercial Biotechnology: An International Analysis. U.S. Congress Office of Technology Assessment. Washington, D.C. 1984. Reprinted and available from the Information Program, North Carolina Biotechnology Center. P.O. Box 13547, Research Triangle Park, NC 27709.

Genetic Engineering and Biotechnology Firms Worldwide 1987-1988. M. Sittig and R. Noyes. Sittig and Noyes Publishers, Kingston, NJ, 1987.

Genetic Engineering News Guide to Biotechnology Companies. Mary Ann Liebert Publishing Co., New York, NY, 1987, 1986.

Information Sources in Biotechnology. A. Crafts-Lighty. Stockton Press, New York, NY; Macmillan Publishers Ltd., London, UK, 1986.

Journals and Newsletters
Abstracts in BioCommerce. (Biweekly) BioCommerce Data Ltd., Slough, UK.

Bio/Technology. (Monthly) Nature Publishing Co., New York, NY.

Bioengineering News. (Weekly) Bioengineering News Pub., Port Angeles, WA.

Biopharm Manufacturing. (Monthly) Aster Pub. Co., Eugene, OR.

Biotechnology Bulletin. (Monthly) IBC Technical Services., London, UK.

Biotechnology News. (Monthly) CTB International Pub. Co., Maplewood, NJ.

Biotechnology Newswatch. (Biweekly) McGraw-Hill, New York, NY.

Chemical and Engineering News. (Weekly) Am. Chemical Soc., Washington, DC.

Chemical Business. (Monthly) Schnell Publishing Co., New York, NY.

European Biotechnology Newsletter. (22 issues/year) Biofutur, S.A., Paris, France.

Genetic Engineering Letter. (Monthly) Environews, Inc., Washington, DC.

Genetic Engineering News. (Monthly) Mary Ann Liebert Pub. Co., New York, NY.

Genetic Technology News. (Monthly) Technical Insights, Inc., Englewood, NJ.

High Technology Business. (Monthly) High Technology Pub. Corp., Boston, MA.

Nature. (Weekly) Macmillan Magazines Ltd., London, UK.

New Biotech. (Monthly) Winter House Scientific Publishers, Winnepeg, Canada.

Science. (Weekly) Am. Assoc. for the Advancement of Science., Washington, DC.

The Scientist. (Biweekly) ISI Corp., Philadelphia, PA.

Scrip World Pharmaceutical News. (Biweekly) PJB Publications Ltd., Surrey, UK.

Trends in Biotechnology. (Monthly) Elsevier Pub. Co., Cambridge, UK.

Newspapers
In addition to the above journals, the following newspapers are scanned daily.

New York Times

Washington Post

Wall Street Journal

INDEX

Index

Index